FRANKINCENSE – GUM OLIBANUM

Botany, Oleoresin, Chemistry, Extraction, Utilization, Propagation, Biotechnology, and Conservation

AAP Focus on Medicinal Plants

FRANKINCENSE – GUM OLIBANUM

Botany, Oleoresin, Chemistry, Extraction, Utilization, Propagation, Biotechnology, and Conservation

T. Pullaiah, PhD
K. Venkata Ratnam, PhD
Mallappa Kumara Swamy, PhD
Lepakshi Md. Bhakshu, PhD

AAP | APPLE ACADEMIC PRESS

First edition published 2024

Apple Academic Press Inc.
1265 Goldenrod Circle, NE,
Palm Bay, FL 32905 USA

760 Laurentian Drive, Unit 19,
Burlington, ON L7N 0A4, CANADA

CRC Press
6000 Broken Sound Parkway NW,
Suite 300, Boca Raton, FL 33487-2742 USA

4 Park Square, Milton Park,
Abingdon, Oxon, OX14 4RN UK

Library and Archives Canada Cataloguing in Publication

Title: Frankincense--gum olibanum : botany, oleoresin, chemistry, extraction, utilization, propagation, biotechnology, and conservation / edited by T. Pullaiah, PhD, K. Venkata Ratnam, PhD, Mallappa Kumara Swamy, PhD, Lepakshi Md. Bhakshu, PhD.

Names: Pullaiah, T., editor. | Ratnam, K. Venkata, editor. | Swamy, Mallappa Kumara, editor. | Bhakshu, Lepakshi Md., editor.

Series: AAP focus on medicinal plants.

Description: First edition. | Series statement: AAP focus on medicinal plants | Includes bibliographical references and index.

Identifiers: Canadiana (print) 20230146414 | Canadiana (ebook) 2023014652X | ISBN 9781774912638 (hardcover) | ISBN 9781774912645 (softcover) | ISBN 9781003348016 (ebook)

Subjects: LCSH: Frankincense. | LCSH: Gums and resins.

Classification: LCC QK495.B8 F73 2023 | DDC 583/.75—dc23

Library of Congress Cataloging-in-Publication Data

CIP data on file with US Library of Congress

ISBN: 978-1-77491-263-8 (hbk)
ISBN: 978-1-77491-264-5 (pbk)
ISBN: 978-1-00334-801-6 (ebk)

AAP FOCUS ON MEDICINAL PLANTS

ABOUT THE SERIES

This new book series, edited by T. Pullaiah, focuses on bioactives and pharmacology of medicinal plants.

For millennia, medicinal plants have been a valuable source of therapeutic agents, and still many of today's drugs are based on plant-derived natural products or their derivatives. Bioactive compounds typically occur in small amounts, and they have more subtle effects than nutrients. Bioactive compounds influence cellular activities that modify the risk of disease and help to alleviate disease symptoms. The bioactive compounds have potentially important health benefits, and these compounds can act as antioxidants, enzyme inhibitors and inducers, inhibitors of receptor activities, and inducers and inhibitors of gene expression among other actions. A wide array of biological activities and potential health benefits of medicinal plants have been reported, which include antiviral, antibacterial, antifungal, antioxidant, anticancer, anti-inflammatory, antidiabetic, hepatoprotective, cardioprotective, nephroprotective properties as well as other protective effects on the liver, kidney, heart, and nervous system.

The volumes aim to be comprehensive desk references on bioactives and pharmacology of all the medicinal plants. They will also be important sourcebooks for the development of new drugs.

Book Series Editor

Prof. T. Pullaiah
Department of Botany
Sri Krishnadevaraya University,
Anantapur 515003, A.P., India
Email: pullaiah.thammineni@gmail.com

Book in the Series

- Bioactives and Pharmacology of Medicinal Plants, 2-volume set
- Biomolecules and Pharmacology of Medicinal Plants, 2-volume set
- Bioactives and Pharmacology of Legumes

- Phytochemistry and Pharmacology of Medicinal Plants, 2-volume set
- Bioactives and Pharmacology of Lamiaceae
- Frankincense – Gum Olibanum: Botany, Oleoresin, Chemistry, Extraction, Utilization, Propagation, Biotechnology, and Conservation
- Phytochemical Composition and Pharmacy of Medicinal Plants, 2-volume set

OTHER BOOKS FROM AAP BY DR. T. PULLAIAH

Ethnobotany of India, 5-volume set:
Editors: T. Pullaiah, PhD, K. V. Krishnamurthy, PhD, and Bir Bahadur, PhD

- Volume 1: Eastern Ghats and Deccan
- Volume 2: Western Ghats and West Coast of Peninsular India
- Volume 3: North-East India and the Andaman and Nicobar Islands
- Volume 4: Western and Central Himalaya
- Volume 5: Indo-Gangetic Region and Central India

Global Biodiversity, 4-volume set:
Editor: T. Pullaiah, PhD

- Volume 1: Selected Countries in Asia
- Volume 2: Selected Countries in Europe
- Volume 3: Selected Countries in Africa
- Volume 4: Selected Countries in the Americas and Australia

Handbook of Research on Herbal Liver Protection: Hepatoprotective Plants
T. Pullaiah, PhD, and Maddi Ramaiah, PhD

Bio-Inspired Technologies for the Modern World: Emerging Trends and Applications
R. Ramakrishna Reddy, PhD, and T. Pullaiah, PhD

Biodiversity of Hotspots–36 volumes
Editor: T. Pullaiah, PhD

ABOUT THE EDITORS

T. Pullaiah, PhD
Former Professor, Department of Botany, Sri Krishnadevaraya University, Andhra Pradesh, India

T. Pullaiah, PhD, is a former Professor at the Department of Botany at Sri Krishnadevaraya University in Andhra Pradesh, India, where he has taught for more than 35 years. He has held several positions at the university, including Dean, Faculty of Biosciences, Head of the Department of Botany, Head of the Department of Biotechnology, and Member of Academic Senate. He was the President of the Indian Botanical Society (2014), President of the Indian Association for Angiosperm Taxonomy (2013), and Fellow of the Andhra Pradesh Academy of Sciences. He was awarded the Panchanan Maheshwari Gold Medal, the Prof. P. C. Trivedi Medal, the Dr. G. Panigrahi Memorial Lecture Award of the Indian Botanical Society, and Prof. Y. D. Tyagi Gold Medal of the Indian Association for Angiosperm Taxonomy, and the Best Teacher Award from Government of Andhra Pradesh. Under his guidance, 54 students obtained their doctoral degrees. He has authored 52 books, edited 23 books, and published over 330 research papers. His books include *Advances in Cell and Molecular Diagnostics* (published by Elsevier), *Camptothecin*, and *Camptothecin producing Plants* (Elsevier) *Ethnobotany of India* (5 volumes published by Apple Academic Press), *Global Biodiversity* (4 volumes, Apple Academic Press), *Red Sanders: Silviculture and Conservation* (Springer), *Genetically Modified Crops* (2 volumes, Springer), *Monograph on Brachystelma and Ceropegia in India* (CRC Press), *Flora of Andhra Pradesh* (5 volumes), *Flora of Eastern Ghats* (7 volumes), *Flora of Telangana* (3 volumes), *Encyclopedia of World Medicinal Plants* (7 volumes, 2nd edition), and *Encyclopedia of Herbal Antioxidants* (3 volumes). He was also a member of the Species Survival Commission of the International Union for Conservation of Nature (IUCN). Professor Pullaiah received his PhD from Andhra University, India, attended Moscow State University, Russia, and worked as Post-Doctoral Fellow during 1976–1978.

K. Venkata Ratnam, PhD
Assistant Professor, Department of Botany, Rayalaseema University, Kurnool, India

K. Venkata Ratnam, PhD, is an Assistant Professor in the Department of Botany at Rayalaseema University, Kurnool, India. He has 12 years teaching (both under- and postgraduate) and 18 years of research experience in ethnobotany and biodiversity conservation studies. He has published 36 research papers in national and international journals, presented 17 research papers in international and national conferences, and participated at 18 national and international symposia. He received two best poster paper awards at national-level symposia. Dr. Ratnam received postdoctoral and research associate fellowships from the University of Hyderabad, Telangana, India. He has completed his master's degree in Botany at Sri Venkateswara University, Tirupati, and MPhil and PhD from the Department of Botany at Sri Krishnadevaraya University, Anantapur, India.

Mallappa Kumara Swamy, PhD
Professor, Department of Biotechnology, East West First Grade College, Bengaluru, India

Mallappa Kumara Swamy, PhD, is a Professor and Head of the Department of Biotechnology at East West First Grade College, Bengaluru, India. Dr. Swamy has authored 65 research publications in peer-reviewed journals and 53 book chapters. Recently, he has edited 10 books published by Springer Nature Singapore Pte Ltd.; one book by CRC Press, USA; one book by Studium Press, India; and one edited book volume by Elsevier. Dr. Swamy also serves as an editorial board member and reviewer for several high-impact factor international journals. His current research is focused on cell and tissue culture technology, phytochemistry, phytocompounds isolation, production and their biological evaluation, and nanotechnology for medical applications. He has more than 19 years of teaching and research experience in the fields of plant biotechnology, secondary metabolite production, phytochemistry, and bioactive studies. He completed his postdoctoral research at the Department of Crop Science, Faculty of Agriculture, Universiti Putra Malaysia (UPM), Serdang, Malaysia. Before that, he had worked as an Associate Professor and Head of the Department of Biotechnology, Padmashree Institute of Management and Sciences, Bangalore University, Bengaluru, India. He received his PhD (Biotechnology) from Acharya Nagarjuna University, Guntur, India.

Lepakshi Md. Bhakshu, PhD
Assistant Professor, Department of Botany, PVKN Govt. College (A), Chittoor, India

Lepakshi Md. Bhakshu, PhD, is an Assistant Professor in the Department of Botany at PVKN Govt. College (A), Chittoor, India. He has published 24 research publications in national and international journals, presented eight research papers in international and national conferences, and participated at 20 national and international symposia. He received two best poster paper awards at national-level symposia. He worked as a project fellow at the Environment Protection Training and Research Institute (EPTRI), Hyderabad, and was a co-author of the Proceedings of the National Seminar on the Conservation of Eastern Ghatsm, published by the EPTRI. He has 15 years of teaching and 20 years of research experience in ethnobotany, ethnopharmacology, phytochemical studies, and biodiversity conservation. He is recipient of a postdoctoral fellowship and the Dr. DS Kothari National Fellowship (UGC) from the University of Hyderabad, Telangana, India. Dr. Bhakshu completed his master's degree in Botany at Sri Venkateswara University, Tirupati, and PhD at the Department of Botany in 2002, Sri Krishnadevaraya University, Anathapuramu, India.

CONTENTS

CONTRIBUTORS

Abdel Rahman Al-Tawaha
Department of Biological Sciences, Al-Hussein Bin Talal University, Ma'an, Jordan

Gaber El-Saber Batiha
Department of Pharmacology and Therapeutics, Faculty of Veterinary Medicine, Damanhur University, Damanhur – 22511, AlBeheira, Egypt

Lepakshi Md. Bhakshu
Department of Botany, PVKN Government College (Autonomous), Chittoor – 517002, Andhra Pradesh, India, E-mail: lmbakshu@gmail.com

Ercan Bursal
Department of Biochemistry, Mus Alparslan University, Turkey

Abhijit Dey
Department of Life Sciences, Presidency University, 86/1 College Street, Kolkata – 700073, West Bengal, India, E-mail: abhijit.dbs@presiuniv.ac.in

Padmanabh Dwivedi
Department of Plant Physiology, Banaras Hindu University, Banaras, Uttar Pradesh, India

Mimosa Ghorai
Department of Life Sciences, Presidency University, 86/1 College Street, Kolkata – 700073, West Bengal, India

Vartika Jain
Department of Botany, Government Meera Girls' College, Udaipur – 313001, Rajasthan, India

Snigdha Majumder
School of Biotechnology, Presidency University (2nd Campus), Kolkata – 700156, West Bengal, India

Tabarak Malik
Department of Biochemistry, College of Medicine and Health Science, School of Medicine, University of Gondar, Ethiopia

Samapika Nandy
Department of Life Sciences, Presidency University, 86/1 College Street, Kolkata – 700073, West Bengal, India

Shreya Nath
School of Biotechnology, Presidency University (2nd Campus), Kolkata – 700156, West Bengal, India

Potshangbam Nongdam
Department of Biotechnology, Manipur University, Canchipur, Imphal – 795003, Manipur, India

Devendra Kumar Pandey
Department of Biotechnology, Lovely Professional University, Phagwara, Punjab, India

Subhabrata Paul
School of Biotechnology, Presidency University (2nd Campus), Kolkata – 700156, West Bengal, India

T. Pullaiah
Department of Botany, Sri Krishnadevaraya University, Anantapur – 515003, Andhra Pradesh, India, E-mail: pullaiah.thammineni@gmail.com

Boregowda Purushotham
Department of Biotechnology, East-West First Grade College, Bangalore, Karnataka, India

K. Venkata Ratnam
Department of Botany, Rayalaseema University, Kurnool – 518007, Andhra Pradesh, India, E-mail: drvenkatapkd@gmail.com

Rupa Sanyal
Department of Botany, Bhairab Ganguly College, Kolkata, West Bengal, India

Mahipal S. Shekhawat
Department of Plant Biology and Biotechnology, Kanchi Mamunivar Government Institute for Postgraduate Studies and Research, Pondicherry, Tamil Nadu, India

Mallappa Kumara Swamy
Department of Biotechnology, East-West First Grade College, Bangalore, Karnataka, India

Champa Keya Tudu
Department of Life Sciences, Presidency University, 86/1 College Street, Kolkata – 700073, West Bengal, India

ABBREVIATIONS

2,4-D	2,4-dichlorophenoxy acetic acid
5-HETE	5-hydroxy-eicosa-tetra-enoic acid
AC	axial canals
ACE	angiotensin-converting enzyme
AChE	acetylcholinesterase
AD	Alzheimer's disease
AELTR	area enclosure with long tapping resting period
AEMTR	area enclosure with medium tapping resting period
AFLP	amplified fragment length polymorphism
AHP	analytical hierarchy process
AKBA	acetyl-11-keto-β-boswellic acid
ALP	alkaline phosphatase
ALT	alanine aminotransferase
AOM	azoxymethane
APN	aminopeptidase N
APTT	activated partial thromboplastin time
AST	aspartate aminotransferase
AβBA	acetyl-β-boswellic acid
BA	6-benzyladenine
BAP	6-benzylaminopurine
BAs	Boswellic acids
BE	boswellin
BLASTn	basic local alignment search tool
BRT	Biligiri Rangaswamy Temple Tiger Reserve
BSE	*B. serrata* extract
BSTFA	N,O-bis (trimethylsilyl) trifluoroacetamide
CAPS	cleaved amplified polymorphic sequences
CAR	conditioned-avoidance response
CEMB	cyano enone of methyl boswellates
CFA	complete Freund's adjuvant
CI	combination index
CITES	convention on international trade on endangered species
C-KβBA	keto-β-BA

COVID-19	coronavirus disease
COX-2	cyclooxygenase-2
CWS	Cauvery Wildlife Sanctuary
DBH	diameter at breadth height
DMBA	7,12-dimethylbenz[a]anthracene
DMN	dimethyl nitrosamine
DOX	doxorubicin
DPPH	2,2-diphenyl-1-picrylhydrazyl
DT	dhamtari
DW	dry weight
EA	ethyl acetate
EAC	Ehrlich's ascites carcinoma
EBV-EA	Epstein-Barr virus early antigen
EGFR	epidermal growth factor receptor
EN	endangered
ESR	erythrocyte sedimentation rate
FCA	Freund's complete adjuvant
FEV	forced expiratory volume
FGNTR	free grazing with no tapping resting period
FGRT	free grazing with a rotational tapping
FRAP	ferric ion reducing antioxidant power
FVC	forced vital capacity
GATA3	GATA-binding protein-3
GBH	girth at breast height
GC-FID	GC with flame-ionization detection
GC-MS	gas chromatography-mass spectrometry
GFA	global frankincense alliance
GPx	glutathione peroxidase
GSH	glutathione
GST	glutathione-s-transferase
HCV	hepatitis C-virus
HO-1	heme oxygenase-1
HSV	herpes simplex virus
IAA	indole-3-acetic acid
IBA	indole-3-butyric acid
ISSR	inter-simple sequence repeat
ITS	internal transcribed region
IUCN	international union for conservation of nature
KBA	11-keto-β-boswellic acid

LC	least concern
LPO	lipid peroxidation
LPS	lipopolysaccharide
MALDI-TOF MS	matrix-assisted laser desorption ionization time-of-flight mass spectrometry
MBC	minimum bactericidal concentration
MCAO	middle cerebral artery occlusion
MDA	malondialdehyde
MEGA	molecular evolutionary genetics analysis
MIC	minimum inhibition concentration
MLD-STZ	multiple low dose-streptozotocin
MMH	Malai Mahadeshwara Wildlife Sanctuary
MOP	myeloperoxidase
MRSA	methicillin-resistant *Staphylococcus aureus*
MS	Murashige and Skoog
NAA	naphthalene acetic acid
NAFLD	non-alcoholic fatty liver disease
NEP	neutral endopeptidase
NF-κB	nuclear factor-kappa B
NGS	next-generation sequencing
NO	nitric oxide
NP	Narayanpur
OB	outer bark
OTUs	operational taxonomic units
OVA	ovalbumin
PAE	post-antibiotic effect
pAECs	porcine aortic endothelial cells
PBMCs	peripheral blood mononuclear cells
PC	parenchyma
PC	pyruvate carboxylase
PCR	polymerase chain reaction
PD	Parkinson's disease
PEFR	peak expiratory flow rate
PEPCK	phosphoenol-pyruvate carboxykinase
PGPRs	plant growth promoting regulators
PLSR	partial least square regression technique
PMNL	polymorphonuclear leukocytes
POBA	3-α-propionyloxy-β-BA

pSTAT6	phosphorylated signal transducer and activator of transcription-6
PT	prothrombin time
PTZ	pentylenetetrazole
PVP	polyvinylpyrrolidone
RAPD	random amplified polymorphic DNA
RCBD	randomized complete block style
RFLP	restriction fragment length polymorphisms
ROS	reactive oxygen species
RRC	radial resin canal
SC	sclerenchyma
SCR	secondary-conditioned response
SEM	scanning electron microscope
SFE	supercritical fluid extraction
SNPs	single nucleotide polymorphisms
SOD	superoxide dismutase
SPME	solid-phase micro-extraction
SSCP	single-strand conformation polymorphism
SSLP	simple sequence length polymorphism
TBARS	thiobarbituric acid reactive substances
TCM	traditional Chinese medicine
TGF-b1	transforming growth factor-b1
THLE	tonic hind limb extension
TMZ	temozolomide
TNBC	triple-negative breast cancer
TNF	tumor necrosis factor
TPA	12-O-tetradecanoylphorbol-13-acetate
TSGs	tumor suppressor genes
UV-C	ultraviolet light
VEGF	vascular endothelial growth factor
VRE	vancomycin-resistant *Enterococcus faecium*
VRSA	vancomycin-resistant *S. aureus* isolates
VU	vulnerable
WFL	wood fiber length
WFW	wood fiber width
XBP 1	x-box-binding protein 1
α-BA	α-boswellic acid
β-BA	β-boswellic acid

PREFACE

Frankincense is the oleoresin obtained from the *Boswellia* species. Its main application is in incense (sharp lingering sweet smell) products. *B. sacra* is the main source of frankincense, which is commonly found in the countries of Oman, Yemen, Ethiopia, and Somalia. However, B. *papyrifra* is another chief source of Frankincense from Africa. Indian frankincense is obtained from B. *serrata. Boswellia,* belonging to the family Burseraceae, contains 24 species found in Africa and Asia. These, including a few other species, remained in the market for centuries as frankincense. The increasing human population enjoyed it, and the growing demand resulted in the search for other sources of frankincense. The global demand and market lobby began to be a threat for the existence of the *Boswellia* species. This was felt as a threat to the communities who relied on it as their livelihood and to the export revenue of the frankincense-growing developing countries.

Hence, there is an urgent need for the assessment of the species yielding frankincense. There is a need for the evaluation of scientific methods of extraction and propagation of the trees, in addition a review of its utilization. Also, the biotechnological applications of the species of *Boswellia* require a detailed review. This book discusses the significance, botany, pharmacology, chemistry, biotechnology, and uses of the *Boswellia* species and frankincense to readers.

Readers will learn about the distribution and availability of species that yields frankincense. Foresters and tribal groups can use the information presented here to develop scientific methods for extraction and propagation of this species and for increasing the livelihoods of people living on the extraction of the oleoresin. Biotechnological methods like tissue culture can be used for enhancing the productivity of the oleoresin. This volume aims to be a desktop reference for frankincense and *Boswellia* species.

Dr. N. Venugopal, an ethnobotanist and expert in science and sustainable extraction from gum olibanum and gum karaya in Andhra Pradesh, was requested by the Madhya Pradesh Government (India) to train gum collectors of Olibanum (Salai Guggul in Hindi). During this period, he came to know that the Convention on International Trade in Endangered

Species of Wild Fauna and Flora (CITES) wanted to include species of *Boswellia* under the CTES category. They elicited opinions from different stakeholders. As a part of that, Dr. Venugopal asked the researchers and field botanists in India (including the first author of this book) to send representation to CITES with their observations. Most of us sent representation saying that if *Boswellia* is included in the CITES list, thousands of local people will lose their livelihoods.

Hence, we thought of bringing a book on all aspects of gum olibanum/frankincense. In fact, Dr. Venugopal proposed this book. Due to his busy schedule with gum training, he could not participate in the writing of this book. However, he was kind enough to share his knowledge, experience, and photographs with us. The authors are grateful to him for his concerns for the well-being of gum-collecting tribes in India. We extend our gratitude for sharing his photographs and information.

We also thank several scientists/botanists for sharing their literature. Dr. Majeed of Sami/Sabinsa was kind enough to share his valuable publications on *Boswellia*. Dr. Lalithamba shared literature and information on the species of *Boswellia*. Dr. A. Madhusudhana Reddy shared photographs of Indian *Boswellia*, and we thank him for the same. We thank all the journal editors, publishers, and authors for permitting us to use their figures in our book, and they are acknowledged at appropriate places.

Readers are requested to bring our notice any deficiencies or mistakes in this volume so that they can be rectified in the future.

—Editors

CHAPTER 1

FRANKINCENSE (OLIBANUM): AN AROMATIC RESIN WITH ECONOMIC IMPORTANCE

MALLAPPA KUMARA SWAMY,[1] T. PULLAIAH,[2*] and ABHIJIT DEY[3]

[1]*Department of Biotechnology, East-West First Grade College, Bangalore, Karnataka, India*

[2]*Department of Botany, Sri Krishnadevaraya University, Anantapur – 515003, Andhra Pradesh, India*

[3]*Department of Life Sciences, Presidency University, Kolkata, West Bengal, India*

Corresponding author. E-mail: pullaiah.thammineni@gmail.com

ABSTRACT

Frankincense, also branded as olibanum, is a precious fragrant resin, mainly extracted from the tree species of the *Boswellia* genus in the family Burseraceae. The resins of these species have significant commercial importance as unprocessed material of incense and balm. Recent investigations show the diminishing population of *Boswellia* species partially credited to the over-utilization. The International Union for Conservation of Nature (IUCN) has listed *Boswellia* species in danger of extinction. The resins obtained from this species constitute several bioactive compounds and exhibit numerous pharmacological properties. This chapter briefly introduces about the significance and uses of *Boswellia* species and frankincense to the readers.

Frankincense – Gum Olibanum: Botany, Chemistry, Extraction, Utilization, Propagation, Biotechnology, and Conservation, T. Pullaiah, K. Venkata Ratnam, Mallappa Kumara Swamy, Lepakshi Md. Bhakshu (Eds.)

1.1 INTRODUCTION

Frankincense is a precious fragrant resin, mainly extracted from the tree species of the *Boswellia* genus in the family Burseraceae. It is also branded as olibanum. This resin is secreted when tree bark is injured, and it is an aromatic, semi-solid, yellow-brown substance. 'Frankincense' name is a derivative of the Olden French franc encens, denoting 'high-quality incense.' The word 'Franc' in Old French means 'noble, pure.' Though, it is named as 'Frankincense,' the name does not refer to the 'Franks.' It is widely used in the manufacturing of perfumery goods. This resin finds its applications worldwide in several ethnic and sacred ceremonials, not only in ancient times but also in the present day. Frankincense trade has been documented for the past 5,000 years, and it has braced pleasure and wealth in many civilizations and the countryside (https://en.wikipedia.org/wiki/Frankincense).

In Arabic, the resins produced by *Boswellia* species are known as 'loban' or 'luban,' which represents the pure silvery resin exudate of a tree, and connects it through the 'milk' (laben) of a tree. Thus, in the ancient trade, the names 'libanum' and 'olibanum' were used. The Hebrew name is 'lebona,' the Greek is 'libanos' or 'libanotos' and the Latin is 'tus.' The early Egyptian designation was 'netersent.' 'Salai guggal' is the name used for Frankincense in India. In Somalia, 'moxor' is represented for frankincense of *B. carterii*, while 'jagcaar' and 'boido' names are given for frankincense of *B. frereana* and *B. papyrifera*, respectively (Ben-Yehoshua et al., 2012; Coder, 2011; Al-Harrasi et al., 2019).

The main application of frankincense was incense (sharp lingering sweet smell). Also known as olibanum, this resin and resin-derived essential oils and alcohol extracts are widely traded internationally and are assimilated into a wide-ranging home care, healthcare, aromatherapy, diet supplementing, and cosmetic products. *B. sacra* was the main source of frankincense found in Oman, Yemen, and Somalia. *B. papyrifera* later became the main source for frankincense from Africa. Indian frankincense is obtained from *B. serrata*. *Boswellia*, belonging to the family Burseraceae, contains 24 species found in Africa and Asia (Thulin, 2020). These, including a few other species, remained in the market for centuries as frankincense. The increasing human population enjoyed it and the growing demand fell into search for another source of frankincense. The global demand and market lobby started threatening the frankincense existence. The Convention on International Trade in Endangered Species of Wild Fauna and Flora (CITES) in the year 2021 wanted to keep the trade under CITES list. This is felt as a threat to the livelihood of the relying communities and export revenue of the developing countries.

Hence there is an urgent need for the assessment of the species yielding frankincense. There is also a necessity for scientific methods of extraction and propagation of the trees. A review of the utilization is also necessary. Biotechnological applications of the species of *Boswellia* are needed for the improvement of the yield of frankincense. This book is an answer to all these.

1.2 ABOUT *BOSWELLIA* SPECIES

The genus *Boswellia*, which belongs to the Burseraceae family, was named in honor of Johann Boswell (1719–1780). More than 24 *Boswellia* species have been identified, and their distribution is observed in the regions of Burkina Faso, Mali, Oman, and Tanzania (Thulin, 2020). Most of these species are distributed in the Horn of Africa province. About 11 species are highly concentrated in the Socotra archipelago of Yemen.

There are about more than 600 species belonging to 17 genera in the family Burseraceae. These species include both shrubs and trees, often spiny. Most of them possess gum resins, latex, or essential oils that are intensely aromatic. Burseraceae family is widely spread in the tropical and sub-tropical counties, which is the foremost portions of the vegetation in dry lowland East-African provinces. The resins of these species are having a significant commercial importance as unprocessed material of incense and balm (Moussaieff and Mechoulam, 2009).

There are currently 24 recognized species. Certain species have inadequate environmental range. For example, more than 10 species are prevalent in the regions of Socotra in the Red Sea between Yemen and the Horn of Africa. Yet together, *B. serrata, B. sacra, B. papyrifera, B. dalzielii, B. neglecta*, and *B. rivae* are found to occur in 17 nations, from India, through the Arabian Peninsula, the greater Horn of Africa to West Africa, with some individual species ranges spanning more than 2,000 kms. *Boswellia* species have a habit of growing in dry to semi-dry environments, supported either by orographic moisture (*B. sacra* and *B. frereana*) and/or seasonal rain (*B. serrata, B. papyrifera, B. dalzielii, B. neglecta,* and *B. rivae*).

The major *Boswellia* species in trade include *B. sacra* occurring in South Arabia and *B. frereana* in the Horn of Africa, found in Nubia (an ancient region in northeastern Africa), *B. serrata* (Indian frankincense), found in India, *B. papyrifera* occurring in Ethiopia, Eritrea, and Sudan. *Boswellia* Roxb. ex Colber genus mainly occurs in North-East Africa, where nearly 75% of the species are endemic. The natural habitat of *Boswellia* species includes both dry lands (such as Ethiopia, Sudan, Somalia, Oman, and Yemen) and humid habitations (represented by South-Asian countries, including India). In Oman, the chemical composition of resins from *B. sacra* shows variations, and is attributed to the topographical locality (hilly and coastal zones) of trees. This difference reflects the commercial scores of the frankincense extracted from this tree (Al-Harrasi et al., 2019).

Recent investigations show the diminishing population of *Boswellia* species, partially attributed to the over-exploitation. The seeds produced by heavily taped *Boswellia* trees show only 16% of the germination rate, while more than 80% germination rate can be observed from seeds of trees that had not been tapped. Further, grazing, forest fire, and natural attacks by the longhorn beetle have reduced *Boswellia* species population. Another threat is the clearing of frankincense woodlands to commercial agriculture (Dejenea et al., 2013; https://en.wikipedia.org/wiki/Frankincense).

In 1998, the International Union for Conservation of Nature (IUCN) has cautioned about the possibility of 'near threat' status to *B. sacra,* which is the major frankincense species sources. CITES do not cover *Boswellia* species under the protection; however, authorities claim that these species meet the measures for protection. A study in 2006 by an ecologist belonging to Wageningen University has claimed that, *B. papyrifera* species in Eritrea were seen to be very rare by the late-1990s. Further, another study in 2019 had anticipated nearly 50% decline in *B. papyrifera* population within the next 20 years. *B. papyrifera,* occurring majorly in Ethiopia, Sudan, and Eritrea is the source for more than two-thirds of the overall production of frankincense. All *Boswellia* species are under threat due to habitat loss and overexploitation. Harvesting and marketing the *Boswellia* tree's resin is one of the chief sources of revenue for the dwellers, thus, causing overtapping (https://en.wikipedia.org/wiki/Frankincense).

1.3 USES

The use of frankincense for spiritual and therapeutic determinations dates back to early civilizations. A wide-ranging bioactive compounds, belonging to diverse chemical classes are recognized in this fragrant resin, and their pharmacological activities are attributed to the harmonizing properties exerted by these phytochemicals. The major phytochemicals isolated are sabinene, α-thujene, p-cymene, α-gurjunene, α-cubenene, δ-elemene, α-ylangene, α-copaene, β-caryophyllene, isocaryophyllene, trans-β-farnesene, α-humulene, aromadendrene, α-amorphene, germacrene d, α-elemene, cembrene γ-cadinene, α-cadinol, δ-cadinene, elemol, α-eudesmol, guaiol, etc. The phytochemical profile of the frankincense differs considerably amongst different *Boswellia* species. Monoterpenes are rich in some species, while in others, diterpenes and triterpenes are found. The aromatic essential oils of *Boswellia* species are reported to possess about 300 chemical compounds (Mertens et al., 2009). Likewise, the resin is reported to contain nearly 100 terpenes (Al-Harrasi et al., 2018). Further, the plant origin, topographical place, harvesting time and other ecological aspects give together to the complete chemo-composition, and thus the pharmacological activities of frankincense.

The clinical investigations have proven that frankincense resin is effective against several disease conditions, including rheumatoid arthritis, osteoarthritis, asthma, irritable bowel diseases, inflammatory bowel diseases, diabetes mellitus and relapsing-remitting multiple sclerosis (Kavitha et al., 2007; Ernst, 2008; Hamidpour et al., 2013; Stürner et al., 2018). The essential oil extracted from this resin shows antimicrobial activities, and neuroprotective activity (Ameen et al., 2017; Rajabian et al., 2020). Extensive investigations on frankincense and its effect on the human physiology could be the right way to discover a probable novel alternative natural medicine for controlling, curing, or preventing several kinds of health issues, including neurological problems like Parkinson's and Alzheimer's diseases (ADs).

These multipurpose species of *Boswellia* provide ecological and economic benefits. *Boswellia* species raise well in regions that are inappropriate for customary farming, and help in stabilizing soil and give shadow and safeguard (Hassan, 2012). They are resilient to drought and remain growing in the bordering zones, even under threatening and erratic environments (Abdoul-Latif et al., 2012). Also, they promote other tree species growth, and appear to repel fire better in comparison to any other allied species (Orwa et al., 2009). *Megaderma frons* and *Cuculus canorus* are the major connected animals (Ackroyd and Harvey, 2016). Camels are seen to eat their fruits. Also, the succulent undeveloped stems, fruitlets, and seeds are given to reared animals. Flowers serve as a source for intrinsic honeybees. *Boswellia* species are also cherished as an income source by people active in bordering territories where other plants cannot grow (Judd et al., 2008; Hassan, 2012; Bantihun and Tesema, 2018).

In addition, it appears that nurseries in the United States are considering to grow *Boswellia* in a large scale for landscape gardeners. *B. sacra* is being grown as bonsai and for xeriscaping (Todd's Tropicals, 2017; https://www.youtube.com/watch?v=li5i7hvJnE4), though its seeds are apparently rare to obtain and, by some reasons, hard to grow (Tortorello, 2011; Enlightenment Garden, 2016; https://www.youtube.com/watch?v=_p0lqauHrZQ).

1.4 EXTRACTION AND COMMERCIAL IMPORTANCE

A careful incision into the barks of *Boswellia* species, i.e., when the external bark piece is skinned out results in the oozing of the milk-like constituent that solidifies with the exposure to air results in frankincense. As a defensive mechanism, frankincense is secreted by the trees' bark, which is likely to protect trees by healing the wound and keeping away from the insect attack. Also, this exudate may probably help in reducing loss of water from the plant. In general, the highest quality frankincense is extracted from the second incision onwards. However, inappropriate incisions, i.e., wider and deeper or excessive incisions may cause trees to be more prone to insects' attack and in due course, could destroy the trees.

Though, the modern medicine has resulted in the reduced pharmacological application of frankincense as a medicine or drug, its societal and religious importance have been continued till date. The mid-20th century has witnessed the vanishing of the word 'Olibanum' from pharmacopeias. During the end of the last century, several advancements in science have taken place with the inclusion of pre-clinical and clinical investigations. With these scientific expansions, frankincense has re-claimed its significance. Several investigations have proven the capabilities of frankincense in treating different diseases, such as ulcerative colitis, rheumatoid arthritis, bronchial asthma, multiple sclerosis, etc. Further, *Boswellia* species solvent extracts or their isolated/purified phytocompounds, particularly diterpenes and triterpenes have proven with greater pharmacological activities, especially anticancer and anti-inflammatory activities. Also, studies have shown that frankincense constitute a rich content of boswellic acids (BAs), which are mainly responsible for different biological properties. Also, incensole and incensole acetates (cembranoids), display remarkable anti-inflammatory and anti-depression properties by activating ion channels in the brain, and thus alleviating depression or anxiety. Frankincense oil is also reported in the usage for migraine relief. This natural oleoresin products are perfect to apply externally on the skin. They are used as an additive in natural perfume and cosmetic products. It reduces swelling, and is a very strong antioxidant. Numerous cosmetic frankincense-based products have appeared in the market. Also, frankincense-based fragrances have been developed for different applications. Frankincense-based products, including its essential oil, predominantly exhibit relaxing, stabilizing, and harmonizing effects, and thus, used in aromatherapy (Al-Harrasi et al., 2019).

The Global Frankincense Alliance (GFA) platform was formed by concerned personalities to address the mounting apprehension on the *Boswellia* species. At the same time, an informal working group on *Boswellia* within the CITES initiated the process to obtain correct facts on the status, distribution, and condition of the trees, and also the impact of these factors influencing the worldwide business. GFA has conducted an online workshop during March 15th, 17th, 18th, 2021 to take stock of the situation of *Boswellia* trees. The increased demand for oleoresin is causing increased worry about the enduring sustainability of the *Boswellia* species. Moreover, forest fire, animal grazing and regular pathogen attacks have significantly reduced the population of these species. The challenges may differ between species and between locations within the range where a species occurs. In some regions and for a few species, a limited factual data is available. A few reports have reported the widespread well-growing *B. neglecta* and *B. rivae* species in the Somali region (Ethiopia) and North parts of Kenya. Also, the naturally regenerating species of some species of *Boswellia* are reported in Sudan, Oman, and Ethiopia. At the same time, there are well-documented reports of declining populations and a dearth of restoration in Eritrea,

Ethiopia, and Sudan. Also, a declining population of *B. serrata* is seen in India. Because of coronavirus disease (COVID-19) pandemic situations, the 2nd International Conference on Frankincense and Medicinal Plants (Oman) was deferred up until 2022. GFA conducted an open virtual workshop, i.e., 'The Future of Frankincense' focused on gathering personalities interested in *Boswellia* species, to manage present-day investigation requisites, in addition to crafting roadmaps for the imperative and crucial accomplishments. More than 300 individuals had filled in the pre-workshop survey and between 96 and 106 people attended each day (GFA, 2021).

There is very little documented information on the communities and harvesters themselves. According to one study, over a quarter of a million people are estimated to be dependent on frankincense for over 50% of their income in the East Golis/Saanag region of Somaliland alone.

1.5 CONCLUSION

During the last few years, the ecosphere of frankincense has experienced dramatic and profound changes. There has been a worldwide boom in its demand, and thus greatly stressed on the Frankincense supply chain. Further, the organization and layout of the marketplace have experienced major transformations. Frankincense has been used in both customary and contemporary natural medication for treating a wide variety of health problems with minimal or devoid side effects. Collecting precise data on the present status, mortality, regeneration rates, and health of the diverse *Boswellia* species and population has constantly arisen as the chief investigation priority. In view of these aspects, this book is written as a monograph on *Boswellia* species, i.e., frankincense and its products. It covers comprehensive information on various aspects of *Boswellia* species, including habitat, taxonomy, morphology, wood anatomy, distribution, phytochemical aspects, pharmacological properties, customary uses, propagation, genetic diversity studies, biotechnological applications, conservation, trade and commerce, and other gray areas of research. Appropriate references are given under each chapter. This book will serve as a reference and field guide to scientists, botanists, traders, and ecologists.

KEYWORDS

- **aroma**
- ***Boswellia***
- **frankincense**
- **habitat**
- **olibanum**
- **perfume**
- **volatile oil**

REFERENCES

Abdoul-Latif, F. M., Obame, L. C., Bassole, I. H. N., & Dicko, M. H., (2012). Antimicrobial activities of essential oil and methanol extract of *Boswellia sacra* Flueck. and *Boswellia papyrifera* (Del.) Hochst from Djibouti. *Intern. J. Management Modern Sci. Technologies, 1*(1), 1–10.

Ackroyd, H., & Harvey, D., (2016). *Conflicted Seeds + Spirit.* Exhibition of new works by Ackroyd & Harvey at the University of Cambridge. https://www.conflictedseeds.com (accessed on 10 October 2022).

Al-Harrasi, A., Khan, A. L., Asaf, S., & Al-Rawahi, A., (2019). Frankincense and human civilization: A historical review. In: *Biology of Genus Boswellia* (pp. 1–9). Springer, Cham, USA.

Al-Harrasi, A., Rehman, N. U., Khan, A. L., Al-Broumi, M., Al-Amri, I., Hussain, J., & Csuk, R., (2018). Chemical, molecular and structural studies of *Boswellia* species: β-boswellic aldehyde and 3-epi-11β-dihydroxy BA as precursors in biosynthesis of boswellic acids. *PLoS One, 13*(6), e0198666. https://doi.org/10.1371/journal.pone.0198666.

Ameen, A. M., Elkazaz, A. Y., Mohammad, H. M., & Barakat, B. M., (2017). Anti-inflammatory and neuroprotective activity of boswellic acids in rotenone parkinsonian rats. *Can. J. Physiol. Pharmacol., 95(7), 819–829.* doi: 10.1139/cjpp-2016-0158.

Bantihum, A., & Tesema, T., (2018). Regeneration status of *Acacia polyacantha* and *Boswellia papyrifera* species in Shimelegir Forest, Jawi District, Ethiopia. *J. Nat. Sci. Res., 8*(14), 27–36.

Ben-Yehoshua, S., Borowitz, C., & Ondrej, H. L., (2012). Frankincense, myrrh, and balm of Gilead: Ancient spices of Southern Arabia and Judea. *Hort. Rev., 39*(1), 3–66.

Coder, K. D., (2011). *Frankincense & Myrrh: A Gift of Tree History.* University of Georgia, Georgia, USA.

Dejenea, T., Lemenih, M., & Bongers, F., (2013). Manage or convert *Boswellia* woodlands? Can frankincense production payoff? *J. Arid Environ., 89*, 77–83.

Ernst, E., (2008). Frankincense: Systematic review. *BMJ, 337*, a2813. doi: https://doi.org/10.1136/bmj.a2813.

Global Frankincense Alliance, (2021). https://www.penhanetwork.org/wp-content/uploads/2021/05/GFA-FOF-Main-outcomes-and-Road-Map-LOW-11-pages.pdf (accessed on 10 October 2022).

Hamidpour, R., Hamidpour, S., Hamidpour, M., & Shahlari, M., (2013). Frankincense (乳香 Rǔ Xiāng; *Boswellia* species): From the selection of traditional applications to the novel phytotherapy for the prevention and treatment of serious diseases. *J. Trad. Compl. Med., 3*(4), 221–226.

Hassan, A. M. M., (2012). Distribution *Boswellia sacra* in Dhofar mountains, sultanate of Oman: Economic value and environmental role. *J. Life Sci., 6*, 632–636.

Judd, W. S., Campbell, C. S., Kellogg, E. A., Stevens, P. F., & Donoghue, M. J., (2008). *Plant Systematics: A Phylogenetic Approach* (3rd edn.). Sinauer Associates, Inc., Sunderland, Massachusetts, USA.

Kavitha, J. V., Rossario, J. F., Chandran, J., Anbu, P., & Bakkiyanathan, (2007). Hypoglycemic and other related effects of *Boswellia glabra* in alloxan-induced diabetic rats. *Indian J. Physiol. Pharmacol., 51*, 29–39.

Mertens, M., Buettner, A., & Kirchhoff, E., (2009). The volatile constituents of frankincense: A review. *Flav. Frag. J., 24*(6), 279–300.

Moussaieff, A., & Mechoulam, R., (2009). *Boswellia* resin: From religious ceremonies to medical uses; a review of *in-vitro, in-vivo* and clinical trials. *J. Pharm. Pharmacol., 61*(10), 1281–1293.

Orwa, C., Jamnadass, R. H., Kindt, R., Mutua, A., & Simons, A., (2009). *Boswellia serrata*. In: *Agroforestry Database: A Tree Reference and Selection Guide, Version 4.0*. World Agroforestry Centre, Nairobi, Kenya. https://www.cabdirect.org/cabdirect/abstract/20177200441 (accessed on 10 October 2022).

Rajabian, A., Sadeghnia, H., Fanoudi, S., & Hosseini, A., (2020). Genus Boswellia as a new candidate for neurodegenerative disorders. *Iranian J. Basic Med. Sci., 23(3), 277–286.* doi: 10.22038/ijbms.2020.35288.8419.

Stürner, K. H., Stellmann, J. P., Dörr, J., Paul, F., Friede, T., Schammler, S., Reinhardt, S., et al., (2018). A standardized frankincense extract reduces disease activity in relapsing-remitting multiple sclerosis (the SABA phase IIa trial). *J. Neurology, Neurosurgery & Psychiatry, 89*(4), 330–338.

Thulin, M., (2020). The genus *Boswellia* (Burseraceae). The frankincense trees. *Botanic Symbols of Upsalien* (Vol. 39, pp. 1–149). Works from the Botanical Department in Uppsala.

Tortorello, M., (2011). *Turns Out You Can Grow Frankincense in the U.S. 2011.* The Bulletin-Home/News. https://www.bendbulletin.com/news/1425262-151/turns-out-you-can-grow-frankincense-in-the (accessed on 10 October 2022).

CHAPTER 2

BOTANY OF *BOSWELLIA*

T. PULLAIAH

Department of Botany, Sri Krishnadevaraya University, Anantapur – 515003, Andhra Pradesh, India, E-mail: pullaiah.thammineni@gmail.com

ABSTRACT

Boswellia is a genus of about 24 species distributed in tropical Africa, southern Arabia, and India. In this chapter, the species of the *Boswellia* genus are enumerated systematically. The systematic enumeration includes citation of the species, brief description, distribution, flowering and fruiting period, and IUCN status. Local names, trade names, and English names are also given.

2.1 INTRODUCTION

Burseraceae is a family of about 18 genera and 540 species. The genus *Boswellia* Roxb. in Colebrooke (1807, p. 379) belongs to this Burseraceae family and comprises of about 24 species distributed in tropical Africa, southern Arabia, and India (Mabberley, 2017). In Tropical Africa, it occurs in Eritrea, Ethiopia, Kenya, Somalia, Sudan, and Tanzania. In southern Arabia, it occurs in Yemen, Oman, and Socotra. Around 40 scientific plant names of species ranked for the genus *Boswellia* are given in 'The Plant List,' of which 30 are accepted names of species. Thulin (2020), who revised the genus, recognized 24 species in the genus *Boswellia*. Around 11 species of *Boswellia* are endemic to Socotra Island of Yemen.

Frankincense – Gum Olibanum: Botany, Chemistry, Extraction, Utilization, Propagation, Biotechnology, and Conservation, T. Pullaiah, K. Venkata Ratnam, Mallappa Kumara Swamy, Lepakshi Md. Bhakshu (Eds.)

There are three centers of distribution of *Boswellia* species. The first center of distribution is the African-Arabian region, where the following species occur. *B. dalzielii, B. frereana, B. globosa, B. microphylla, B. neglecta, B. occulta, B. ogadensis, B. papyrifera, B. pirottae, B. rivae*, and *B. sacra*. The second center of distribution is Socotra Island in Yemen, where the following species are endemic to this Island: *B. ameero, B. asplenifolia, B. bullata, B. dioscoridis, B. elongata, B. "hesperia," B. nana, B. papoviana, B. samahensis, B. scopulorum*, and *B. socotrana*. The third center of distribution is India, where two species – *B. ovalifoliolata* and *B. serrata* occur.

Gum olibanum is being extracted from different species of *Boswellia*. Since there is a controversy regarding the accepted names, there is a necessity for taking stock of the taxonomy of *Boswellia* species. This chapter is an attempt to clear this riddle.

2.2 TAXONOMY

The name *Boswellia carteri* Birdwood appears in many papers on Pharmacology and Phytochemistry. It is a synonym of *Boswellia sacra* Flück. *Boswellia elegans* Engl., *B. hildebrandtii* Engl., *B. holstii* Engl. and *B. multifoliolata* Engl. are treated as valid names in 'The Plant List' but they are treated as synonyms of *Boswellia neglecta* S. Moore, by Thulin (2020) also in this book. *Boswellia boranensis* Engl. is treated here and by Thulin (2020) as a synonym of *Boswellia rivae* Engl. *Boswellia chariensis* Guillaumin, *B. occidentalis* Engl. and *B. odorata* Hutch. treated as valid names in The Plant List are synonyms of *Boswellia papyrifera* (Caill. ex Delile) Hochst. *Boswellia bricchettii* (Chiov.) Chiov. treated as a valid name in The Plant List is a synonym of *Lannea obovata* (Oliv.) Engl. (https://www.gbif.org/species/5636159). *Boswellia madagascariensis* Capuron treated as a valid name in The Plant List is a synonym of *Ambilobea madagascariensis* (Capuron) Thulin, Beier, and Razafirn. (https://www.gbif.org/species/5636128). *Boswellia ruspoliana* Engl. is a synonym of *Boswellia rivae* Engl. (https://powo.science.kew.org/taxon/127064-1).

In the following pages, the species of the genus *Boswellia* are enumerated systematically. The systematic enumeration includes citation of the species, brief description, distribution, flowering and fruiting period, and IUCN status.

2.2.1 *BOSWELLIA* ROXB. EX COLEBR.

- *Boswellia ameero* Balf. f., Proc. Roy. Soc. Edinburgh 11:505, 1882; Thulin, Symb. Bot. Upsal. 39:88, 2020.

It is a large tree, 5–8 m high, trunk gray, with flaking yellowish to brown color bark. Leaves densely crowded at shoot-apices, imparipinnate compound, leaflets 11–17 in number, 14–25 cm long, surface with tomentose to sparsely pubescent texture; leaflets 1.5–10 × 1–7 cm, broadly oblong to ovate, margin variable crenate to serrate, apex obtuse or rounded. Racemes short stemmed with dense with bright pink flowers. Calyx cup-shaped, petals 7, pink or reddish, glabrous; stamens 10. Capsules large, 4–5-locular, seeds–5 (Figures 2.1 and 2.2).

Common in dry and semideciduous woodlands on limestone or granite at altitudes from 120 m to 1,050 m above m.s.l. Socotra Island, endemic. Fl.& Fr.: February–April.

IUCN Category: Vulnerable (Miller, 2004a; IUCN, 2011).

FIGURE 2.1 *Boswellia ameroo* (Image 1).

Source: Image courtesy: Joël Lodé, with permission.

FIGURE 2.2 *Boswellia ameroo* (Image 2).

Source: Image courtesy: Joël Lodé, with permission.

- *Boswellia asplenifolia* (Balf. f.) Thulin, Symb. Bot. Upsal. 39: 128. 2020. *Odina asplenifolia* Balf. f., Proc. Roy. Soc. Edinburgh 11: 508. 1882. *Lannea asplenifolia* (Balf. f.) Engl., in Engl. Jahrb. 24: 499. 1898.

  Deciduous tree, to 8 m high; bark gray or reddish-brown, smooth to rough, sometimes flaking. Young branchlets reddish brown or grayish. Leaves densely crowded at shoot apices, imparipinnate, 7–23-foliolate; leaflets opposite or subopposite, 0.15–1 × 0.15–0.5 cm, lanceolate or elliptic-oblong, glabrous or sometimes sparsely puberulous. Flowers cream or pale yellow in racemes or panicles; bract and bracteoles 0.5–1 mm long; calyx cup-shaped; petals elliptic; stamens 10. Fruits sub-globose or ovoid, 3-locular, reddish brown when ripe.

  Coastal plains and rocky slopes on limestone rocks in Socotra. Fl.: December–January.

IUCN category: Endangered C1 (Thulin, 2020).

- *Boswellia bullata* Thulin, Kew Bull. 56: 983. 2001 and Symb. Bot. Upsal. 39: 98. 2020.

It is a small tree with 4 m height; bark pale brown to light gray with outer papery flaking layers. Leaves densely crowded at the shoot apex, pinnately compound, large, 7–18 cm long; leaflets 11–19 in number, 1–4 × 0.5–1.5 cm, white-felted beneath; leaflets densely hirsute on upper surface, margins irregularly serrate–crenate. Inflorescence panicle with many creamish-red flowers; bracts and bracteoles ovate to lanceolate, calyx cup-shaped, stamens 10. Fruit 4–5-celled capsule, pyriform, glabrous (Figure 2.3).

Cliff-dwelling tree. Endemic to Socotra Island in Yemen. Fl. and Fr.: February–April.

IUCN Category: Vulnerable (Miller, 2004b).

FIGURE 2.3 *Boswellia bullata.*

Source: Image courtesy: Joël Lodé, with permission.

- *Boswellia dalzielii* Hutch., Bull. Misc. Inform. Kew 1910: 137. 1910; Thulin, Symb. Bot. Upsal. 39: 42. 2020.

Deciduous tree, to 13 m high, with slender, ascending branches. Trunk gray or yellow, with exfoliation. Bark papery, pale, peeling, and ragged. Leaves 20–45 cm long, imparipinnate, 11–21-foliolate, leaflets 5–12.5 × 1–3 cm, oblong-lanceolate, glabrous. Flowers white, fragrant, in many-flowered racemes. Fruit 3-locular.

Locally abundant in wooded Sudano-Sahelian Savanah, Benin, Burkina Faso, Northern Ivory Coast to Nigeria and into Cameroon. It is known as the frankincense tree, and is found mainly in the Savannah region of West Africa; locally abundant in northern Nigeria, Cameroun, and Ubangi-Shari (Murray, 1995). Fl.: March.

Locally, the plant has different vernacular names, i.e., the Bini called it Eban-Arukhoe, Fulani called it Januhi, while the Hausa called it Ararrabi, Basamu, and Hanu.

IUCN category: Least concern (Thulin, 2020).

- *Boswellia dioscoridis* Thulin, Kew Bull. 56: 984. 2001 & Symb. Bot. Upsal. 39: 120. 2020.

Medium-sized tree up to 7 m tall, bark flaking, yellowish brown. Leaves compound, large, crowded at the apex of secondary branches, glabrous, leaflets 11–21, oval shape, leaflets finely pubescent along the veins on the upper surface, rather deeply or irregularly serrate or crenate margins. Panicles or racemes with many cream-colored glabrous flowers, petals pink, and glabrous outside. Fruits 3–5-celled, 7–11 × 5–9 mm, narrow at both ends (Figure 2.4).

On limestone cliffs, in dry semideciduous forests. Endemic to southern and Eastern Socotra Island in Yemen. Fl. & Fr.: February–April.

IUCN Category: Vulnerable (Miller, 2004c).

FIGURE 2.4 *Boswellia dioscoridis*.

Source: Image courtesy: Joël Lodé, with permission.

- *Boswellia elongata* Balf. f., Proc. Roy. Soc. Edinburgh 11:505. 1882; Thulin, Symb. Bot. Upsal. 39: 94. 2020.

Medium-sized tree, to 8 m height, trunk 0.8–1 m in diameter; bark flaking, yellowish, pale brown, or bronze. Leaves pinnate, large, 15–30 cm long, crowded at shoot-apices, glabrous, leaflets 11–21, oval shape, leaflets 3–10 × 0.7–3.5 cm, finely pubescent along the veins on upper surface, rather deeply or irregularly serrate or crenate margins. Flowers pale cream color, in racemes or panicles. Fruit capsule, seeds 5, rarely 4 (Figure 2.5).

Dry forests and dry shrubland of subtropical and tropical regions in Socotra Island in Yemen. Endemic. Fl. & Fr.: February–April.

IUCN Category: Vulnerable (Miller, 2004d).

FIGURE 2.5 *Boswellia elongata.*

Source: Image courtesy: Joël Lodé, with permission.

➢ *Boswellia frereana* Birdw., Trans. Linn. Soc. London 27: 146. 1870; Thulin, Symb. Bot. Upsal. 39: 66. 2020.

Medium-sized tree up to 8 m high, resin copious, milky, yellow after drying. Leaves compound, oblanceolate in outline, 10–30 cm long, leaflets 9–15-foliolate. Flowers in dense thyrses, 10–30 cm long, pedicels 1–3(–4) mm long; bracts 1.5–6 mm long. Fruit (5–)6(–8)-celled, 5.5–9 × 3–7 mm, pear-shaped, glabrous; stones trigonous, often slightly winged. The resin is copious, milky, and yellowish.

Endemic Northern Somalia, also cultivated in Yemen (Figure 2.15). This species often grows on steep vertical slopes to a height of 750 m above sea level and frankincense obtained is known as *jagcaar* and *meydi* resin in Somalia (Svoboda et al., 2001; Al-Harrasi et al., 2019). Fl. & Fr.: November–December. Locally known as Dhidin, Maydi, and Yigaar tree. It is known as Elemi frankincense tree and King of Frankincense.

Chromosome number: 2n = 22 (Thulin & Warfa, 1987).

IUCN Category: Vulnerable c1 (Thulin, 2020).

➢ *Boswellia globosa* Thulin, Nordic J. Bot. 24: 273. 2006 publ. 2007 & Symb. Bot. Upsal. 39: 84. 2020.

Small tree, 2.5–4 m tall. Bark purplish black and smooth. Young shoots are grayish to purplish brown, longitudinally ridged. Leaves pinnately compound, 11–21-foliolate, 2.5–6 cm long, rachis densely pubescent with spreading glandular hairs; leaflets subsessile, opposite, or sub-opposite, 3–6.5 × 1.5–4 mm, elliptic to obovate, apex obtuse, base cuneate to rounded, asymmetric, margin entire, densely pubescent with spreading hairs on both surfaces. Flowers in few-flowered racemes; bracts triangular, fruit sub-globose, winged, and glabrous.

It is endemic to northern Somalia. Among rocky limestone boulders in semidesert bushland.

IUCN Category: Vulnerable D1 (Thulin, 2020).

➢ *Boswellia "hesperia"* Thulin, Symb. Bot. Upsal. 39: 118. 2020.

Dwarf tree. Leaves simple, crowded at shoot apices.

Rare in Socotra on limestone rocks.

IUCN Category: Critically Endangered (Thulin, 2020).

Note: Thulin (2020) provisionally proposed this species but not yet confirmed.

➢ *Boswellia microphylla* Chiov., Ann. Bot. (Rome) 13: 404. 1915; Thulin, Symb. Bot. Upsal. 39: 74. 2020.

Small tree, up to 6 m high; bark dark gray to bluish, not peeling. Leaves 0.5–4.5 cm long, imparipinnate, 5–17-foliolate, leaflets 0.2–1.3 × 0.1–0.8 cm, elliptic to obovate, glabrous to sparsely pubescent. Flowers white or cream-colored in racemes; calyx-cup-shaped, petals 2–3 mm long. Fruit rounded, truncate at apex, 3-locular.

On gravely soil, with limestone at 400–1,300 m a.s.l. in Ethiopia, Somalia, and Kenya. Fl. & Fr.: December–January. Vern. Names: Beeyo-Madow, Bei Bei, Chewder, Mirafur, Moguli, Muqlo.

IUCN Category: Least concern (Thulin, 2020).

➤ *Boswellia nana* Hepper, Hooker's Ic. PI., 37, t.3676. 1971; Thulin, Symb. Bot. Upsal. 39: 114. 2020.

It is a prostrate or creeping shrub; bark flaking with yellowish-red color. Bark covered with crustose lichens. Leaves simple, less than 4 cm long, stalk shorter than 3 cm, obovate, leaf margin serrates but never pinnate. Flowers crowded in unbranched inflorescence, petals pink or reddish, glabrous outside. Fruits 3–4-locular, purplish brown (Figure 2.6).

Growing on limestone cliffs, and areas of flat limestone pavement at an altitude of 300–550 m a.s.l. Endemic to Northeast part of Socotra Island in Yemen. Fl. & Fr.: February–April.

IUCN Category: Vulnerable (Miller, 2004e); Critically Endangered D (Thulin, 2020).

***Note*:** Thulin and Al-Gifri (1998) regarded *B. nana* as a reduced form of *B. popoviana.* However, it differs from that species in that it has smaller, more reduced panicles of red, not yellowish flowers, and the undersurface of its leaves are glabrous and not white-felted (Al-Haraasi et al., 2017).

FIGURE 2.6 *Boswellia nana.*

Source: Image courtesy: Joël Lodé, with permission.

- *Boswellia neglecta* S. Moore, J. Bot. 15: 67. 1877; Thulin, Symb. Bot. Upsal. 39: 78. 2020. *B. hildebrandtii* Engl., Bot. Jahrb. Syst., 15: 98. 1893. *B. multifoliolata* Engl., Ann. Bot. (Rome) 7: 16. 1907. *B. holstii* Engl., Veg. Erde 9 (III 1): 786. 1915. *B. elegans* Engl., Bot. Jahrb. Syst. 34: 314. 1904.

  Deciduous tree, up to 8 m high. Leaves imparipinnate, leaflets 5–45. Flowers white or greenish-white, in panicles up to 6(–10) cm long; pedicels 1–8 mm long. Calyx c. 1.5 mm long; petals ovate, c. 2–3 mm long, glabrous. Fruit 3(–4)-celled, pear-shaped to sub-globose, sometimes winged, glabrous.

  B. neglecta S. Moore (*B. hildebrandtii* Engl. *B. multifoliolata* Engl.), distributed in Ethiopia (Bale, Gamo Gofa, Hararghe, and Sidamo) and neighboring countries such as Kenya, Somalia, Tanzania, and Uganda (Vollesen, 1989) (Figure 2.15). The aromatic gum resins widely used as incense known as "Dakara" (Oromifa), "Borena Etan" (Amharic) or "Borena type olibanum" (Mokria et al., 2017). Fl. & Fr.: November–December.

  Note: In the Plant List *Boswellia hildebrandtii* Engl., *B. multifoliolata* Engl., *B. elegans* and *B. holstii* are treated as accepted names while in most floras they are treated as synonyms of *Boswellia neglecta.*

  IUCN category: Least concern (Thulin, 2020).

- *Boswellia occulta* Thulin, DeCarlo, & S. P. Johnson, Phytotaxa, 394(3):219, 2019; Thulin, Symb. Bot. Upsal., 39: 62, 2020.

  Small tree, up to 5 m tall, base of trunk swollen; bark flaking; resin copious. Leaves simple, elliptic-oblong, petiole 5–20 mm long. Racemes 10–20 flowered; petals 5, white, stamens 10, inserted at the base of and between the lobes of the disk. Fruits 4–5-loculed, 6–8 × 4.0–4.5 mm, pyriform, glabrous.

  Ceel Afweyn District of Somaliland (north-western Somalia) (Thulin et al., 2019). Fl. & Fr.: October–December.

  IUCN category: Endangered B2a (Thulin, 2020).

- *Boswellia ogadensis* Vollesen, Kew Bull. 40: 39. 1985; Thulin, Symb. Bot. Upsal. 39: 48. 2020.

  Tree, up to 4 m tall, with smooth flaking bark. Leaves imparipinnate, 5–9-foliolate, leaflets ovate-elliptic, 1–4 × 0.6–4 cm. Flowers yellowish, in panicles; calyx cup-shaped, stamens 10. Fruit pyriform, 3-locular.

  Southern Ethiopia. Stony hills on Gypsum at 280–350 m a.s.l. Fl. & Fr.: May.

  IUCN Category: Critically Endangered (Alemu et al., 2018).

- *Boswellia ovalifoliolata* N.P. Balakr. & A. N. Henry, J. Bombay Nat. Hist. Soc. 58: 546. 1961; Chitra & A. N. Henry in Hajra et al., Fl. India 4: 432. 1997; Pullaiah & Sandhya Rani, Trees Andhra Pradesh 147. 1999.

  Medium-sized trees, up to 10 m high, bark papery, ash-colored. Leaves crowded at ends of branches, 9–20 cm long, alternate, imparipinnate, exstipulate; leaflets opposite or alternate, sessile, 9–13 in number, ovate-oblong, 1.6–7.4 × 1–5.2 cm, base inequilateral and rounded, margin entire or slightly wavy, apex obtuse and retuse, glabrous, glaucous beneath, veins reddish. Flowers large, greenish in axillary panicles, panicles larger than leaves, up to 30 cm; petals and sepals completely glabrous, petals smaller, 5 × 3 mm, obovate-oblong. Drupe 3-gonous, containing three pyrenes, 1.5 × 0.8 cm, bony, winged; seeds compressed (Figures 2.7 and 2.8).

  Endemic to foothills of Seshachalam hill ranges in Chittoor, Kadapa, Kurnool, Nellore, and Prakasam districts of Andhra Pradesh, India, up to an altitude of about 600–900 m (Pullaiah et al., 2018). Vern.: Guggilam.

  Fl. & Fr.: February-April.

  IUCN Category: Vulnerable A2cd, NB1ab (I, ii, iii) (Saha et al., 2015).

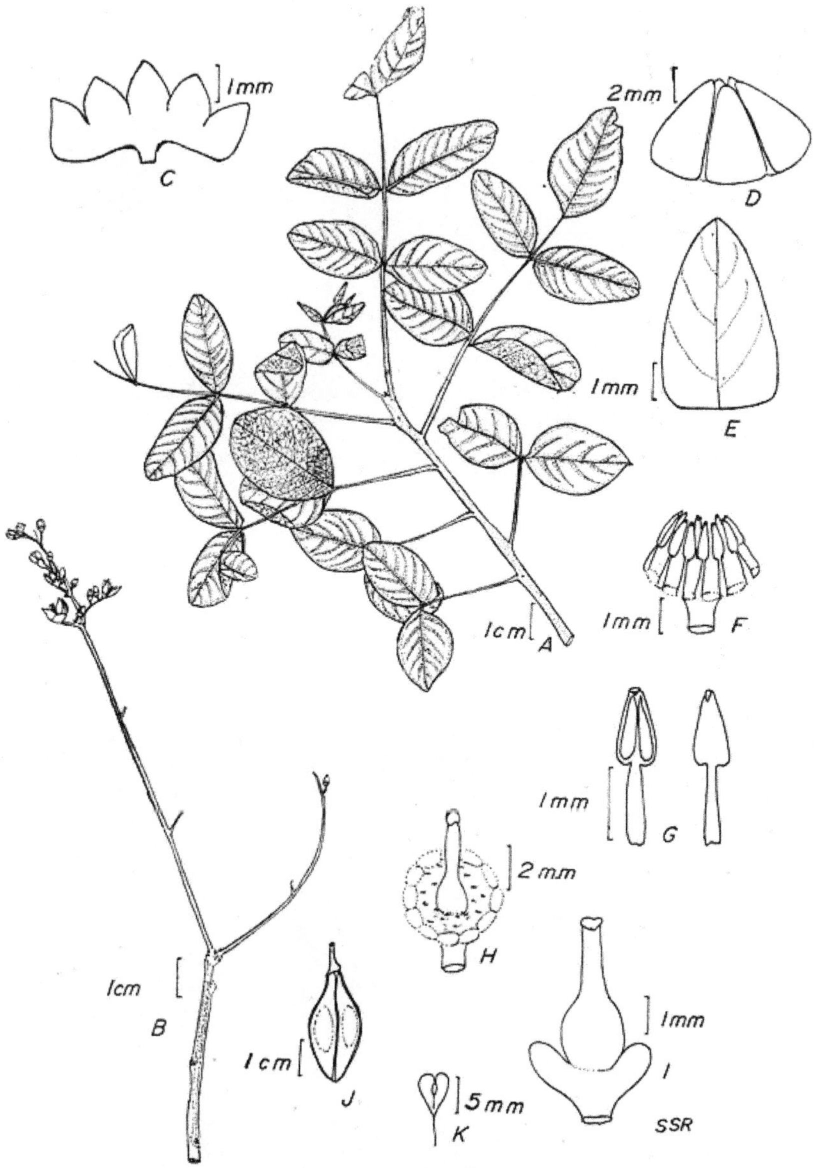

FIGURE 2.7 *Boswellia ovalifoliolata* N. P. Balakr & A. N. Henry. (A) Twig; (B) inflorescence branch; (C) calyx; (D) corolla; (E) petal; (F) androecium; (G) stamens; (H) pistil with disc; (I) pistil; (J) capsule; and (K) seed.

FIGURE 2.8 *Boswellia ovalifoliolata.* (A) Habit; (B) bark; (C) leaves; (D) inflorescence; and (E) fruits.

Source: Images by Dr. A. Madhusudana Reddy, with permission.

- *Boswellia papyrifera* (Caill. ex Delile) Hochst., Flora 26: 81. 1843; Thulin, Symb. Bot. Upsal. 39: 36. 2020. *Amyris papyrifera* Caill. ex Delile, *Voy. Méroé 4 (prepr.): 99 1826. B. odorata* Hutch., Bull. Misc. Inform. Kew 1910:138, 1910. *B. chariensis* Guillaumin, Bull. Soc. Bot. France 55(8): 33, 1908. *B. occidentalis* Engl., Bot. Jahrb. Syst. 46. 292, 1911.

  Deciduous tree up to 12 m tall, bark smooth, whitish to pale brown, peeling in flakes, young branches pubescent. Leaves pinnately compound, large, 20–40 cm long, arranged on long stalks, leaflets 11–29, subsessile, narrowly ovate-oblong, margin waved or toothed. Inflorescence panicle with sweet-scented flowers, petals 5, white to pink, stamens 10. Fruit capsule, red in color with 2 cm long, seeds three tapered.

  Stony hill sides; in highland tropical dry forests at 1,200–1,500 m a.s.l. in Ethiopia, Nigeria, Cameroon, Central African Republic, Chad, Sudan, Uganda, and Eritrea (Figure 2.15). Fl. & Fr.: October to February. Common name: Elephant tree, Sudanese Frankincense. Frankincense obtained from *B. papyrifera* is known as *boido* in Somalia (Al-Harrasi et al., 2019).

  IUCN Category: Vulnerable A2cd + 3 cd (Thulin, 2020).

  ***Note*:** *B. papyrifera* shows a lot of phenological variations because of environmental conditions, like rainfall (Lemenih and Kassa, 2011).

- *Boswellia pirottae* Chiov., Ann. Bot. (Rome) 9: 53. 1911; Thulin, Symb. Bot. Upsal. 39: 46. 2020.

  Deciduous tree, up to 10 m tall with gray or black bark, not peeling. Leaves imparipinnate, 15–27-foliolate, leaflets 3–8 × 0.8–2.5 cm, oblong-lanceolate, glaucous. Flowers white, in many-flowered panicles, calyx cup-shaped, stamens 10. Fruit 3-locular, seeds ovate, not winged.

  Native to NW and W Central Ethiopia. Endemic to Ethiopia. Vern. Names: Tantaani (Oromo), Yeitan Zaf (Amharic).

  Uses: The gum resin is collected and locally marketed and used in church ceremonies (Thulin, 2020).

IUCN Category: Vulnerable C1 (Awas et al., 2018). Fewer than 10,000 individuals are seen, with numbers declining. It is also threatened by the building of a dam.

- *Boswellia popoviana* Hepper, Hooker's Ic. Pl. 37: t.3677. 1971; Thulin, Symb. Bot. Upsal. 39: 110. 2020.

Deciduous tree, to 5 m high, in crevices and holes of steep limestone cliff faces, bark flaking, yellowish to reddish. Leaves simple, 3–16 × 1.5–9 cm, crenate or deeply lobed and irregularly crenate-sinuate, hairy beneath. Panicles or racemes many-flowered, found after the end of the foliage period, petals pink or reddish, glabrous outer side. Fruit capsule 9–14 × 4–7 mm (Figures 2.9–2.11).

It is found in limestone-rich cliffs, in dry semi-deciduous forests. Socotra Island. Endemic. Fl. & Fr.: February–April.

IUCN Category: Vulnerable (Miller, 2004f).

FIGURE 2.9 *Boswellia popoviana* (Image 1).

Source: Image courtesy: Joël Lodé, with permission.

FIGURE 2.10 *Boswellia popoviana* (Image 2).

Source: Image courtesy: Joël Lodé with permission.

FIGURE 2.11 *Boswellia popoviana* (Image 3).

Source: Image courtesy: Valerian, CC4. Open source, free to reproduce.

- *Boswellia rivae* Engl., Ann. Bot. (Rome), 7: 16. 1907; Thulin, Symb. Bot. Upsal. 39: 70. 2020. *B. boranensis* Engl. Bot. Jahrb. Syst. 34: 316. 1904. *B. ruspoliana* Engl., Nat. Pflanzenfam. Ed.2. 19a: 423. 1931.

  Small, deciduous tree, up to 7 m tall, often wider than tall. Leaves pinnately compound, up to 18 cm long, the 3–20 mm long petiole included, 11–29-foliolate. Racemes or panicles with few flowers, pubescent, 1.5–6 cm long; bracts elliptic-oblong, pedicels 3–10 mm long. Flowers pink, pubescent outside, petals 5, filaments densely pilose. Fruit 3(–4)-celled, 17–27 × 7–9 mm, angular, pear-shaped, pubescent.

  It is found in wooded grasslands in red sandy to stony hills at 250 to 80 m a.s.l. altitude. Ethiopia, Somalia, and Kenya. Fl. & Fr.: January–May. Common name: Quara, Black incense.

  IUCN Category: Least concern (Thulin, 2020).

- *Boswellia sacra* Flück., Lehrb. Pharmak. 31. 1867; Thulin, Symb. Bot. Upsal. 39: 52. 2020. *B. carteri* Birdw. Trans. Linn. Soc. London 27: 143. 1870.

  Medium-sized tree up to 8 m tall. Leaves compound, 11–19-foliolate. Racemes or panicles 26 cm long, flowers white. Fruit 3–4-celled, narrowly, or broadly pear-shaped, glabrous.

  Found more often on cliffs in dry mountainous region and desert woodland in North-eastern Tropical Africa, Somalia to Yemen (Hadramaut and Mahrah), and Oman (Dhofar) (Figure 2.15). Flowers in winter and may be extending to April. Oleo-gum-resin obtained is translucent, deep yellow, reddish, or pale white and also known as *beeyo* (Svoboda et al., 2001).

  Chromosome number 2n = 22 (Thulin and Warfa, 1987).

  IUCN Category: Least concern (IUCN, 2011).

- *Boswellia samahensis* Thulin & Scholte, Symb. Bot. Upsal. 39: 106. 2020.

  Dwarf tree, up to 1.5 m high; with smooth grayish bark, flaking. Leaves imparipinnate 11–19-foliolate; leaflets lanceolate-oblong, 1–3.5 × 0.4–1.5 cm, densely hirsute with spreading hairs. Flowers reddish, in racemes, densely pubescent, calyx cup-shaped, stamens 10. Fruit subglobose to pyriform, seeds trigonous, narrowly winged.

Limestone escarpment in Socotra, Yemen. Fl. & Fr.: March-April.

IUCN category: Critically Endangered C2A (Thulin, 2020).

- *Boswellia scopulorum* Thulin, Symb. Bot. Upsal. 39: 102. 2020.

  Small tree, up to 5 m high; with smooth, light gray bark which sometimes flaking. Leaves imparipinnate, 3–15-foliolate. Flowers red, in panicles. Fruits subglobose.

  North-eastern Socotra on limestone boulders and cliffs. Fl. & Fr.: February–March.

  IUCN category: Critically Endangered D (Thulin, 2020).

- *Boswellia serrata* Roxb. ex Colebr., Asiat. Res. 9. 379. t. 5. 1807; Bennet in Hook. f., Fl. Brit. India 1: 528. 1875; Chitra & A. N. Henry in Hajra et al., Fl. India 4: 432. 1997. Thulin, Symb. Bot. Upsal. 39: 28. 2020. *B. glabra* Roxb., Pl. Coromandel t. 207. 1811. *B. serrata* Roxb. var. *glabra* (Roxb.) Bennet in Hook. f., Fl. Brit. India 1: 528. 1875.

  Large, deciduous trees, up to 15 m high, bark papery, exfoliating. Leaves apically clustered, imparipinnate, leaflets 8–23 pairs, variable in size, shape, and degree of pubescence, oblong to lanceolate, 3–8 × 1–3 cm, glabrous, rusty below, base cuneate, oblique, margin dentate-serrate, apex acute. Flowers bisexual, pale pink or white in axillary panicles; calyx tube broadly campanulate, calyx-lobes triangular-ovate, pubescent, persistent; petals 5, ovate-oblong, shortly clawed, inflexed at apex, pink or white, pubescent outside, except margin; stamens 10. Drupe trigonous, greenish-yellow, pyrenes 3, each 1-seeded; seeds heart-shaped (Figures 2.12–2.15).

  Common in rocky areas of dry forests of Central and Peninsular India and Pakistan. In India, it occurs in the states of Andhra Pradesh, Assam, Bihar, Chhattisgarh, Gujarat, Jharkhand, Karnataka, Madhya Pradesh, Maharashtra, Orissa, Punjab, Rajasthan, Tamil Nadu, Telangana, Uttar Pradesh, and West Bengal. Fl. & Fr.: March-August.

  It is known as Kunduru, Sallaki in Sanskrit, Salai, Madi, Salga, Salhe, Sali, and Salar in Hindi, Anduga in Telugu, Kuurdur in Arabic, Luban, Salai in Bengali, Gugal, Saleda, and Dhup in Gujarati, Parangisambrani, Kungli, Kundrikam, Gugulu, and Moranda in Tamil, and The Indian Olibanum tree and Indian Frankincense tree in English. Trade name is Salai.

Chromosome number: 2n = 44 (Ram et al., 2014).

IUCN Category: Least concern.

FIGURE 2.12. *Boswellia serrata*. (A) Habit; (B) bark; (C) leaves; (D) inflorescence; and (E) fruits.

Source: Images courtesy of Dr. A. Madhusudana Reddy, with permission.

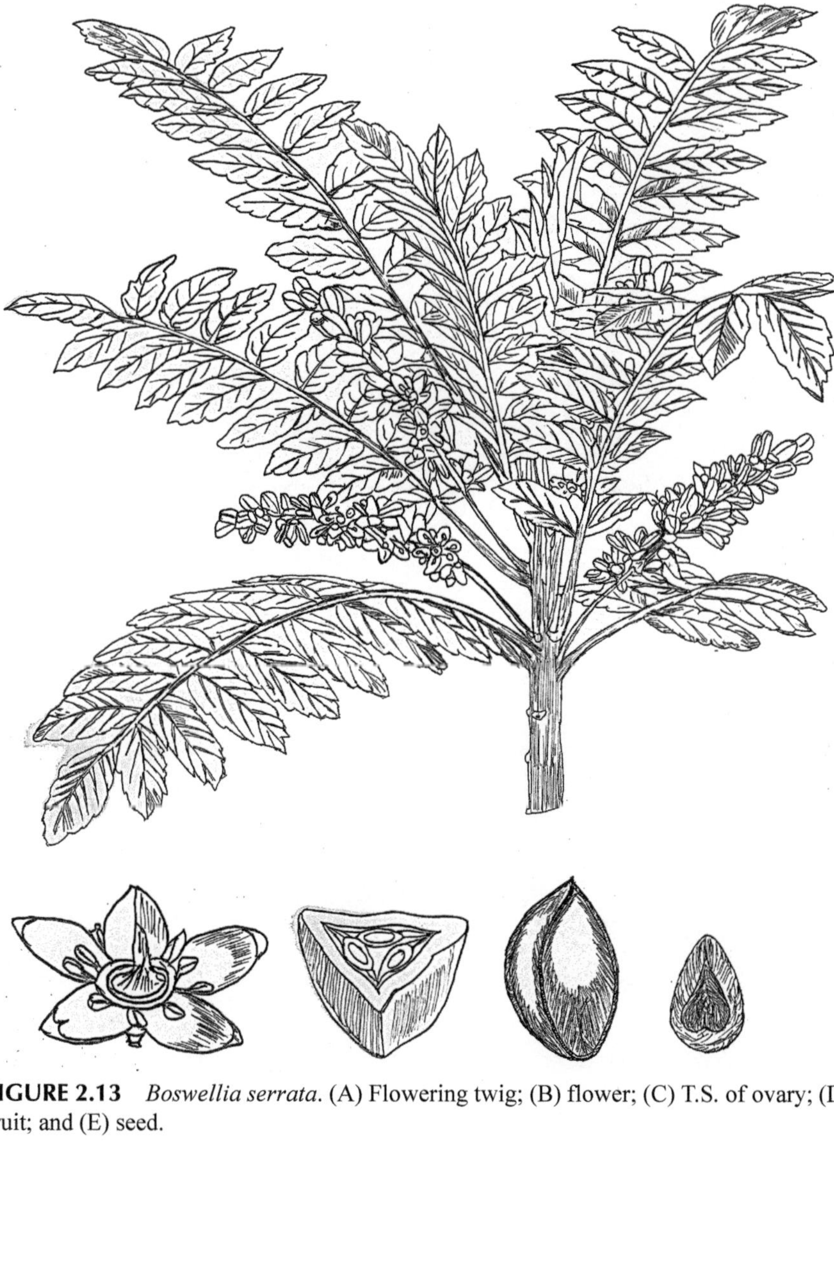

FIGURE 2.13 *Boswellia serrata*. (A) Flowering twig; (B) flower; (C) T.S. of ovary; (D) fruit; and (E) seed.

FIGURE 2.14 *Boswellia socotrana.*

Source: Image courtesy: Joël Lodé, with permission.

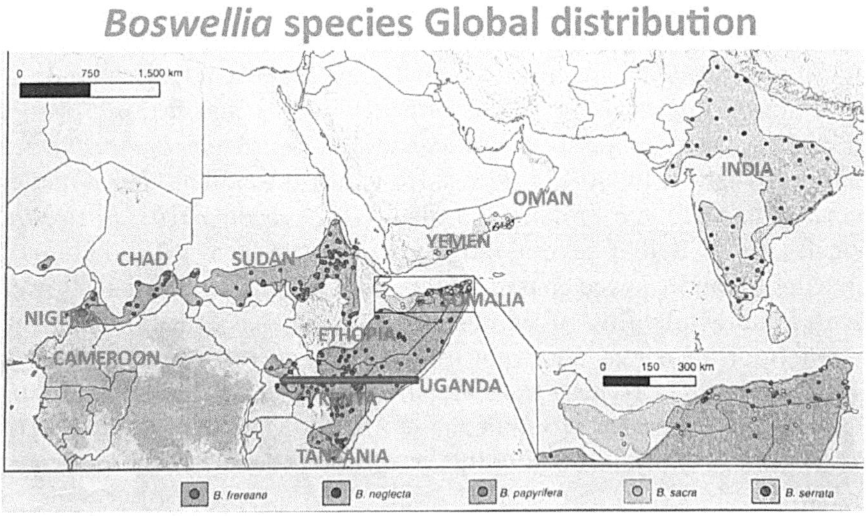

FIGURE 2.15 Distribution of *Boswellia* species.

Source: Adapted from Bongers et al., 2019.

➢ *Boswellia socotrana* Balf. f., Proc. Roy. Soc. Edinburgh 11: 505. 1882; Thulin, Symb. Bot. Upsal. 39: 124. 2020.

Medium-sized tree up to 8 m tall, stem bark gray to brownish. Leaves compound, rachis winged, resembling a fern frond, 3–10 × 0.5–1 cm, leaflets 7–31, 1.5–10 × 1.5–7. Racemes many-flowered, flowers creamy to yellowish. Petals glabrous on both sides. Fruit capsule, 3-celled, 6–8 × 4.5–6 mm, broadly ovoid or sub-globose (Figure 2.14).

Subtropical or tropical dry forests and subtropical or tropical dry shrubland of Socotra Island. Ground dwelling (not on cliffs). Endemic and is with a limited area of occupancy. Fl. & Fr.: February–April.

IUCN Category: Vulnerable (Miller, 2004g).

2.3 REPRODUCTIVE BIOLOGY

Sunnichan et al. (2005); and Raju et al. (2012) investigated the reproductive biology of *Boswellia serrata* and *B. ovalifoliolata*, respectively. Flowers in all the species of *Boswellia,* including *B. serrata* and *B. ovalifoliolata* are bisexual and actinomorphic. Flowers of *B. ovalifoliolata* are mildly odoriferous and weakly protandrous. In almost all the species of *Boswellia* flowering and fruiting occurs in the dry season. Most of the species of the genus *Boswellia* are leafless when in bloom. In *Boswellia sacra* the pollen size is large, and the shape is sub-prolate to prolate (Daly et al., 2010). *Boswellia* species are limited by several intrinsic biological factors. Limitation of cross-pollination, space constraints for seed production from all flower ovules and availability of limited resources in rocky, dry litter of the forest floor appear to constrain higher fruit set in *Boswellia* species (Raju et al., 2012). These trees are strictly self-incompatible and thus depends on outcrossing (Sunnichan et al., 2005; Raju et al., 2012). In *B. serrata* 85% of the fresh pollen grains are viable and the pollen to ovule ratio is 3348:1 (Sunnichan et al., 2005). Flowers of *B. serrata* are pollinated by honey bees *Apis dorsata* (Asian honey bee) and *A. cerana* var. *indica* (Indian honey bee). The flowers offer both pollen and nectar. Both entomophily and ornithophily are seen in *Boswellia ovalifoliolata.* In this species, flowers are pollinated by bees (*Apis dorsata, A. cerana*, A. *florea, Trigona iridipennis, Xylocopa latipes,* and *X. pubescens*), wasps (*Eumenes petiolata, E. conica*), flies and butterflies (*Catopsilia pomona, Junonia lemonias, Acraea violae* and *Danaus chrysippus*) (Figures 2.16 and 2.17). Pollinating birds here include *Nectarinia asiatica* (Sun bird) and *N. zeylonica* (Raju et al., 2012). Floral characteristics suggest that entomophily is the principal mode but sunbirds also pollinate. Mature fruits dehisce and disseminate their papery winged, lightweight seeds with the aid of wind. Following rainfall seed germination takes place, but further growth depends on soil water and nutritional status. Seedling recruitment success rate is highly limited. This could be due to nutrient-poor soil and water stress (Raju et al., 2012).

FIGURE 2.16 *Boswellia ovalifoliolata.* (A) Tree; (B) flowering inflorescence; (C) position of stamens; (D) pollen grain; (E) & (F) pistil; (G) *Apis dorsata*; (H) *Trigona iridipennis*; (I) *Ceratina* sp.; (J) *Xylocopa latipes*; (K) *Xylocopa pubescens*; (L) *Eumenes conica*; (M) *Eumenes petiolate*; (N) *Eumenes* sp.; (O) *Rhynchium* sp.; (P) *Hyperalonia* sp.; and (Q) *Catopsilia pomona.*

Source: Reprinted from Raju et al. (2012), with permission; https://creativecommons.org/licenses/by/4.0/

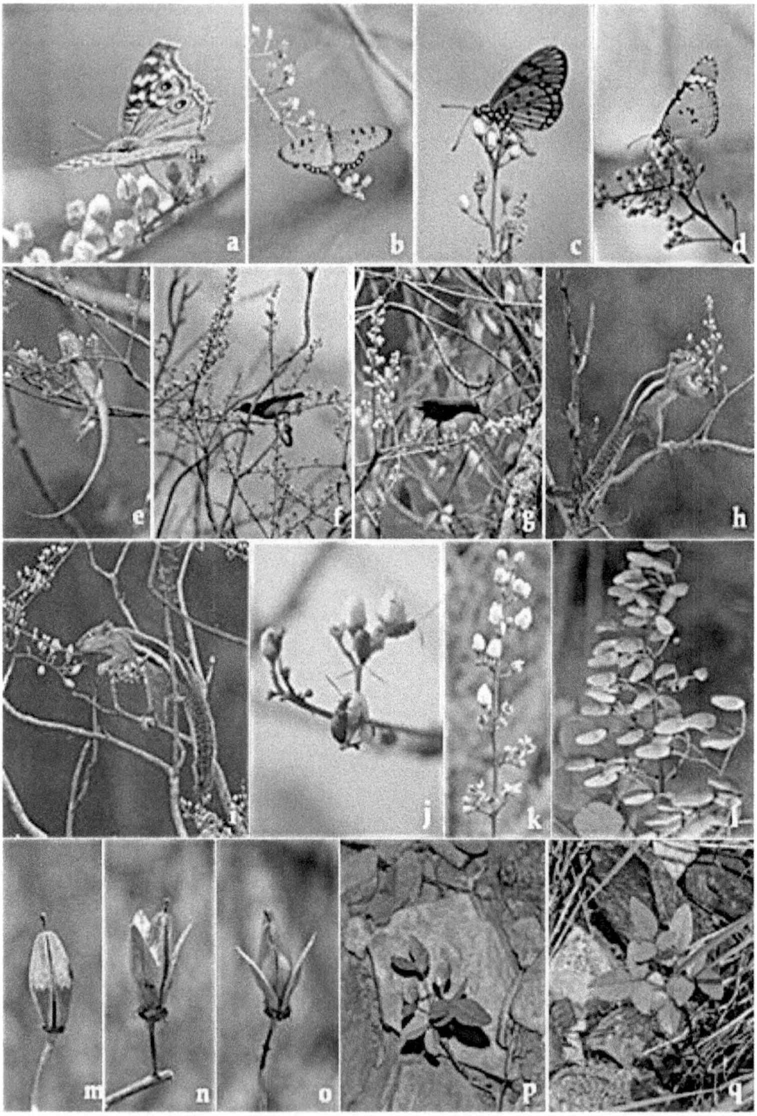

FIGURE 2.17 *Boswellia ovalifoliolata*. (A) *Junonia lemonias*; (B) & (C) *Acraea violae*; (D) *Danaus chrysippus*; (E) *Calotes versicolor*; (F) & (G) *Nectarinia asiatica* (male and female); (H) & (I) *Funambulus palmarum* (flower and fruit feeding); (J) & (K) weevil feeding on buds and flowers; (L) fruit set; (M) maturing fruit; (N) & (O) fruit dehiscence; (P) & (Q) healthy seedlings.

Source: Reprinted from Raju et al. (2012), with permission; https://creativecommons.org/licenses/by/4.0/

Toon et al. (2006) in *B. papyrifera* have shown that intensive tapping for gum causes the trees to divert too much carbohydrate into resin at the expense of reproductive organs, such as flowers, fruit, and seeds. Due to this, the trees produce fewer smaller fruits with seeds of lower weight and reduced vitality than non-tapped trees. At the stand level, non-tapped trees produced three times as many healthy and filled seeds as tapped trees. Germination success was > 80% in stands with non-tapped trees and < 16% for those with tapped trees (Toon et al., 2006).

In *B. sacra* nectariferous ring of flowers change colors with significant ecological role through 3 phases; Yellow, Orange-Red, and dark red phases. Yellow flowers are visited by a large number of different pollinators – bees, wasps, ants, little flies, and butterflies – starting from half an hour after sunrise until the sunset. Red flowers are not visited, orange flowers are visited mostly by walking insects. Fruits usually begin to ripe in March. Adult mortality rate is low and even lower if the tree is over-harvested for frankincense. Adult *Boswellia* mortality is 6–7% per year independent from sap tapping (https://ayublab.com/frankincense).

2.4 WOOD ANATOMY

Mokria et al. (2017) studied periodicity of wood formation and leaf phenology and to assess the growth dynamics of *B. neglecta. B. neglecta* forms two growth rings per year. Larger vessels are seen at the beginning of each growing season and smaller vessels formed later in the growing season. Seasonality in cambial activity matches with a bimodal leaf phenological pattern. The mean annual radial growth rate of *B. neglecta* trees is 2.5 mm. *B. neglecta* forms distinct growth rings. The occurrence of tangentially arranged smaller solitary vessels and thicker cell-wall thickness are wood anatomical features depicting growth rings boundaries. The vessel patterns across growth ring indicate that *B. neglecta* is a diffuse-porous tree, with larger vessels in the beginning of the growth ring (earlywood) and successively smaller vessels toward the end of the growth ring (latewood). Vessels are mostly solitarily and only sporadically occurred in clusters. The size of the earlywood and latewood vessels range from 4 to 62 × 10^3 μm^2 (mean = 20 ± 11 SD × 10^3 μm^2) and 1.6 to 30 × 10^3 μm^2 (mean = 7 ± 4 SD × 10^3 μm^2), respectively. The average vessel size of the earlywood and latewood is significantly different ($t_{(1633)} = 25.8$, $p < 0.0001$).

2.5 CONCLUSION

For a genus of 24 species, studies on botanical aspects, as seen above, are too scanty. For example, chromosome numbers and chromosome morphology are very limited. There is a huge scope in this field. Similarly, the studies on reproductive biology are carried out only in two species, and there is a necessity for such studies on the other 22 species. Even in those two species where such studies have been carried out, information is scanty. Wood anatomy of such an important genus yielding gum from its trunk is completely lacking except for stray mentioning. Studies on the ecology of these desert species is a must to know about its propagation and existence.

KEYWORDS

- ***Boswellia***
- **Burseraceae**
- **endemism**
- **frankincense**
- **gum olibanum**
- **reproductive biology**
- **Socotra Island**

REFERENCES

Alemu, S., Alemu, S., Atnafu, H., Awas, T., Belay, B., Demissew, S., Luke, W. R. Q., et al., (2018). *Boswellia ogadensis*: *The IUCN Red List of Threatened Species*. e.T34385A128140745. https://dx.doi.org/10.2305/IUCN.UK.2018-2.RLTS.T34385A128140745.en (accessed on 10 October 2022).

Al-Harrasi, A., Khan, A. L., Asaf, S., & Al-Rawahi, A., (2019). Taxonomy, distribution, and ecology of *Boswellia*. In: Al-Harrasi, et al., (eds.), *Biology of Genus Boswellia* (pp. 11–34). Springer.

Awas, T., Belay, B., Demissew, S., Nemomissa, S., Mekbib, E., Atnafu, H., Alemu, S., & Alemu, S., (2018). *Boswellia pirottae: The IUCN Red List of Threatened*

Species. e.T34394A128137387. https://dx.doi.org/10.2305/IUCN.UK.2018-2.RLTS.T34394A128137387.en.

Birdwood, G., (1870). On the genus *Boswellia*, with descriptions and figures of three new species. *Trans. Linn. Soc. London, 27*, 111–148. https://doi.org/10.1111/j.1096-3642.1870.tb00205.x.

Bongers, F., Groenendijk, P., Bekele, T. et al. Frankincense in peril. Nat Sustain 2, 602–610 (2019). https://doi.org/10.1038/s41893-019-0322-2

Bruno, A. M., Lavranos, J. J., & James, G. J., (2000). Frankincense on Soqotra Island (*Boswellia*, Burseraceae; Yemen). *Cactus and Succulent Journal, 72*, 265–278.

Daly, D., Harley, M., Martínez-Habibe, M., & Weeks, A., (2010). Burseraceae flowering plants. *Eudicots* (pp. 76–104). Springer.

Gillett, J. B., (1991). *Flora of Tropical East Africa: Burseraceae*, 1–94.

Hedberg, I., & Edwards, S., (1990). *Flora of Ethiopia and Eritrea* 1989; publ. 1990 (Vol. 3, pp. 1–659). The National Herbarium, Addis Ababa University, Ethiopia & The Department of Systematic Botany, Uppsala.

Hepper, F. N., (1971). *Boswellia nana* and *B. popoviana*. *Hooker's Icones Plantarum Ser. 2, 7*(4), 3676, 3677.

IUCN, (2011). *IUCN Red List of Threatened Species*. https://www.iucnrelist.org (accessed on 10 October 2022).

Lemenih, M., & Kassa, H., (2011). *Management Guide for Sustainable Production of Frankincense, A Manual for Extension Workers and Companies Managing Dry Forests for Resin Production and Marketing* (p. 5). Chapter 4 – Flowering, seed production and propagation.

Mabberley, D. J., (2017). *Mabberley's Plant-Book* (4th edn.). Cambridge University Press, UK. https://doi.org/10.1017/9781316335581.

Miller, A., (2004a). *Boswellia ameero*: *The IUCN Red List of Threatened Species*. e.T30414A9546504. http://dx.doi.org/10.2305/IUCN.UK.2004.RLTS.T30414A9546504.en.

Miller, A., (2004b). *Boswellia bullata*: *The IUCN Red List of Threatened Species 2004*. e.T44812A10950015. https://dx.doi.org/10.2305/IUCN.UK.2004.RLTS.T44812A10950015.en (accessed on 10 October 2022).

Miller, A., (2004c). *Boswellia dioscoridis: The IUCN Red List of Threatened Species 2004*. e.T44813A10950138. http://dx.doi.org/10.2305/IUCN.UK.2004.RLTS.T44813A10950138.en (accessed on 10 October 2022).

Miller, A., (2004d). *Boswellia elongata: The IUCN Red List of Threatened Species 2004*. e.T30415A9546667. http://dx.doi.org/10.2305/IUCN.UK.2004.RLTS.T30415A9546667.en (accessed on 10 October 2022).

Miller, A., (2004e). *Boswellia nana: The IUCN Red List of Threatened Species 2004*. e.T44814A10950258. http://dx.doi.org/10.2305/IUCN.UK.2004.RLTS.T44814A10950258.en (accessed on 10 October 2022).

Miller, A., (2004f). *Boswellia popoviana: The IUCN Red List of Threatened Species 2004*. e.T37866A10082448. http://dx.doi.org/10.2305/IUCN.UK.2004.RLTS.T37866A10082448.en (accessed on 10 October 2022).

Miller, A., (2004g). *Boswellia socotrana: The IUCN Red List of Threatened Species 2004*. e.T30416A9546843. http://dx.doi.org/10.2305/IUCN.UK.2004.RLTS.T30416A9546843.en (accessed on 10 October 2022).

Mokria, M., Tolera, M., Sterck, F. J., Gebrekirstos, A., Bongers, F., Decuyper, M., & Sass-Klaassen, U., (2017). The frankincense tree *Boswellia neglecta* reveals high potential for restoration of woodlands in the Horn of Africa. *Forest Ecology and Management, 385*, 16–24.

Murray, M., (1995). *The Healing Power of Herbs* (pp. 162–171). Prima Publishing, Rocklin, California, United States of America.

Pullaiah, T., Chennaiah, E., & Sandhya, R. S., (2008). *Flora of Andhra Pradesh* (2nd edn., Vol. 1, pp. 311, 312). Scientific Publishers, Jodhpur, India.

Raju, A. J. S., Lakshmi, P. V., Ramana, K. V., & Chandra, P. H., (2012). Entomophily, ornithophily and anemochory in the self-incompatible *Boswellia ovalifoliolata* Bal. & Henry (Burseraceae), an endemic and endangered medicinally important tree species. *J. Threatened Taxa, 4*(7), 2673–2684.

Ram, H., Sharma, S. K., Kumar, A., & Rama, R. S., (2014). Male meiotic studies in some *Boswellia serrata* (Roxb.) Ex. from the Aravali region of Rajasthan, India. *Cytologia, 79*, 41–47.

Saha, D., Ved, D., Ravikumar, K., & Haridasan, K., (2015). *Boswellia ovalifoliolata: The IUCN Red List of Threatened Species 2015*. e.T50126567A50131280. http://dx.doi.org/10.2305/IUCN.UK.2015-2.RLTS.T50126567A50131280.en.

Sunnichan, V. G., Mohan, R. H. Y., & Shivana, K. R., (2005). Reproductive biology of *Boswellia serrata. Bot. J. Linn. Soc., 147*, 73–82.

Svoboda, K. P., Hampson, J. B., & Hall, L., (2001). *Boswellia* from Somalia, a source of high quality frankincense. *Med. Plant Conserv., 7*, 16–19.

Thulin, M., & Al-Gifri, A., (1996). The frankincense trees (*Boswellia* spp.) on Soqotra. In: Dumont, H. J., (ed.), *Proceedings of the First International Symposium on Soqotra Island: Present and Future* (pp. 107–113). United Nations Publications, New York.

Thulin, M., & Warfa, A. M., (1987). The frankincense trees (*Boswellia* spp., *Burseraceae*) of Northern Somalia and southern Arabia. *Kew Bulletin, 42*, 487–500. https://doi.org/10.2307/411006.

Thulin, M., (2000). *Burseraceae*. In: Thulin, M., (ed.), *Flora of Somalia* (Vol. 2, pp. 183–228). Royal Botanic Gardens, Kew, UK.

Thulin, M., (2020). *The Genus Boswellia (Burseraceae): The Frankincense Trees* (Vol. 39, pp. 1–149). Botanic symbols of Upsalien, Works from the Botanical Department in Uppsala.

Thulin, M., Decarlo, A., & Johnson, S. P., (2019). *Boswellia occulta* (*Burseraceae*), a new species of frankincense tree from Somalia (Somaliland). *Phytotaxa, 394*(3), 219–224.

Toon, R., Woldeselassie, O., Marius, W., & Frans, B., (2006). The effect of tapping for frankincense on sexual reproduction in *Boswellia papyrifera. J. Appl. Ecol., 43*, 1188–1195.

Vollesen, K., (1989). *Burseraceae*. In: Hedberg, I., & Edwards, S., (eds.), *Flora of Ethiopia* (Vol. 3, pp. 442–478). Addis Ababa, Ethiopia.

CHAPTER 3

EXTRACTION OF OLEOGUM RESIN (FRANKINCENSE/OLIBANUM) FROM *BOSWELLIA* SPECIES

LEPAKSHI MD. BHAKSHU[1*] and K. VENKATA RATNAM[2]

[1]*Department of Botany, PVKN Government College, Chittoor – 517002, Andhra Pradesh, India*

[2]*Department of Botany, Rayalaseema University, Kurnool – 518007, Andhra Pradesh, India*

[*]*Corresponding author. E-mail: lmbakshu@gmail.com*

ABSTRACT

Gums, resins, essential oils, secondary metabolites, mucilage, latex, etc., are important plant by-products serving humans with their unique charac teristics through a variety of biological and physicochemical properties. The oleo resins possess both gum and aromatic properties, and thus are used as an excellent source of incense. They are produced from the tree species belonging to the genus *Boswellia* of Burseraceae family. The oleo gum is exuded from the trees due to the natural or deliberate incisions or wounds on the stem bark. The gum secretion when dried for about 15 to 20 days on the stem bark forms into a specific rounded or amorphous crystalline form. Due to its aromatic nature, the gum has attained a commercial value, and became an important commodity, impacting on the global economy. Generally, the rural/tribal people are involved in the collection of gum during February to May in a year. The over-extraction and improper tapping methods and demand in the international markets

Frankincense – Gum Olibanum: Botany, Chemistry, Extraction, Utilization, Propagation, Biotechnology, and Conservation, T. Pullaiah, K. Venkata Ratnam, Mallappa Kumara Swamy, Lepakshi Md. Bhakshu (Eds.)

has led to its over-exploitation, causing decreased population of *Boswellia* species, apart from the other threats, such as grazing, extreme environment, insect attack and infestation, low seed setting, and its success in the fields to grow as mature trees. The major species, such as *B. dalzielii, B. frereana, B. neglecta, B. sacra* (syn. *B. carterii*), *B. papyrifera, B. occulta, B. serrata*, etc., are facing high risks, due to over tapping and improper methods of extraction. Many of these species have entered into endangered (EN) status, in addition to some of them being reported as endemic in nature. Hence, the knowledge on the gum extraction methods serve to save the life of these trees. This chapter discusses on the gum collection protocols, tools, and tapping methods adapted to the *Boswellia* species for an efficient and quality tapping.

3.1 INTRODUCTION

The use of frankincense incorporates long-records in human civilization (Van Beek, 1960). The flavorer oil content with exceptional scent has created it appropriate to burn frankincense as an incense in temple rituals and to use as a base for perfumes and for healthful functions since ancient times (Groom, 1981). The primary recorded observation of frankincense was 15th century B.C. in Egypt. It remarked 22 instances within the Bible, 16 instances for religious worship, twofold as a tribute of honor, as shortly as a writing of merchandise, and three instances as a fancied from the royal gardens of male monarch (Tucker, 1986). These presents are currently not the simplest of the most effective value on the time, but to boot symbolized Christ's existence: gold for royalty, frankincense for divinity and myrrh for suffering. Incense has been associated with religious ceremonies everyplace within the world, after you contemplate that point old. The Greeks and Romans embarked on to use frankincense for sacrifice in the sixth and second centuries B.C. Incense was recognized as a technique of propitiating the gods and as a cleaner and wise chemical, with those horrific smells of primitive town existence may be disguised (Groom, 1981). The employment of frankincense for ritual functions in African country dates once more at the smallest amount to the Aksumite Empire, ca. 500 B.C. (Goettsch, 1986). Frankincense was delivered into church ceremonies at the beginning of Christianity. Even after you contemplate that, the hardened resin burns withinside the Roman Catholic and Orthodox Churches at some purpose of the planet (Yadav & Gupta, 2018). The genus

Boswellia belongs to the family Burseraceae, that possess distinguished vertical schizogenous resin ducts (containing triterpenoid compounds and ethereal oils) withinside the bark, and produces perfumed oils and gum resins which might be in charge of its healthful properties (Murthy et al., 2016).

Gum-resins are a mixture of gum and resin substances and show the characteristics of both of them and possess volatile substances as minor portion and became aromatic oleo resins. Most popular gum-resin used globally are, asafetida, myrrh, frankincense, salai-guggul, etc. The resins collected from *B. rivae*, *B. ogadensis*, and *B. neglecta* occurring in the drylands of Ethiopia are traded as frankincense (Lemenih and Teketay, 2003; Lemenih et al., 2003). However, there is no specific method described for the gum resin extraction, and it is collected using the conventional methods or naturally dried exudates. The species of *Boswellia* were appreciated for its fragrant resin products, known as frankincense, also referred as olibanum, and it can be obtained from the trees after reaching about 10 years of age (Gonzalez, 2020). However, the gum extraction methods have not been studied for the species except *B. ameero, B. dioscorides, B. elongata, B. nana, B. pirottae, B. ovalifoliolata*, etc.

The resin yield of an individual tree on average differs across the globe and among species. It may also depend on the trees which are being tapped regularly or allowed to rest. The recent estimate of its production potential is scanty. In Sudanese species (*B. papyrifera*) reported the yield an average of 175 gram/tree (during the first tapping). A population of *B. papyrifera* in the Nuba Mountains of Sudan was recorded to yield 60–80 kg/hectare (Alemu et al., 2012). As per the reported estimate, yield for a year an average of three kilos of gum-resin per each tree (Eshete, 2002; Hassan, 2012). Based on this, the annual production of gum-resin in Oman, reported at ~230 tons/year (Hassan, 2012).

Whereas, the estimation of its production in Ethiopia, in 1981 as 23,000 tons (Coppen, 1995). Likewise, in a study of Eshete (2002) the annual production olibanum in the Amhara (Ethiopia) alone predicted about 2,03,975 tons annually (based on 3 kg/tree average). This difference by a 10-fold has been attributed data on the census or yields of an area of sample selection in 1981. The production volume also varies by region within a range in a country. Considering this information, the furthermore sections will focus on the occurrence and extraction approaches for frankincense from different species of *Boswellia.*

3.2 OCCURRENCE AND EXTRACTION OF GUM-RESIN

The bark of the *Boswellia* tree is considered as a reservoir of gum resin secretions. When the reservoirs are injured, punctured, and penetrated, a milky juice is excreted onto the bark. When the milky juice becomes exposed to air, it begins to harden, producing the oleogum resin (Tucker, 1986; Murthy et al., 2016), and typically yields essential oils between 5% and 15% (Mertens et al., 2009).

Frankincense is an aromatic oleo gum resin created by dry paleotropical trees of genus *Boswellia*, typically characterized by papery, scaly, and exfoliating bark with compound leaves, and characterized by the production of aromatic organic compound. Ecologically, the resins are a unit, thought to be vital players as deterrents to herbivory, infection, attack by boring beetles, and different sorts of harm. As a product, it has been listed and utilized in ancient medical systems for 5,000 years, creating it one in all the oldest commodities within the world. There are a unit concerning 24 members of this genus, distributed across Africa, Arabia, and South Asia; mostly used regionally, however solely a number of species turn out oleoresins in that area and listed in vital volumes on the world markets especially *B. papyrifera*, *B. serrata*, *B. sacra* (syn. *B. carterii*) and *B. frereana*. These resins are used as incense as well as medication (Marshall, 2003; Johnson et al., 2019). *B. neglecta* is referred as black frankincense for its dark color after drying of secretions (tears) and distributed in certain countries such as Kenya, Ethiopia, Tanzania, Somalia, and Uganda (Mokria et al., 2017).

The *Boswellia* trees possessed a brown papery bark in which inner part shows light and dark brownish outer and inner layer with gummy nature (Thulin and Warfa, 1987). A notch is created by making partial scar on the bark of old aged trees, and the secretion flows and converted into a resin like substance which were collected after complete drying. This gum resin hardens into an amorphous substance to an orange-brown gum which is named as frankincense or olibanum (Salai guggul in Hindi) which consists of specific aromatic fragrance. The gum resin is reported to possess the mixture of essential oil 5–15%, gum 6–30% (mixture of polysaccharides) and 60–85% resins with a mixture of terpenes (Hamidpour et al., 2013). The Frankincense oil obtained through steam distillation possesses a spicy, woody, and specific odor. The essence obtained from the gums of *Boswellia* species were extracted using traditional methods (steam distillation and Soxhlet extraction using low polar solvents routinely used for oil extraction). The traditional techniques are not cost effective, and recently, supercritical fluid extraction (SFE), an environmentally friendly procedure has been adopted to minimize the constraints with quick and higher quality products were standardized by Boutin and Badens (2009).

Traditionally, frankincense from *Boswellia* vegetation is produced through synthetic wounding of the timber, a method known as tapping. Tapping is completed at some stage in the dry season. However, it is not always clear whether or not the dry season is selected out of comfort for tappers or due to the fact it is far dictated via way of means of the body structure of the tree. The conventional tapping includes barely shaving the outside layer of the bark and forming a round wound of approximately 1–2 cm high, 1–1.5 cm huge and 0.5–1.0 cm deep. Usually, three tapping spots are made on every facet of the tree, beginning at approximately 0.5 m from the bottom of the stem (Figure 3.8). Basing the tree size and health there may be more than three tapping positions were being fixed (Lemenish and Kassa, 2011). The process of tapping or scratching became a regular and cyclic, which can be repeated by 15–20-day interval and after the older scratches refreshed at their upper margins through slight carving by 2 cm down and continued until the rainy season. Thus, a tree is tapped 8–12 instances in a year, every wound can also additionally achieve a width of approximately 10 cm or extra (Lemenish and Kassa, 2011).

Though the yield of the gum might be noticed to increase with proportional to scratching spots, there may be adverse effect on the tree health and also may interfere with the reproductive cycle and seed setting besides affecting the quality of gum since the tears size may become small and gum appears as fine dust and not attract the buyers or exporters. Hence it is recommended to restrict the tapping cycles and number of scratches (six tapping spots on a tree with 30 cm Diameter at Breadth Hight) to ensure the quality product (Lemenish and Kassa, 2011).

Another crucial detail in coping with the tapping regime is renewing the wound at the permitted intervals. If wounds are not renewed at the best intervals, the vintage wounds can also additionally heal completely. Wounding after the older wounds heal results in the tree to postpone exuding the resin. This is sort of equal to beginning the entire method of tapping throughout again, with a view to bring about decreased ordinary annual yield and affect the benefits. The extraction of oleogum from the trees of *Boswellia* species are discussed below.

3.3 *BOSWELLIA DALZIELII* HUTCH.

This aromatic tree growing in Sudano-Sahelin grasslands with rocky, shallow, and dry soils (Ouédraogo et al., 2006). The people of the eastern area of Burkina Faso termed it as "Tree Man" which is a major frankincense tree found north Segbana. This tree possesses a characteristic rounded clear crown and may reach 13 m height with yellowish-gray stem bark peels as stripes, and it secretes white sap with fragrance when get scratched (Ouédraogo et al., 2006).

In a study, 20 oleoresin samples of *B. dalzielii* were collected from the source mature stands in specified locations across the western and central Burkina Faso. The analyzed samples were randomly taken in dry seasons from various aerial parts of stem or branches through wounds. The essential oils from the resin samples (Figures 3.1 and 3.2(A)–(C)) of Burkina Faso are compared and found similarities with the components of samples obtained from Nigeria especially rich with α-pinene. Moreover, the samples of Nigeria, significantly possessed a rich amount of sesquiterpenes with a chemotype spiked with myrcene indicating the presence of a specific insect which may induce its synthesis (DeCarlo et al., 2019).

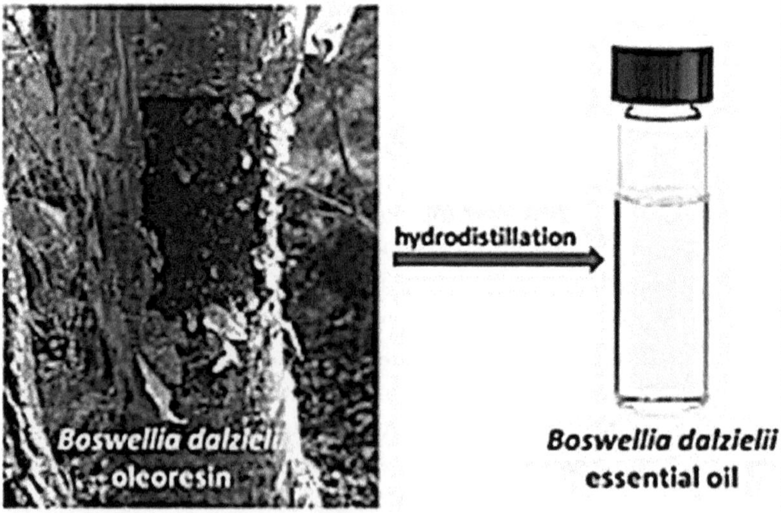

FIGURE 3.1 Tears of *B. dalzielii* (oleo resin) and essential oil of oleo resin.

Source: DeCarlo et al. (2019).

FIGURE 3.2 (A) & (B) habit; and (C) oleo gum of *B. dalzielii.*

Source: Reprinted from DeCarlo et al., 2019. © 2019 by the authors. Licensee MDPI, Basel, Switzerland. https://creativecommons.org/licenses/by/4.0/

3.4 *BOSWELLIA FREREANA* BIRDW.

The most expensive olibanum of Somalian species, *B. frereana.* is natively known as "jagcaar." The resin is called 'lobãn majdi' or 'maydi' in the markets (Basar, 2005). The specific extraction methods were not mentioned and may be following the traditional tapping methods.

The resin extracted from the stem, is considered as a high-quality frankincense (Uphof, 1959). Most resin is obtained by making artificial scratches on the bark of the tree. And the milky ooze exuded has been hardened through the open exposure of these droplets or 'tears' and can be easily removed and gathered by the gum tappers.

3.5 *BOSWELLIA NEGLECTA* S. MOORE

B. neglecta (=*B. hildebrandtii*) is the source of frankincense, and economically viable important resource gum resin (Figure 3.3) of Kenya and locally called Lobathin and it is also used as chewing gum. *B. neglecta* is very common and occurs in most drylands in *Acacia-Commiphora* bushland, particularly on red sandy soil over basement complex and on lava flows, between 200 m and 1,350 m elevations with 250–600 mm rainfall. The local names of this gum include, Hancha (Borana, Gabra), Dakar (Orma), Dakhara, Kinodo (Kamba), Lkinoo, Lecholoo (Samburu), Magafur or Mathefur in Somali (Gachathi and Muga, 2002).

FIGURE 3.3 (A) *B. neglecta* habit; (B) twig with fruits; and (C) frankincense.

Source: Reprinted from Gachathi and Muga (2002).

3.6 *BOSWELLIA OCCULTA* THULIN

It is a novel species (Figure 3.4(A)–(E)) identified in 2019 (Thulin et al., 2019) and regarded as one of the frankincense species and the details on the methods of collection and extraction of gum are not discussed and regarded to collect in traditional ways (tapping) as described for other *Boswellia* species as far as by scientific reports yet to be published from the type location. However, the chemical characterization was discussed by Ayubova et al. (2019).

FIGURE 3.4 *B. occulta* tapping method for gum extraction.

Source: Thulin et al. (2019) from the Journal Phytotaxa. ©Magnolia Press; reproduced with permission from the copyright holder.

Moreover, no one emphasized on the extraction methods apart from the tapping, since the *B. occulta* populations were located far from the regular gum (Figure 3.4(A)–(E)) yielding stands (Gonzalez, 2020). Ayubova et al. (2019) discussed about the different grades during the characterization of essential oils of oleo gum resin of *B. occulta* (Figure 3.5).

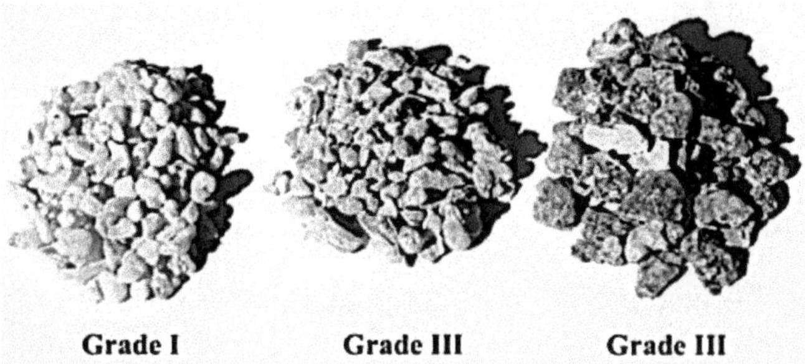

FIGURE 3.5 Different grades of gum resin of *B. occulta*.

Source: Reprinted with permission from Ayubova et al. (2019). © 2019 Elsevier.

3.7 *BOSWELLIA PAPYRIFERA* (DEL.) HOCHST

The important tree taxa in Ethiopia such as *Boswellia, Commiphora*, and *Acacia* stands major resource for the valuable gums and growing in different micro-climatic areas and the gum may be harvested from the natural stands (Vollesen, 1989; Lemenih and Teketay, 2003; Wubalem et al., 2002). The collection of gum or incense products may provide a best off-form source for employment. Around 7,000 laborers were benefited during the gum extractions as a seasonal employment (western Tigray) in a year and of which women are 31%. A taper can collect about 10–15 quintals of gum resin in a year and may earn about income of 100 to 150 US$ (Aregawi, 1997) and expected to raise its demand throughout the world. In Ethiopia, enterprises involved in frankincense international trade, are recently focusing more on ensuring the quality by offering high prices to collectors of export grades (Wubalem et al., 2002). In Ethiopia the frankincense source is *B. papyrifera* which oozes a white resin during injury of the stem bark and metamorphosed as tear/club/pear shaped product. The product varies according to the season or tree size and produce 100 and 1,000 grams from a tree in a year (Wubalem et al., 2002).

The popular oleo gum resin is extracted from *B. papyrifera* is the main source of frankincense in Ethiopia which occupied revenue more than 70% of income generated through the oleogum business. Since there is an ever-increasing demand poses a continued risk on this species which have uncertainty of its existence and facing risk. In addition to the commercial demand certain anthropogenic cues such as deforestation, fuel wood, domestic grazing of cattle, urbanization, expansion of agricultural land, etc., threaten the survival of these trees. Even certain environmental issues also being noticed for its survival, such as forest fires, natural pests/ diseases, etc., also enhanced the risk for the expansion of its natural populations in Ethiopia (Tadesse et al., 2020).

3.7.1 TAPPING

Techniques of tapping or scratching and harvesting of frankincense remained really the equal given that historic instances (Groom, 1981). Tapping of *B. papyrifera* is provided with the help of scratching as a thin layer of the bark with a scalpel-like handy device referred to as Mengaff in Ethiopia. A tree can be tapped 8–12 instances all through the dry intervals of the year. The first 3–4 tapings are undertaken at 21–30 days, and this interval may be fastened within 10–15 days since the temperature rises during the tapping periods. The secretion at the injuries became dried and turned to hard globular to pear-shaped during the exposure to natural air (Verghese, 1988). Few kilograms of frankincense are amassed from a tree according to year (Coppen, 1995). The quantity may range relying at the diameter of the tree, productiveness, and the season. The species can retain the capacity of incense generation which may last for as much as 50–60 years of its life (Tilahun, 1997; Gebrehiwot et al., 2003).

3.7.2 SORTING AND GRADING OF FRANKINCENSE

The grading might be performed after collection and preliminary processing of gum and sorted based on the quality parameters such as color, size, purity, and made into five following categories. The First grade designated with white colored granules of 6 mm or more; and the second grade for the 4–6 mm, the third grade is 2–4 mm whereas the fourth grade has brown or black and the fifth grade has powdery appearance with < 2 mm diameter. *B. papyrifera* is a primary tree species which denoted for gum resin production in Ethiopia especially from Tigray and Amhara are major regions.

B. papyrifera population in the country is highly affected in its natural habitat, and there may be an impaired domestication. The major challenge associated with *B. papyrifera* is hampered natural regeneration. The majority of the gum extraction, harvesting, and processing is through traditional methods of tapping. Hence, the Governmental and non-governmental organizations are actively involved in frankincense production and commercialization activities throughout the country adopting intensive training and the sustainable methods for its conservation.

The secretion of gum from the trees has been evoked by the seasonal pattern hence the tapping and gathering of frankincense is carried out accordingly from middle of September to June on the onset of dry season. The tapping technique usually carried through the scratching or shaving stem bark in thin, reaching 2 mm depth with a 4–8 mm width. The Mengaff (Figure 3.7) a locally designed tool makes suitable circular incision on the bark externally has been widely used (Girmay, 2000). The second tapping, may be performed after a gap period of 30–40 days a moderate expansion of wound which has made during the first attempt and the process can be continued until the wound reaches a width of 4 cm in a 3–4 months duration. The maximum injuries for the tapping may be done on three to four facets reaching to 6–8 tapping/scratching spots are designed on suitable trees (Girmay, 2000). The fine oleo gum has been collected after the drying within two or three weeks of injury/wounding may initiate the oozing of an exudate which metamorphosed as frankincense (Figure 3.6(A)–(C); Tadesse et al., 2007). The wound is re-scratched after the gathering of gum to enable the next cycle of collection and may be repeated with an interval of 15 to 21 days until the onset of the rainy season.

FIGURE 3.6 (A) *B. papyrifera* habit; (B) gum tapping extraction; and (C) fresh gum oozing tears.

Source: Reprinted with permission from Tadesse et al. (2007).

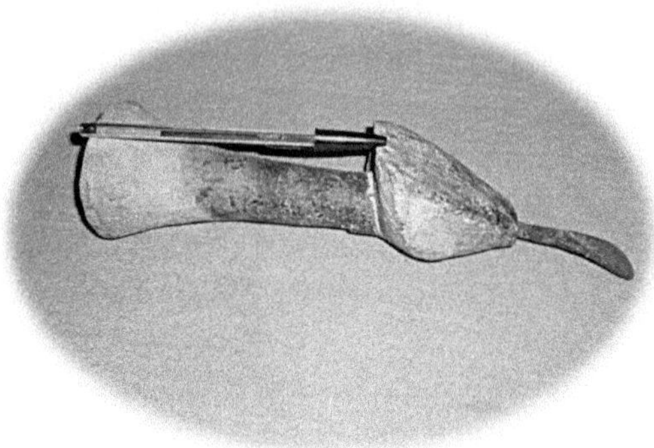

FIGURE 3.7 Mangef/Mengaff, a hand tool used for gum tapping on the trees of *B. papyrifera*.

Source: Reprinted from Ali et al., 2009. © 2009 Taylor & Francis.)

The onset of leaves in the first week of June initiates the process of photosynthesis which is necessary to the plant's life, and the tapping and collection have been stopped by the extractors up to October. An average of 10–12 quintals of dried gum has been collected by the tappers, and an average of 500 grams of gum from an individual tree in a cycle of the favorable season. The entire harvest is stored temporarily and transported to a center where further sorting, grading, and categorization are performed (Girmay, 2000). The first two grades were sold through the international markets, and others were sold or utilized for domestic purposes or incense material for ceremonies, sacred places or used in coffee, etc. (Tadesse et al., 2007).

In an investigation, Ali et al. (2009) conducted a bi-factorial experiment designed as RCBD (randomized complete block style) within 162 trees. The experiments were performed within different stem diameters and using scratching tools such as Sonki and Mengaff on two sides or four sides, and the study revealed the best resin-yielding tree as per the diameter of the stem at DBH with more than 20 cm.

The investigation carried out by Khamis (2001) at Jebel Mara stated that the minimum suitable DBH of the tree that could be tapped was 12 cm. In this area, the producers categorized the gum-tapping trees based on the nature of the bark, i.e., hard bark, soft bark, or medium. The total gum production of each gum-yielding tree per season is highest in the Jebel Mara region (2.8 kg/tree/year) than in Kordofan State (2.2 kg/tree/year) (Khamis, 2001). This may be due to differences in gum-tapping methods and variations in species' metabolic activities (Harsh et al., 2003).

Based on the above information, Ali et al. (2009) concluded the following observations:

- The enhanced tapping attempts are not affecting the yield of gum production, and 2–4 incisions were recommended, which can minimize injuries and the span of life of the tree.
- Tapping positions at different heights of tree stem base, i.e., 150, 100, and 50 cm, showed notable variations in gum resin quantity. Tapping at 150 cm stem base is recommended to get more amount of gum resin.
- Of the two tools used to gum tapping, i.e., Mengaff and Sonki, Mengaff was recommended tool for gum tapping due to having a wide blade and it allows one to scratch bigger piece of bark than Sonki.
- On average, *B. papyrifera* yields (> 250 g per tree) higher amount of gum resin than *Acacia senegal* (250 g per tree).

The study of Girma et al. (2015) revealed the collection and sampling of the gum resin from the *B. papirifera* depends the age and season using the tapped gum randomly from the woods. The gum resin, collected from model trees, was analyzed, and explained based on the occurrence of marker compounds such as boswellic acid and incensole acetate which are correlated with the trunk diameter at breast height (gbh) and maximum secondary metabolites and recommended the tapping for sustainability. Moreover, the relationship of seed viability and the maximum carbon source found on the standing trunks were also demonstrated. The gum extracts from the trees by the group of scientists using general tapping method using sickle and knife and used in the investigation.

Frankincense is produced by making wounds with 2.5 cm^2 surface area and 1 mm depth using a traditional axe at several places of the stem bark, and subsequently, the secreted gum will be collected from the wounds of *B. papyrifera* tree bark. The tapping practice will be repeated 8–12 cycles in the dry season. At every tapping cycle the old gum resin is removed by enlarging the opened wound. As per Ogbazghi (2001); and Gebrehiwot et al. (2003) trees were tapped with 6–12 spots on the stem. Due to high market value of frankincense, currently, 27 tapping wounds are made in one tree in commercially important areas (Kebede, 2010). The yield of frankincense has been declined ninth tapping cycle from each tree (Eshete et al., 2012a). Depending on size of the tree, productivity site and season, frankincense yield was varied between 41 g/tree/season and 3,000 g/tree/

season. Eshete et al. (2012a, b); and Tilahun et al. (2011) reported that a considerable amount of frankincense yield was observed at initial tapping rounds. However, it ultimately reduced towards the end of the dry season.

3.8 *BOSWELLIA SACRA* FLÜCK.

B. sacra (olibanum-tree or frankincense) became a source of frankincense, which is harvested after drying of the resinous sap. It is known by the Arabians as "maghrayt d'sheehaz" and the resin it produces as "luban dhakar" and believed to be a native species of Arabian-peninsula and horn of Africa (Basar, 2005). The resin is harvested from the incisions deeply into the tree stands and the wounding leads the tree to ooze a milky substance which can seal the wound and prevents the infection and helps in the healing process.

Frankincense or olibanum from *B. sacra* (=*B. carterii*) (Figure 3.8) is referred to in Bible as a source of incense (https://www.fao.org/3/Y4496E/Y4496E29.htm). The gum is collected from the mature stems using the traditional tapping method (Tadesse et al., 2007) and in natives of Somalia, called it as "moxor" and the gum resin is called "lobãn dakar" or more commonly as "beeyo" quality (Basar, 2005). *B. sacra* is valid name and *B. carterii*, is a synonym. Hence the literature on the gum extraction methods was studied together with the *B. sacra*.

FIGURE 3.8 (A) Habit of *B. sacra*; (B) tapping pattern; (C), (D), and (E) graded samples of oleo gum obtained from stem bark tapping.

The frankincense essential oil of the *B. sacra* gum resin was processed in a custom-made hydro distiller in Salalah (Oman) and its gives frankincense essential oil (8–13% w/w) through hydro-distillation (Lin et al., 2013; Ni et al., 2012). The essential oil obtained from the oleo gum is chemically highly diversified and it also source of new phytochemicals as reported by Yamei et al. (2021) is also subject for the new frontiers in research.

The grading of the gum resin of *B. sacra* (Figure 3.9), performed and classified and graded as explained here. The first grade is Hougari Royal or Najdi from deserts of the north of Dhofar State, Oman, collected during monsoon (rainy) area with very little rain. Hougari Superior grade (Figure 3.9(C)), with a green tinge, commonly used by the indigenous people through water to get relief from respiratory illness. Hougari Regular (Figure 3.9(D)), the popular frankincense, is used as incense. Hougari Sha'bigrade (Figure 3.9(E)) the lowest graded gum resin is obtained from the aged and dying stands of the coasts and terrestrial areas during the monsoons (Lin et al., 2013).

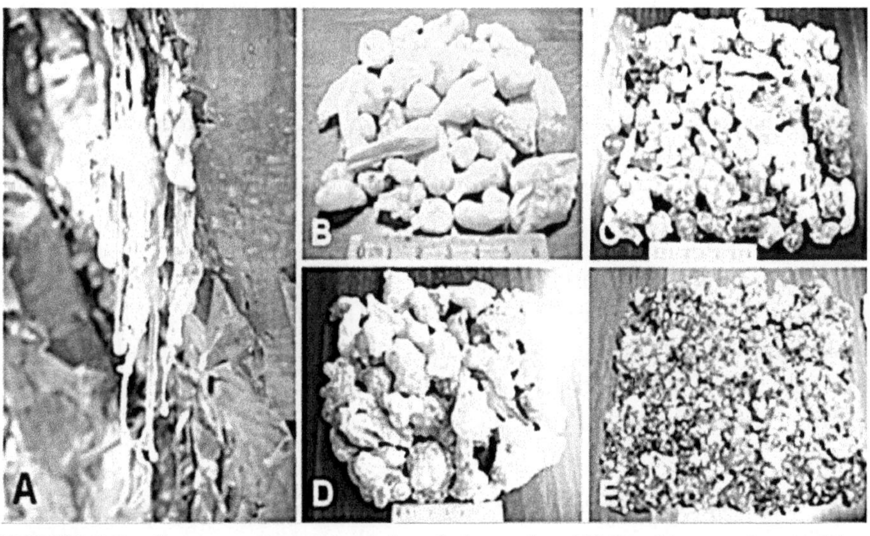

FIGURE 3.9 Gum secretion with grades of oleo resins. (A) Frankincense tree incision and tears; (B)–(E) gum resin grades obtained from *B. sacra*.

Source: Reprinted from Lin et al., 2013. https://creativecommons.org/licenses/by/4.0/

Khan et al. (2018) studied the molecular mechanism and involvement of phytohormones in relation to secretion of oleo gum from stem bark and yields regulated by Gibberellic acids, Jasmonic acid through modulation and expression of stress-related genes or enzymes during wounding. In another study by Rashan et al. (2021), oleoresins were collected from verified *B. sacra* trees during 2016 from the Dhofar region (Oman), is a good source of essential oil and having more phytochemical diversity (Yemini et al., 2021).

The oleoresins obtained through scarification (Figure 3.8) were gathered by conventional methods and graded as follows:

- Houjari was collected from Jebel Samhan (east of Salalah). These trees growing at higher altitudes with cool temperatures compared to the Najdi or Sahli cultivars, in addition, this area also receives considerably higher downpour (600–700 mm/year).
- The gum resin grade which is designated as Sahli were obtained from Al Magseel (Shabi valley) region which west of Salalah persistent with desert climate, with hotter temperature and low precipitation (110–360 mm/year), which is better than the Wadi Doka.
- The sample, Najdi (least graded), was supplied from Wadi Doka, on the plains of north of Salalah, receives desert climate, with lower precipitation (<100 mm/year) and steep variations in temperature.

Three different oleoresins of different grades were obtained from different regions of Dhofar, near Salalah (Rashan et al., 2021). In both cases, the extraction of oleoresin has followed the traditional methods from the plants of different origin with eco-climatic conditions, which also justified the grading quality of the gum.

3.9 *BOSWELLIA SERRATA* ROXB. (SALAI GUGGUL OR SALPHA TREE)

Boswellia serrata Roxb., commonly known as 'Salai Guggal,' is an important and multi-use tree species. The plant is a good source of oleo-gum-resin, which shows medicinal properties.

3.9.1 HARVESTING/COLLECTION OF GUM RESIN

Generally, trees at breast height with 90 cm girth are selected for gum tapping. Trees with <90 cm girth should not be tapped, because they may adversely affect their growth and development. A thin scratch about 30 cm in length and 20 cm width is made on the trunk of the tree at a height of 0.75 meters from the ground level to expose the resin canal and ducts to the environment. The scratch is freshened regularly after 4–5 days. After two weeks of scratching, they start to collect gum resin from scratched trees. The gum tapper scraps the tree bark in a half-circular ridge on the scratch and the gum is collected in a given interval. The scratch is widened by another round of scratch, and it widened accordingly (Vikaspedia, 2020).

Fortunately, oleoresin producing ducts in *B. serrata* are present in the superficial layers of the outer cortex of the bowl. The cambium which plays a key role in skin recovery is not affected in the tapping process is a good sign. The trees under tapping are recovering perfectly well and are re-tapped several times in the Salai lifetime.

3.9.1.1 COLLECTION/HARVESTING PERIOD

Generally, tapping and collection of gum resin begins in the month of November, and it closes at the beginning of monsoon. Collection of gum-resin may start after one month of scratching.

Consequent refreshing of the scratch and collection of gum resin from trees may be done every two weeks' time period.

3.9.1.2 YIELD

Average estimated gum resin yield to be 1 kg/tree/year. However, mature and healthy tree yields 2 to 2.5 kg gum resin/year. The production capacity of gum resin found to be varied according to locality, season, and size of the tree. In general, trees with larger girth, and vigorous growth yields more gum resin than trees with lower girth, growing in moist localities with very short height, bark with black.

India is the main distributor of *B. serrata* gum resin (Brendler et al., 2018). In the year 2008–2013, highest amount of gum resin (530 tons) was collected from Sheopur district of Madhya Pradesh state (Bhattacharya and Hayat, 2004), whereas only 99.8 tons from Gujarat and 27 tons from Andhra Pradesh, respectively (Yogi et al., 2014). Brendler et al. (2018) provided a synthesis of frankincense production in India.

3.9.2 PROCESSING AND VALUE ADDITION

Semi-solid substance secreted in scratched portion is collected, and impurities were cleaned manually. The collected crude gum is kept in a bamboo basket to flow out fluid portion. It is about 8 to 10% of the original crude gum used in varnishes and paints. The leftover solid part of the gum resin is dried completely and treated with powders of soapstone to make it brittle. Then, the gum is broken into small pieces and graded based on color and impurities.

Generally, in the market four grades of gums available. They are i) Superfine quality: Gum in light yellow color and translucent in appearance ii) Quality I: Gum in brownish yellow and translucent in appearance iii) Quality II. Gum in brown color and translucent in appearance vi) Quality III. Gum in dark brown color and opaque in nature.

3.9.2.1 BIO-ACTIVE CONSTITUENTS

The Salai gum contains terpenoids, essential oils, and gum. The terpenoid portion consists mixture of boswellic acids (BAs), which have been identified as pharmacologically active principles of gum resin of *Boswellia.*

3.9.3 PROPERTIES

- Depending on the collection season and presence of impurities, the gum appears transparent golden brown to greenish brown color.
- It consists volatile oil 8–10%, gum 30–35%, moisture 10–11%, insoluble matter 4–5%
- Penta cyclic boswellic acids are the chief components of the gum resin, which have been reported as potent anti-arthritic, anti-inflammatory, and anti-atherosclerotic properties.
- It dissolves in organic solvents.

- It shows softening point (65–72°C) and melting point from 73–78°C.
- It burns readily and diffuses a pleasant odor.
- The resin portion has an acid value of 1.87, ester value of 63.14, specific gravity of ointments 0.91, iodine value of around 200 and saponification value of 65.
- The gum portion consists carbohydrates like pentose and arabinose.

3.9.4 INDUSTRIAL APPLICATIONS

- Because of its unique aromatic nature, Salai gum resin is used as incense in traditional systems of medicine.
- In Ayurveda, the Indian Traditional medicinal system, the gum is used to cure arthritis and asthma. It is a more effective drug like phenylbutazone and other drugs used to reduce inflammatory condition.
- In indigenous medicine, the gum is used for rheumatism, astringent, and nervous disease.

3.9.5 COMMERCIAL GUM TAPPING METHODS IN SALAI-GUGGUL

For commercial purpose, tapping of gum resin from trees is done by making deep incision or peeling of stem or by blazing (Nair, 2000). Unscientific tapping of gum resin, such as making deep incision at one place repeatedly and overexploitation of gum resin leads to the death of trees. Rijkers et al. (2006) reported that frequent gum collection from *B. papyrifera* affects flowering fruiting conditions and production of viable seeds. Different researchers have made several attempts to establish standardized methods for sustainable yield of Salai gum resin (Mishra et al., 2012; Shiva, 2008; ICAR, 2014).

In this connection, Gufic Bioscience Ltd. Navsari, Gujarat, carried out a field experiment in *Boswellia* plantation of about 110 acers, located at Gothada village of Vadodara, Gujarat, to standardize gum tapping methods. About 30 trees of 14 years old were selected for their experiment.

The method of tapping is most important since it involves the life of the tree and the reproductive cycle. In this connection, the Indian Council of Agriculture Research (ICAR, 2014) developed and safe tapping methods depends on the age or diameter and health of the selected tree.

The girth at breast height (GBH) of trees selected for gum tapping should be 32 to 35 cm. About 10 kinds of incisions on the trees were made at the GBH with a half depth of its total bark depth (=0.75 cm) in the January (2016) first week. Instruments like chisel and mechanical driller were used for making hole in the bark. The illustrations (Figure 3.10) earmark the types of cuts used for gum tapping on stem bark. They are: (a) square cut; (b) circular cut; (c) spiral shape; (d) irregular cut; (e) V cut; (f) rill cut; (g) hole with 1.5 cm diameter; (h) triangular cut with bark lid; (i) V cut with 1.5 cm diameter hole at bottom of tree, treated with 3 ml of 39% ethephon solution. Three individual trees were selected for each type of cuts. Particularly, for hole method, two holes were made for each tree in opposite directions. The oleo gum resin ooze out after seven days period and was collected in a circular tray and yield was calculated using weighing digital balance. The highest yield of oleo-gum resin was found in hole made on bottom of tree in V-cut method and treated with ethephon. However, the yield was optimum in the case of the hole method treated with ethephon and patched up with clay. This may be due to use of ethephon, a plant growth regulator. A similar type of results was reported in other gum yielding plants (Nair, 2000; ICAR, 2014).

A very less amount of gum yield was observed in trees with irregular cut, triangular cut with bark lid, circular cut, and hole method. During tapping work, they noticed that slant cut yielded more gum resin than straight cut. Mishra et al. (2012) reported that no gum resin yield was observed in *Boswellia* trees with 35 cm girth size. They suggested that two methods, i.e., V-cut method with 1.5 cm width hole, treated with ethephon and cuts patched up with clay suitable methods for gum tapping in *Boswellia* plantations (Mishra et al., 2012).

Sinha et al. (2016) demonstrated the effect of ethephon on the stem bark of Salai guggul (Figure 3.10), which provides the basis for sustainable utilization through high yield in addition to improvement of oleo gum resin. Sinha et al. (2016) also suggested to have more scientific studies in this context to improve the boswellic acid.

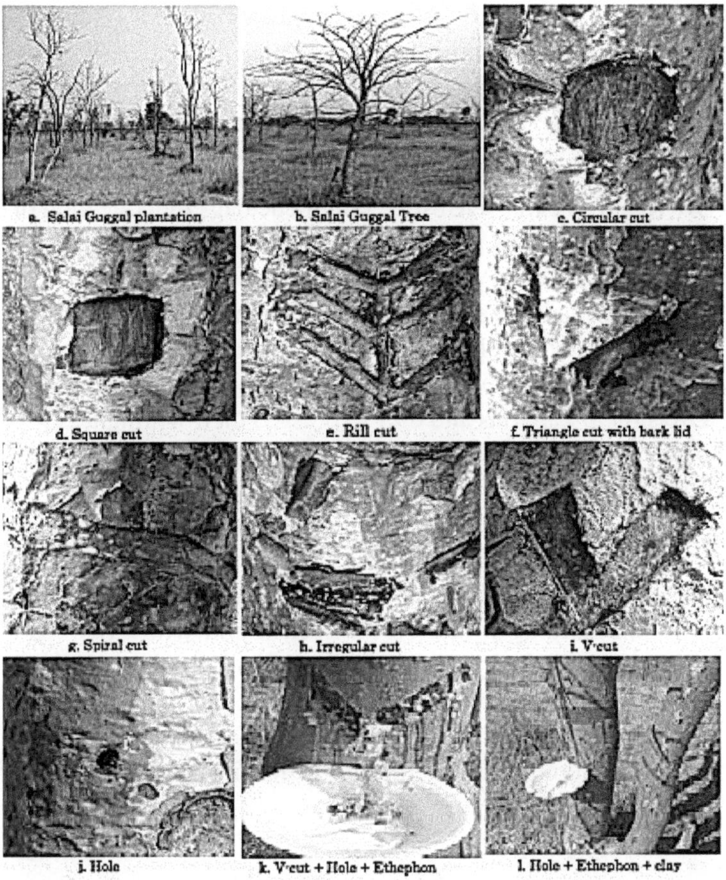

FIGURE 3.10 Pattern of tapping methods and gum extraction of *B. serrata* stem bark.
Source: Reprinted with permission from Sinha et al. (2016).

Nayak and Prajapathi (2020) investigated about the potentiality of gum yielding of natural forest stands of *B. serrata* in mixed stand of Deopur Forest Range in Balodabazar, Chhattisgarh, India. The two tapping techniques were tried for extracting of gum using the mechanical tapping techniques and mechanical along with chemical tapping techniques through making a square shape, V shape, and hole shape cuts and data has been collected in winter and summer seasons. The chemical inducers of gummosis are ethephon and injected in different cuts. Studies with ethephon treatment and mechanical tapping in two consequent seasons such as winter and summer (2018–2019) revealed that 20% ethephon demonstrated better results which yielded 52.94 g/tree and recommended a square type of scratch rather than V or hole type (Nayak and Prajapathi, 2020).

Further, Dr. N. Venugopal, a trainer (Madhya Pradesh, India) of gum extraction from *B. serrata*, made the following observations (personal communication):

1. We found an extensive use of polythene bags for collection of Cheed by the Sangrahaks. The polythene gets thoroughly adhered to the oleoresin and doesn't get out of it. Everyone nowadays is aware of the polythene issue as a universal problem. It's a big quality issue and we need to handle effectively to protect the market. For this, we have shown a Bamboo basket being used by Gond tribes at Chhatarpur Division to Saharias and they expressed their willingness to use it in place of Polythene bags. Therefore, Supply of Bamboo baskets for collection of Cheed Gond will address the quality issue.
2. The Oleoresin collected is admixed with a lot of foreign organic materials. Sorting out these can help improving the quality. They are using a black iron 'Ghantia' for collection of Cheed which is a bell hanging in the cow's neck found with every Sangrahak (Figures 3.11–3.13). This requires an attitudinal change towards the quality aspect. If a stainless-steel vessel is provided to collect the Cheed conveniently in place of 'Ghantia' may be useful in avoiding some amount of impurities into the oleoresin. These amounts to primary value addition.

FIGURE 3.11 *B. serrata* gum tapping tools and oleoresin tapping.

Source: Dr. N. Venugopal, with permission.

FIGURE 3.12 *B. serrata* oleoresin oozing pattern from blazes pattern on stem bark.

Source: Dr. N. Venugopal, with permission.

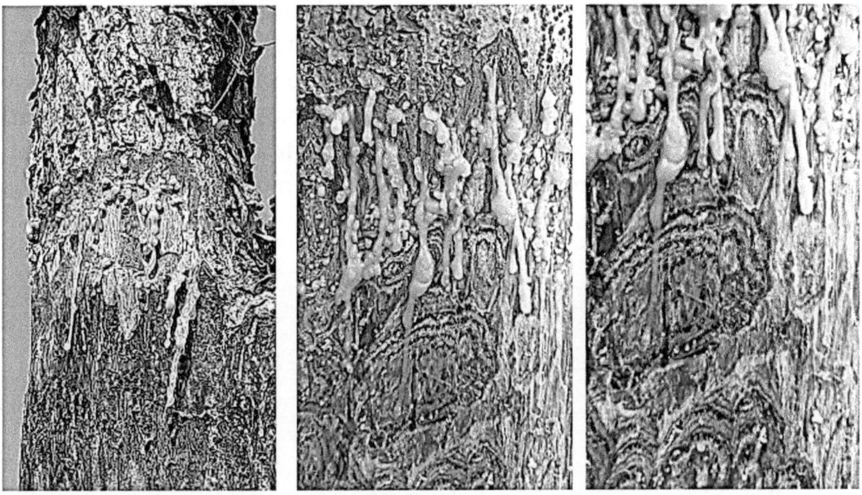

FIGURE 3.13 *B. serrata* oleoresin oozing pattern from blazes.

Source: Dr. N. Venugopal, with permission.

3. We found in many places the trees are having bigger gall like many protrusions on the bole and they are due to the unskilled and indiscriminate blazing by the new generation. This is reducing the trappable area and also loss to the tree health. To handle this problem, we need to work hand in hand and make several follow-ups. We cannot make them stop their unintentional habit easily. To bring in the behavioral and psychological change in the Sangrahaks mindset it demands a polite intermingled

handholding support. This is a time-consuming effort to save the resource (Dr. N. Venugopal, Personal communication).

3.10 CONCLUSION

The immense economic value, religious belief, and medicinal potentiality of oleogum resin of species of *Boswellia* genus have been apprised and leave the plant with more stress for high yields. Many of the species were under the threatened category of the IUCN list seeking conservation and sustainable utilization of natural resources. However, many scientific reports were mentioned on the systematic tapping methods, especially with the species such as *B. dalzielii*, *B. papyrifera*, *B. sacra* (syn. *B. carterii*), *B. serrata*, etc., providing seasonal and altitudinal variations also. In a few countries, the traditional tapping methods and over-tapping of the frankincense and trees were facing disturbance in the reproductive cycle and not able to produce fertile seeds for natural propagation apart from pollination and geographical isolation in a few species. More scientific investigations are to be focused on tapping methods or conservation strategies for its sustainable utilization.

KEYWORDS

- ***Boswellia* species**
- **Burseraceae**
- **dhoop**
- **frankincense**
- **gum extraction**
- **ritual ceremonies incense**
- **supercritical fluid extraction**
- **tapping**

REFERENCES

Alemu, A. A., Pretzsch, J., Mahmoud, T. E. S., & Adam, Y. O., (2012). Commodity chain of frankincense from the dry woodlands of Nuba Mountains, South Kordofan State, Sudan. *Small-Scale Forestry, 11*(3). doi: 10.1007/s11842-011-9189-4.

Ali, A. H., Kamal, E. M. F., & Musa, A., (2009). Effect of position of tapping, tree stem diameter and tapping tools on frankincense yield of *Boswellia papyrifera* in South Kordofan State, Sudan *Forests, Trees and Livelihoods, 19*, 19–26.

Aregawi, B., (1997). *Preliminary survey on forest products utilization and marketing in Tigray* (p. 67). Tigray Bureau of Agricultural Development and Natural Resources, Mekelle, Ethiopia.

Ayubova, M., Zahra, O. G., Moktar, O. G., Hugues, B., & Baldovini, N., (2019). Analytical investigations on *Boswellia occulta* essential oils. *Phytochemistry, 164*, 78–85.

Basar, S., (2005). *Phytochemical Investigations on Boswellia Species: Comparative Studies on the Essential Oils, Pyrolysates and Boswellic Acids of Boswellia Carterii Birdw., Boswellia Serrata Roxb., Boswellia Frereana Birdw., Boswellia Neglecta S. Moore and Boswellia Rivae Engl.* PhD Thesis, Institute of Organic Chemistry, University of Hamburg.

Boutin, O., & Badens, E., (2009). Extraction from oleaginous seeds using supercritical CO_2: Experimental design and products quality. *J. Food Eng., 92*, 396–402.

Brendler, T., Brinckmann, J. A., & Schippmann, U., (2018). Sustainable supply, a foundation for natural product development: The case of Indian frankincense (*Boswellia serrata* Roxb. ex Colebr.). *J. Ethnopharmacol., 225*, 279–286.

Coppen, J. J. W., (1995). *Flavors and Fragrances of Plant Origin.* Food and Agriculture Organization. Rome.

DeCarlo, A., Johnson, S., Ouédrago, A., Dosoky, N. S., & Setzer, W. N., (2019). Chemical composition of the oleogum resin essential oils of *Boswellia dalzielii* from Burkina Faso. *Plants, 8*, *223*, 1–11.

DeCarlo, A., Johnson, S., Okeke-Agulu K.I., Dosoky N.S., Wax, S.J., Owolabi M.S., & Setze, W.N., (2019b) Compositional analysis of the essential oil of Boswellia dalzielii frankincense from West Africa reveals two major chemotypes. Phytochemistry 164, 24-32. https://www.sciencedirect.com/science/article/abs/pii/S0031942219300858

Eshete, A., (2002). *Regeneration Status, Soil Seed Bank and Socio-Economic Importance of Boswellia papyrifera (Del.) Hoschst. in Two Woredas of North Gondar Zone, Northern Ethiopia.* Master's Thesis, Swedish University of Agricultural Sciences and Wondo Genet College of Forestry.

Eshete, A., Sterck, F. J., & Bongers, F., (2012a). Frankincense production is determined by tree size and tapping frequency and intensity. *Forest Ecology and Management, 274*, 136–142.

Eshete, A., Teketay, D., Lemenih, M., & Bongers, F., (2012a). Effects of resin tapping and tree size on the purity, germination and storage behavior of *Boswellia papyrifera* (Del.) Hochst. seeds from Metema District, Northwestern Ethiopia. *Forest. Ecol. Manag., 269*, 31–36.

Gachathi, N. F., & Muga, O. M., (2002). *A Review of Commercial Gums and Gum Resins Resources in Kenya: Taxonomy, Ecology and Distribution* (pp. 1–15).

http://197.248.75.118:8282/jspui/bitstream/123456789/400/1/KEFRI%20Conference%202008%20final.pdf (accessed on 10 October 2022).

Gebrehiwot, K., Muys, B., Haile, M., & Mitloehner, R., (2003). Introducing *Boswellia papyrifera* (Del.) Hochst and its non-timber forest product, frankincense. *Intern. Forestry Review, 5*, 348–353.

Girma, A., Andrew, K. S., De Bie, C. A. J. M., & Bongers, F., (2015). Understanding *Boswellia papyrifera* tree secondary metabolites through bark spectral analysis. *ISPRS J. Photogrammetry and Remote Sensing, 105*, 30–37.

Girmay, F., (2000). *The Status of gum Arabic and Resins in Ethiopia* (pp. 14–22). Report of the Meeting of the Network for Natural Gum and P Resins in Africa (NGARA), Nairobi, Kenya.

Goettsch, E., (1986). Traditional aromatic and perfume plants in Central Ethiopia (a botanical and ethnohistorical survey). *J. Ethiopian Studies, 19*, 81–90.

Gonzalez, K., (2020). *Impacts of Unsustainable Harvesting of Frankincense Producing Boswellia Trees.* Dissertation Paper submitted to Oregon State University.

Groom, N., (1981). *Frankincense and Myrrh: A Study of The Arabian Incense Trade*. Longman, London and New York.

Hamidpour, R., Hamidpour, S. M., Hamidpour, M., & Shahlari, M., (2013). Frankincense (Rǔ Xiang; *Boswellia* species): From the selection of traditional applications to the novel phytotherapy for the prevention and treatment of serious diseases. *J. Trad. Comp. Med., 3*, 221–226.

Harsh, L. N., Khan, H. A., Bohra, M. D., & Tewari, J. C., (2003). Growth performance and gum production of *Acacia senegal* in different landforms in hot arid zone of India. *J. Non-Timber Forest Products, 10*, 1–7.

Hassan, A. M. M., (2012). Distribution *Boswellia sacra* in Dhofar Mountains, Sultanate of Oman: Economic value and environmental role. *J. Life Sci., 6*, 632–636.

ICAR, (2014). Network project on harvesting processing and value addition of natural resins and gums. Annual Report. *Indian Council of Agricultural Research, 2014*, 1–21.

Johnson, S., DeCarlo, A., Satyal, P., Dosoky, S. N., Sorensen, A., & Setzer, N. W., (2019). Organic certification is not enough: The case of the methoxydecane frankincense. *Plants, 88*(8), 1–10.

Kebede, T., (2010). *Current Production Systems of Frankincense from B. papyrifera Tree, its Implications on Sustainable Utilization of the Resource*. MSc Thesis, Mekele University, Ethiopia.

Khamis, M. A., (2001). *Management of Boswellia papyrifera Stands for Resin Production in Jebel Marra Area, W. Sudan, Present Situation and Future Prospects*. M.Sc. thesis, Dresden, Germany.

Khan, A. L., Al-Harrasi, A., Raheem, S., Qari, M. I., Byung-Wook, Y., Yoon-Ha, K., Sang-Mo, K., et al., (2018). Regulation of endogenous phytohormones and essential metabolites in frankincense-producing *Boswellia sacra* under wounding stress. *Acta Physiologiae Plantarum, 40*, 113.

Lemenih, M., & Teketay, D., (2003). Frankincense and myrrh resources of Ethiopia: Its distribution, production, opportunities for dryland development and research needs. *SINET Ethiop. J. Sci., 26*, 63–72.

Lemenih, M., Abebe, T., & Olsson, M., (2003). Gum and resin resources from some *Acacia, Boswellia* and *Commiphora* species and their economic contributions in Liban, South-East. *J. Arid Environ., 55*, 465–482.

Lin, H. K., Suhail, M. M., Fung, K. M., Woolley, C. L., & Young, D. G., (2013). Extraction of biologically active compounds by hydro-distillation of *Boswellia* species gum resins for anti-cancer therapy. *OA Alternative Medicine, 1*(1), 4.

Marshall, S., (2003). Frankincense: Festive pharmacognosy. *Pharmaceut. J., 271*, 862–864.

Mertens, M., Buettner, A., & Kirchhoff, E., (2009). The volatile constituents of frankincense: A review. *Flavor and Fragrance Journal, 24*(6), 279–300.

Mishra, S. P., Behera, N., & Paramanik, T., (2012). Comparative assessment of gum yielding capacities of *Boswellia serrata* Roxb. and *Sterculia urens* Roxb. in relation to their girth sizes. *The Ecoscan (special issue), 1*, 327–330.

Mokria, M., Motuma, T., Frank, J. S., Aster, G., Frans, B., Mathieu, D., & Ute, S., (2017). The frankincense tree *Boswellia neglecta* reveals high potential for restoration of woodlands in the Horn of Africa. *Forest Ecology and Management, 385*, 16–24.

Murthy, K. S. R., Reddy, M. C., Sandhya, R. S., & Pullaiah, T., (2016). Bioactive principles and biological properties of essential oils of *Burseraceae*: A review. *J. Pharmacogn. Phytochem., 5*(2), 247–258.

Nair, B., (2000). Sustainable utilization of gum and resin by improved tapping technique in some species. In: *Harvesting of Nonwood* (p. 293). FAO, Italy. http://www.fao.org/docrep/019/y4496e/y4496e.pdf (accessed on 10 October 2022).

Nayak, A. P., & Prajapati, R. K., (2020). Standardization of sustainable oleo gum resin tapping technique in Salai (*Boswellia serrata* Roxb.) from tropical dry deciduous forest of Chhattisgarh. *Int. J. Curr. Microbiol. App. Sci., 9*(10), 2281–2296.

Ni, X., Suhail, M. M., Yang, Q., Cao, A., Fung, K. M., Postier, R. G., et al., (2012). Frankincense essential oil prepared from hydro-distillation of *Boswellia sacra* gum resins induces human pancreatic cancer cell death in cultures and in a xenograft murine model. *BMC Comp. Altern. Med., 12*, 253.

Ogbazghi, W., (2001). *The Distribution and Regeneration of Boswellia papyrifera (Del.). Hochst. in Eritrea.* PhD Thesis. Wageningen University and Research Centre, The Netherlands.

Ouédraogo, A., Thiombiano, A., Hahn-Hadjali, K., & Guinko, S., (2006). Régénération sexuée de *Boswellia dalzielii* Hutch, un arbre médicinal de grande valeur au Burkina Faso. *Bois. Et. Forêts. Des. Tropics, 289*, 48.

Rashan, L., Alan, W., Manon, H., Nicolas, F., Parag, D., & Ian, E. C., (2021). Chemical composition, antibacterial activity, and antibiotic potentiation of *Boswellia sacra* Flueck. oleoresin extracts from the Dhofar Region of Oman. *Evidence-Based Complementary and Alternative Medicine*, 1–23, Article ID 9918935.

Rijkers, T., Ogbaschi, W., Wessel, M., & Bongers, F., (2006). The effect of tapping for frankincense on sexual reproduction in *Boswellia papyrifera. J. Appl. Ecol., 43*, 1188–1195.

Shiva, A., (2008). Method of sustainable harvesting and value addition for economic uplift and biodiversity conservation. *MFP News, 18*, 21, 22.

Sinha, S. K., Pathak, J. G., Mehta, A. A., & Behera, L. K., (2016). Tapping methods in Salai guggal (*Boswellia serrata* Roxb.) for sustainable yield of oleo-gum resin: A case study. *Int. J. Usuf. Mngt., 17*(2), 13–18.

Tadesse, W., Desalegn, G., & Alia, R., (2007). Natural gum and resin bearing species of Ethiopia and their potential applications. *Investigación Agraria: Sistemasy Recursos Forestales, 16*(3), 211–221.

Tadesse, W., Feleke, S., & Eshete, T., (2004). Comparative study of traditional and new tapping methods on frankincense yield of *Boswellia papyrifera. Ethiopian J. Nat. Resour., 6*, 287–299.

Tadesse, W., Tatek, D., Gizachew, Z., & Getachew, D., (2020). Underutilized natural gum and resin resources in Ethiopia for future directions and commercial utilization. *World J. Agric. Res., 8*(2), 32–38.

Thulin, M., & Warfa, A. M., (1987). The frankincense trees (*Boswellia* spp., Burseraceae) of northern Somalia and southern Arabia. *Kew Bulletin, 42*, 487–500.

Thulin, M., Decarlo, A., & Johnson, S., (2019). *Boswellia occulta* (Burseraceae), a new species of frankincense tree from Somalia (Somaliland). *Phytotaxa., 394*. 219. 10.11646/phytotaxa.394.3.3.

Tilahun, G., (1997). *Boswellia Papyrifera (Del.) Hochst. from Western Tigray: Opportunities, Constraints and Seed Germination Responses.* MSc Thesis, Swedish University of Agricultural Sciences, Skinnskatteberg.

Tilahun, M., Muys, B., Mathijs, E., Kleinn, C., Olschewski, R., & Gebrehiwot, K., (2011). Frankincense yield assessment and modeling in closed and grazed *Boswellia papyrifera* woodlands of Tigray, Northern Ethiopia. *J. Arid Environments, 75*, 695–702.

Tucker, A., (1986). Frankincense and myrrh. *Econ. Bot., 40*(4), 425–433.

Uphof, J. C. Th., (1959). *Dictionary of Economic Plants* (2nd edn., 1968). Lubrecht & Cramer Ltd.

Van, B. G., (1960). Frankincense and myrrh. *Bibli. Archaeol., 23*(3), 7095.

Verghese, J., (1988). Olibanum in focus. *Perfumer and Flavorist, 13*(1), 1–12.

Vikaspedia: Salai, (2020). https://vikaspedia.in/agriculture/post-harvest-technologies/natural-resins-and-gums-of-commercial-importance/introduction (accessed on 10 October 2022).

Vollesen, K., (1989). In: Hedberg, & Edward, S., (eds.), *The Flora of Ethiopia* (Vol. 3, No. I, pp. 442–478). Addis Ababa University: Addis Ababa.

Wubalem, T., Demel, T., Mulugeta, L., & Girmay, F., (2002). *Boswellia* country report for Ethiopia. In: Chikamai, B., (ed.), *Review and Synthesis on the State of Knowledge of Boswellia Species and Commercialization of Frankincense in the Dry Land of Eastern Africa* (pp. 1–33). FAO/EU/FORNESSA Publicatio*n*.

Yadav, S., & Gupta, S. K., (2018). A systematic review on frankincense (*Boswellia* species). *Intern. J. Innovative Res. Technol., 5*(5), 212–224.

Yamei, F., Qiang, Z., & Sun, L., (2021). Five terpenoids from the gum resin of *Boswellia carterii* and their cytotoxicity. *Fitoterapia, 154*, 1–6.

Yogi, R. K., Bhattacharya, A., & Jaiswal, A. K., (2014). *Lac, Plant Resins and Gums Statistics at a Glance 2013* (pp. 1–38). ICAR – Indian Institute of Natural Resins and Gums, Ranchi (Jharkhand), India. Bulletin (Technical) No. 06/2014.

CHAPTER 4

PHARMACOGNOSY OF *BOSWELLIA* SPECIES

LEPAKSHI MD. BHAKSHU[1] and K. VENKATA RATNAM[2]

[1]*Department of Botany, PVKN Government College, Chittoor – 517002, Andhra Pradesh, India*

[2]*Department of Botany, Rayalaseema University, Kurnool – 518007, Andhra Pradesh, India*

**Corresponding author. E-mail: lmbakshu@gmail.com*

ABSTRACT

Frankincense, an aromatic oleo gum resin, which is also popular as olibanum and maximum extent, has been obtained from the four tree species of the taxon *Boswellia*, a rosid Family Burseraceae. The species of *Boswellia* were found to be highly commercial, and hence there might be a chance of adulteration or mixing with inferior gums or plant parts. Hence, pharmacognostic studies help the market men to identify its original source or pure form or its grade. The present discussion paved an understanding of the structural aspects of plant material anatomically and its architecture of gum-secreting organs or orientation of schizogenous canals in the wood and bark. It also helps the gum tappers with the usage of tools for scratching the bark and the extent of the injury on the stem bark for the collection of quality oleo resin aromatic gum without affecting the tree life. The species such as *B. ovalifoliolata, B. papyrifera, B. sacra,* and *B. serrata* are being extensively investigated by scientists globally, and they are discussed in

Frankincense – Gum Olibanum: Botany, Chemistry, Extraction, Utilization, Propagation, Biotechnology, and Conservation, T. Pullaiah, K. Venkata Ratnam, Mallappa Kumara Swamy, Lepakshi Md. Bhakshu (Eds.)

this chapter. However, there are many other *Boswellia* species that are yet to be explored scientifically for their pharmacognostic values.

4.1 INTRODUCTION

Pharmacognosy is a most relevant offshoot of science for the discovery of future medicines and the researchers are encouraged to be mindful of the ethical issues with using natural products as medicines. This science is dedicated to providing proof for correct identification with the help of different technical applications of science to avoid the adulteration of inferior materials or with the parts of allied species. Most of the people of third world countries depend on the traditional systems of medicines in which herbs are an important component and developed as part of the pharmacognosy, play a crucial role in the discovery of safe and effective drugs (Orhan, 2014).

Additionally, the science of medicines has a great demand in the preparation of pure and quality drugs requires high-grade standards and herbal uniquity for safe usage and attention to the risks when used for long periods. In this connection, the present discussion focused on the pharmacognostic reports contributed by global scholars for the correct identification of the oleoresin-obtaining *Boswellia* species, which are essential for the preparation of effective medicines. There are about 24 species of *Boswellia* authenticated and reported to yield oleoresin from its stem bark. However, only certain species were scientifically studied on their pharmacognosy. Since the gum resin of species of *Boswellia* genus were commercially, religiously, and medicinally gained importance and requires pharmacognostic examination for quality control and its quality upgradation.

Among the 24 authenticated species of *Boswellia* (Al-Hassari et al., 2019) some species, such as *B. ameero, B. bullata, B. dalzielii, B. dioscorides, B. elongata, B. frereana, B. nana, B. neglecta, B. occulta, B. pirottae, B. popoviana, B. rivae, B. socotrana*, etc., await the pharmacognostic properties. Thulin (2020) finalized 24 species authenticated so far and described the life forms of these trees.

The pharmacognostic studies reported so far for a few species of *Boswellia* are discussed in furthermore sections.

4.2 *BOSWELLIA OVALIFOLIOLATA* BAL. & HENRY

It is an endemic species restricted to the Seshachalam biosphere of Eastern Ghats, India, gained importance for its medicinal values and aromatic gum resin. The pharmacognostic studies by Sathis Kumar et al. (2011) are given briefly here.

4.2.1 TRANSVERSE SECTION OF LEAF

Histological characters of *B. ovalifoliolata* leaf transverse section showed typical dorsiventral leaf. It includes the following parts as outer epidermis, lower epidermis, and internal tissues. The upper epidermis was single-layered, covered with distinct cuticle. Below the upper epidermis, palisade, and spongy tissues are present. Below the spongy tissue parenchymatous lower epidermis is present. Vascular bundles are enclosed by a layer of pericycle and are embedded in the middle of the spongy parenchyma (PC). The vascular bundles were described as conjoint collateral and open in which xylem and phloem arranged as layers enclosed in bundle sheath.

4.2.2 TRANSVERSE SECTION OF STEM

The transverse section of *B. ovalifoliolata* shoot exhibited three marked zones, from exterior to interior. The epidermis formed as an outermost layer enclosed in the thin cuticle. A collenchymatous tissue below the epidermis, which contain thick cell walls, provides mechanical support. The cortical tissue possesses thin cells of PC articulated in many layers. The inner part of the stem in its cross-section named as endodermis found below the cortex. The endodermis contains the oil glands followed by vascular bundles covering with the pericyclic cell layers. The vascular bundles are with xylem and phloem along with the medullary rays. The central portion constituted the pith composed of parenchymatous cells with intercellular spaces.

4.2.3 MICROSCOPIC STUDIES OF PLANT POWDER

Microscopic studies of the whole plant powder showed the presence of xylem vessels with and without lignified and pitted walls, thin-walled and polygonally arranged cork cells and prism, and square-shaped calcium oxalate crystals. The epidermis is single-layered, and it bears paracytic stomata, and unicellular and glandular trichomes are present.

Along with this type, parenchymatous cells with trichomes and circular or oval-shaped starch grains were also observed. Quantitative estimation of cell components revealed that the leaf power consists of phloem fibers (225–425/50–100 μl/b ratio), starch grains (37.5–75 μ diameter), calcium oxalate crystals l/b ratio 11–18/8–12 mm^2, stomatal number on upper (7) and lower epidermis (12), stomatal index on the upper epidermis (22.58), the lower epidermis (33.33).

4.2.4 PHYSICOCHEMICAL PROPERTIES

Quantitative physicochemical parameters of powdered drug of the whole plant of BO were determined as %w/w. It was revealed that the total amount of ash (5%), ash insoluble in acid (0.9%), water-soluble ash (1.4%), sulfated ash (0.66%), alcohol soluble extractives (17.44), water-soluble extra (25.12%), moisture content (6.6%), crude fiber content (50.5%) and foreign matter (2%), respectively.

4.2.5 PRELIMINARY PHYTOCHEMICAL SCREENING

The powdered drug was estimated for the residues after dissolution in the solvents and the extractives were disclosed as 0.16, 0.66, 2.89, 5.6, and 1.26% for the petroleum ether, chloroform, EA (ethyl acetate), ethanol, and aqueous portions respectively and observed its sticky property. The qualitative phytochemical studies of the petroleum ether extract showed fats and oils; chloroform extract, the flavonoids, proteins and amino acids, fats, and oils; EA extract, the carbohydrates, flavonoids, alkaloids, phenolic compounds, amino acids, proteins, glycosides, fats, and oils. Whereas, the alcohol extract reacted for alkaloids, flavonoids, phenolic compounds, carbohydrates, amino acids, proteins, and glycosides. The aqueous extract showed the presence of phenolic compounds, flavonoids, carbohydrates, amino acids, proteins, and saponins (Sathis et al., 2011).

4.2.6 HISTOCHEMICAL STUDIES

The studies on histology provided information on the localization of groups of secondary metabolites using appropriate chemical reagents or dyes. In this connection, Savithramma et al. (2014) reported the histochemical examinations and discussed them here. The distribution of phytoconstituents, namely, polyphenols, tannins, crystals, and starch grains in leaf, stem, stembark, and roots of *B. ovalifoliolata* stained by using specific chemicals or reagents ($FeCl_3$, toluidine blue reagent, Iodine solution, and HCl). The stained resulted development of bluish black, purple, or blue, bluish green, and dark black colors inferring the occurrence of tannins, polyphenols, starch grains, and crystals, respectively, in various regions like epidermis, endodermis, midrib, cortex, and vascular bundle of leaf (Figure 4.1); stem (Figure 4.2); stem bark (Figure 4.3); and root (Figure 4.4) of *B. ovalifoliolata.*

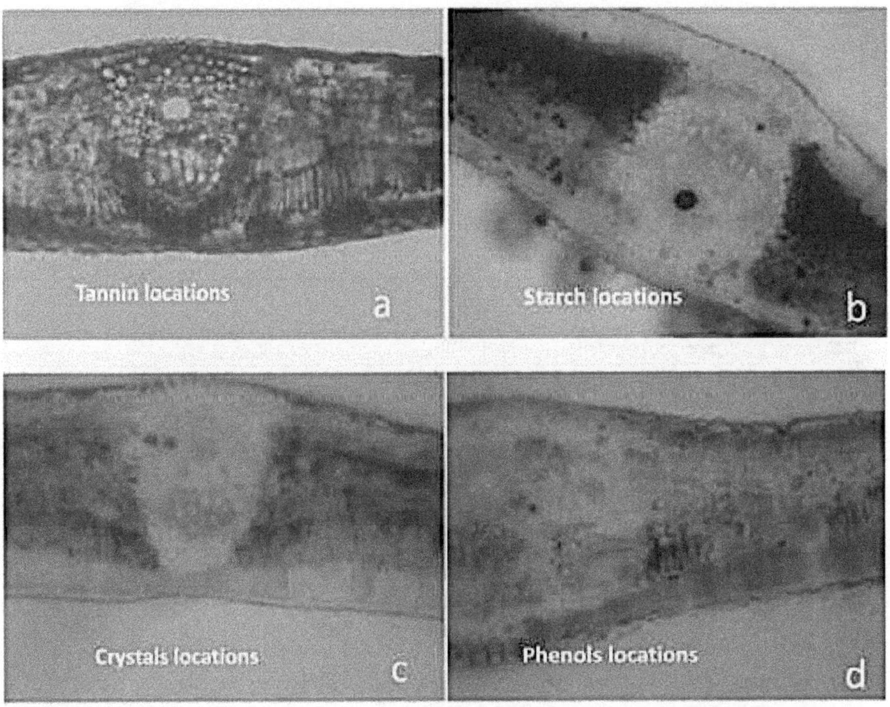

FIGURE 4.1 Histo-localization of components in the leaf of *B. ovalifoliolata*: (a) location of tannins; (b) location of starch; (c) location of crystals; and (d) location of phenols.

Source: Reprinted from Savithramma et al., 2014. © International Journal of Pharmacognosy and Phytochemical Research. Open access

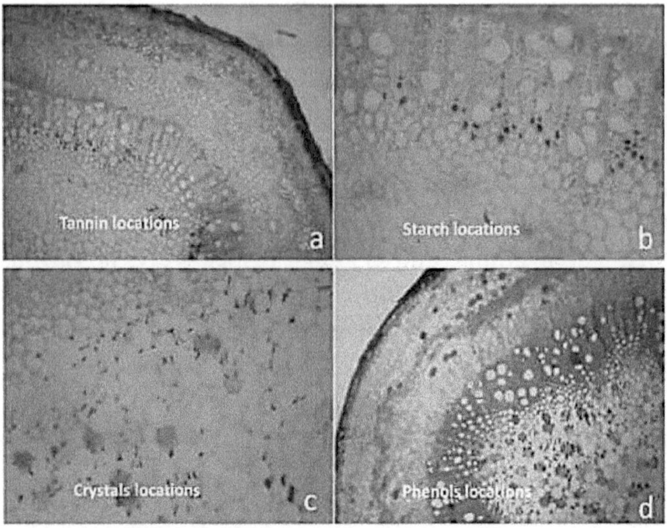

FIGURE 4.2 Histo-localization of components in stem of *B. ovalifoliolata*: (a) location of tannins; (b) location of starch; (c) location of crystals; and (d) location of phenols.

Source: Reprinted from Savithramma et al., 2014. © International Journal of Pharmacognosy and Phytochemical Research. Open access

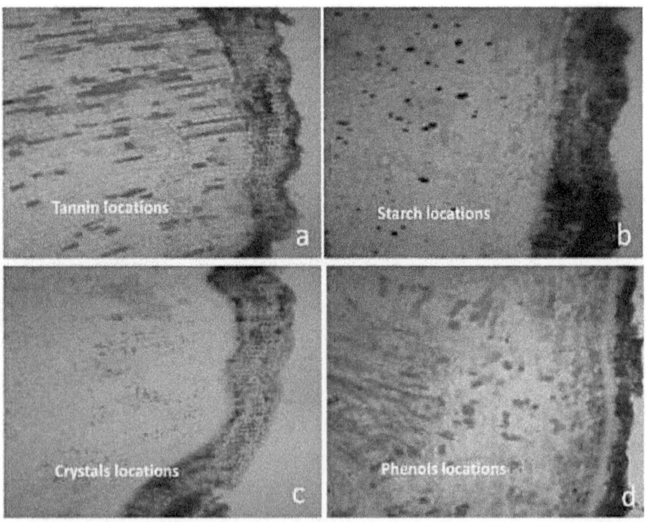

FIGURE 4.3 Histo-localization of components in stembark of *B. ovalifoliolata*: (a) location of tannins; (b) location of starch; (c) location of crystals; and (d) location of phenols.

Source: Reprinted from Savithramma et al., 2014. © International Journal of Pharmacognosy and Phytochemical Research. Open access

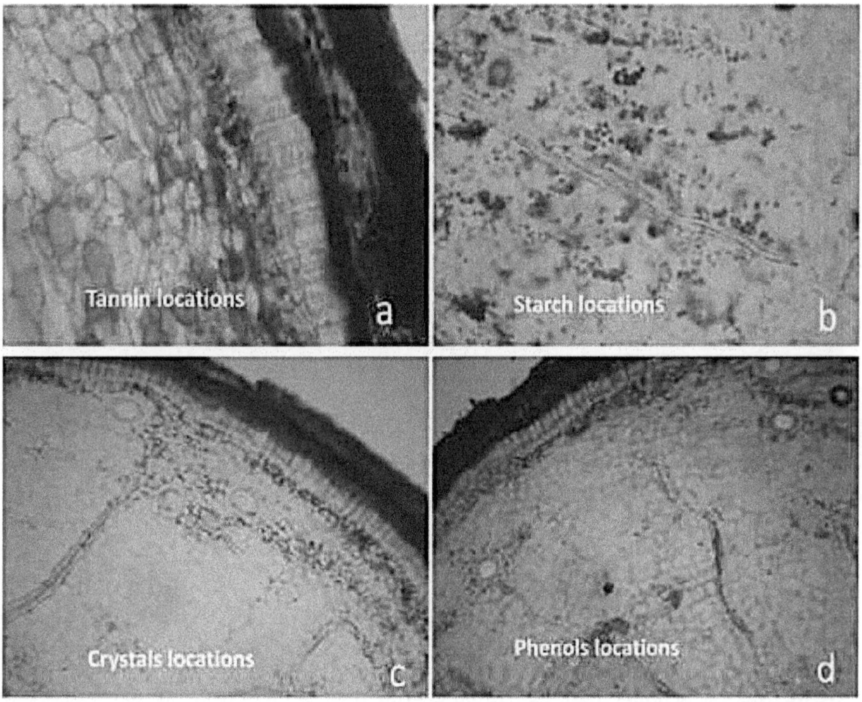

FIGURE 4.4 Histo-localization of components in roots of *B. ovalifoliolata*: (a) location of tannins; (b) location of starch; (c) location of crystals; and (d) location of phenols.

Source: Reprinted from Savithramma et al., 2014. © International Journal of Pharmacognosy and Phytochemical Research. Open access

4.3 *BOSWELLIA PAPYRIFERA* (DEL.) HOCHST.

B. papyrifera is a deciduous tree species, reported for the majorly commercialized and tradable white aromatic incense. It is distributed in Ethiopia, Cameroon, the Central African Republic, Eritrea, Sudan, Nigeria, Chad, and Uganda (northeast). The pharmacognostic study of stem bark was conducted by Tolera et al. (2013) and followed the standard protocols as given below.

The inner bark consists of three zones (Figure 4.5). The intact zone (first zone) shows a portion of the inner bark, which is unaffected by dilatation and located closely with vascular cambium (Figure 4.5(A)). The other portion of the inner bark is affected and dilatated and is divided into partly and highly dilatated zones exhibited structural differences. The partly dilatated zone (Figure 4.5(B)) is close and adjacent to the intact zone. This is not much affected and occupied broader portion made up of sclerenchyma (SC). Whereas, the much-dilatated region (third zone) is dominated by PC (Figure 4.5(B)).

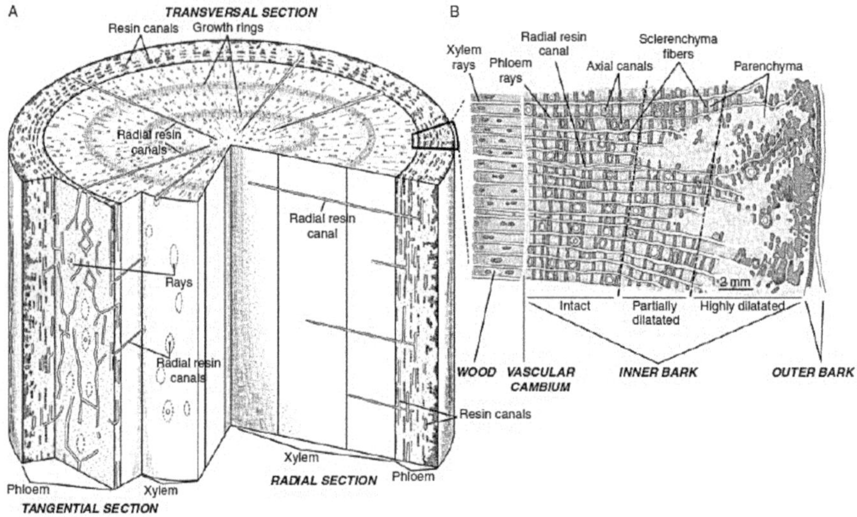

FIGURE 4.5 The inner part of the bark with microscopic structure showing resin secretion units of a *B. papyrifera*: (A) resin secretion structure in the bark and xylem; and (B) resin canals and associated tissues of the bark.

Source: Reprinted with permission from Tolera et al., 2013. © 2013 Oxford University Press.

4.3.1 BARK ANATOMICAL STRUCTURE

The internal part of bark layered around the stem of *B. papyrifera* with 17.2 mm on average occupied externally for a 15% its radius and consists of two prominent portions. The inner broad and outer one with narrow zone originated from the cambium (Figure 4.6(A) and (B)). The outer bark (OB) zone showed thin-walled cells with one or two layers and peels off as thin flakes. Whereas the inner bark consisting multiple tangential layers of SC fibers made up of thick walls and PC made up of thin-walled cells as layers (Figure 4.6(B)). The architecture of the inner bark has been well explained with its detailed cell components of three zones, as shown in Figure 4.6(A)–(C) – can be a landmark in the Pharmacognostic studies of *B. papyrifera.*

4.3.2 RESIN SECRETORY STRUCTURES OF B. PAPYRIFERA

The inner bark is reported as a major source of resin-producing structures with axial as well as radial canals. Resin canals were found in the wood only a few radial canals were in the ray tissue (Figure 4.6(A) and (B)). These canals continue with the inner bark via phloem rays (Figure 4.7(A)) and form a network within the bark. The organization of resin canals were clear in cross-section (Figure 4.6(A)) in which the axial canals (AC) are abundant than the radial canals and the layered with epithelial cells where lignified (inner bark of intact zone), whereas axial resin canal cells were non-lignified (Figure 4.7(C) and (D)). The intact zone part shows mutually interconnected (Figure 4.7(A) and (B)).

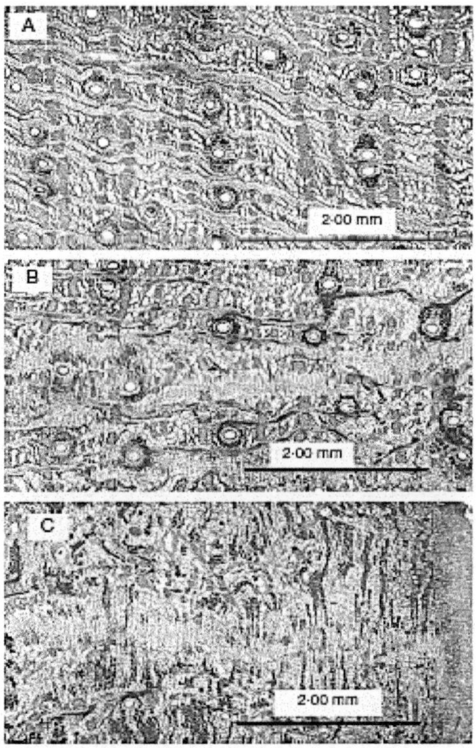

FIGURE 4.6 The zones of bark of transverse section of *B. papyrifera* with tissue ornamentation of resin-secreting parts. (A) The intact zone; (B) the partly dilated; and (C) the highly dilated zone.

Source: Reprinted with permission from Tolera et al., 2013. © 2013 Oxford University Press.

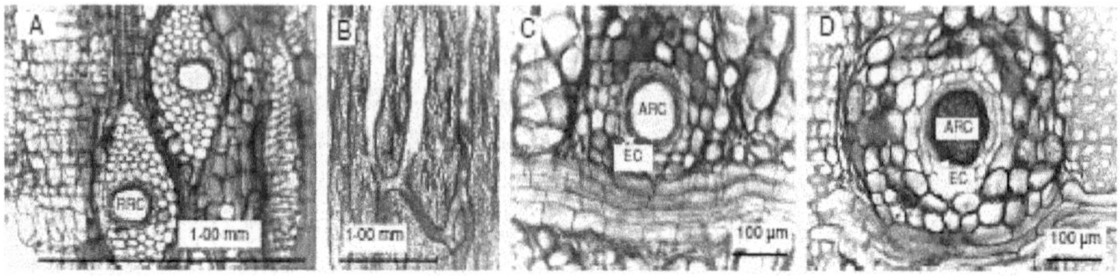

FIGURE 4.7 Architecture of resin canals of bark tissues with the resin-producing structures with different cell components.

Source: Reprinted with permission from Tolera et al., 2013. © 2013 Oxford University Press.

4.3.3 DISTRIBUTION, DENSITY, AND SIZE OF AXIAL RESIN CANALS IN THE INNER BARK

The anatomy of stem bark with intact (39%), partially dilatated (28%), and highly dilatated zones (33%). The details of distribution and proportion of resin canal cells were clearly discussed and evaluated by Tolero et al. (2013) by studying 20 trees with a stem girth up to 232 mm and demonstrated that the diameter varies among the tested trees (Figure 4.7(A) and (B)). It also found that the resin transport radially or axially of resin through the canals from wood and bark and vice versa. Based on the above study, the gum tappers or trainees may improve the gum extractions methods to ease for high yields and sustainable production of frankincense which would be livelihood for poor tribal people and contributes to the nation's economy (Tolera et al., 2013).

The detailed studies denoted that, the interconnected resin-secreting canals facilitate long-distance transport and assumed a well-planned drainage during wounding or tapping. It can be understood based on the studies, that the wounding may not be deep for high yields rather it can induce high stress for healing (Shigo, 1984; Pearce, 1996). The studies of Tadessa et al. (2004) demonstrated enlarged size of wound initially showed high secretion and later declined even number of tapping spots increased (Tilahun et al., 2011; Eshete et al., 2012).

4.4 *BOSWELLIA SACRA* FLUECK. (= *B. CARTERII* BIRDW.)

Al-Harrasi et al. (2018) while studying the chemical and molecular structures, have explained about the internal structure of the bark of stem briefly to understand the extent making of injury on the stem bark (Figure 4.8) helps the tappers to a safe scratching method.

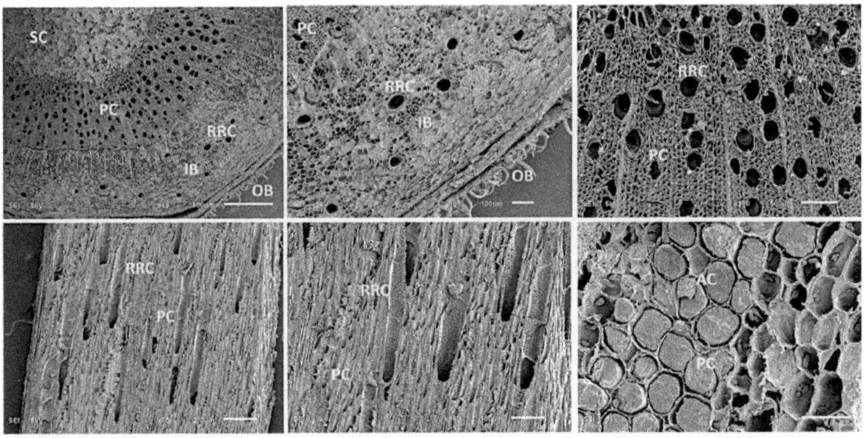

FIGURE 4.8 Transverse and radial anatomy of *B. sacra* stems analyzed using SEM (scanning electron micrographs). The tissues represented as SC (sclerenchyma); PC (parenchyma); AC (axial canals); RRC (radial resin canal); IB (inner bark); OB (outer bark).

Source: Al-Harrasi et al., 2018. © 2018 Al-Harrasi et al. https://creativecommons.org/licenses/by/4.0/

The studies of resin-producing structures evaluated and found to secretion of gum from the cambium, which was evidenced by the ornamentation of internal structure clearly demarked resin canals in the inner bark (Al-Harrasi et al., 2018).

4.5 *BOSWELLIA SERRATA* ROXB.

Oleo-gum resin canal occurrence was reported in the phloem tissues, and the lumen of the ducts were layered with epithelium providing apoplastic passage to resins. Further, these cells are enriched with the dictyosomes, which are the important cellular components for secretions and produce resin through the eccrine and granulocrine secretion into the canals. Even it is also reported that the inner layers of the epithelial cells are possessed with rudimentary plasmodesmata, which infers its secretion mode into the lumen of the canal (Nair & Subrahmanyam, 1998).

Pharmacognostic profile on *B. serrata* gum resin was studied using organoleptic features. The powdered gum resin was whitish yellow in color with hygroscopic nature and with sweet balsamic fragrance. It is with sub-carid, pungent, and sweet taste. A microscopic study of the powder drug indicated that the gum particles appeared like irregular tear shapes with dusty nature. Gum powder is soluble in polar and nonpolar solvents, and is also soluble in mild acids (Gupta et al., 2011). The oleo gum was studied for organoleptic properties such as morphological characteristics and anatomical organization by Senghani and Patel (2013), and reported the internal properties of *B. serrata* leaves and stem bark. The study by McCutcheon (2018) on the adulterations of *Boswellia* gums with *B. sacra* or with Pine resins discussed the standards of the gums and their components as part of the correct identification of gum resin used in the medicinal system, which may impact the biological effects when used.

4.6 CONCLUSION

The pharmacognostic studies of crude drugs are a part of characterization during the formulation of natural medicines in ancient medicines. It became more efficient in the recent past due to the advancement of scientific tools/technologies. So far, few species of *Boswellia* are being reported for pharmacognostic studies, considering their high economic values. Since there are many chances of adulteration with inferior, similar materials, it might be beneficial to take the pharmacognostic properties to the other species to obtain the improvement of non-destructive procedures for gum tapping. Moreover, these studies also find application in the justification of *Boswellia* species, and are helpful in their identification or taxonomical treatments.

KEYWORDS

- ***Boswellia***
- **frankincense**
- **gum olibanum**
- **pharmacognosy**
- **salai guggul**
- **schizogenous canals**

REFERENCES

Al-Harrasi, A., Rehman, N. U., Khan, A. L., Al-Broumi, M., Al-Amri, I., Hussain, J., et al., (2018). Chemical, molecular and structural studies of *Boswellia* species: β-boswellic aldehyde and 3-*epi*-11β-dihydroxy BA as precursors in biosynthesis of boswellic acids. *PLoS One, 13*(6), e0198666.

Eshete, A., Sterck, F. J., & Bongers, F., (2012). Frankincense production is determined by tree size and tapping frequency and intensity. *Forest Ecology and Management, 274*, 136–142.

Gupta, P. K., Harimohan, C., Channappa, H., Vinay, S., Varun, G., & Pankaj, N., (2011). Pharmacognostical and phytochemical evaluation of oleo gum-resin of Shallaki (*Boswellia serrata* Roxb.). *J. Pharmaceut. Biomed. Sci., 6*, 1–5.

McCutcheon, A., (2018). Adulteration of *Boswellia serrata*. Austin, TX: Botanical adulterants prevention program. *Botanical Adulterants Bulletin*, 1–7.

Nair, M. N. B., & Subrahmanyam, S. V., (1998). Ultrastructure of the epithelial cells and oleo-gum resin secretion in *Boswellia serrata* (Burseraceae). *IAWA Journal, 19*, 415–427.

Orhan, I. E., (2014). Pharmacognosy: Science of natural products in drug discovery. *Bioimpacts, 4*(3), 109, 110. doi: 10.15171/bi.2014.001.

Pearce, R. B., (1996). Antimicrobial defenses in the wood of living trees. *New Phytologist, 132*, 203–233.

Sathis, K. D., Srisutherson, N., Reddy, B. P. K., Vinitha, S., Yadhagiri, R. T., & David, B., (2011). Pharmacognostic studies on *Boswellia ovalifoliolata*. *J. Pharmacy Res., 4*(5), 1374–1375.

Savithramma, N., Linga, R. M., & Venkateswarlu, P., (2014). Histochemical studies of *Boswellia ovalifoliolata* Bal. & Henry: An endemic endangered & threatened medicinal plant of Seshachalam hill range of Eastern Ghats of India. *Intern. J. Pharmacogn. Phytochem. Res., 6*(1), 1–6.

Senghani, M. K., & Patel, P. M., (2013). Pharmacognostic and phytochemical study of oleo gum resin from *Boswellia serrata*. *Res. J. Pharmacogn. Phytochem., 5*(5), 244–250.

Shigo, A. L., (1984). Compartmentalization: A conceptual framework for understanding how trees grow and defend themselves. *Ann. Rev. Phytopathol., 22*, 189–214.

Tadesse, W., Feleke, S., & Eshete, T., (2004). Comparative study of traditional and new tapping methods on frankincense yield of *Boswellia papyrifera*. *Ethiopian J. Nat. Resources, 6*, 287–299.

Thulin, M., (2020). *The Genus Boswellia* (*Burseraceae*)*: The Frankincense Trees* (Vol. 39, pp. 1–149). Botanic symbols of Upsalien, Works from the Botanical Department in Uppsala.

Tilahun, M., Muys, B., Mathijs, E., Kleinn, C., Olschewski, R., & Gebrehiwot, K., (2011). Frankincense yield assessment and modelling in closed and grazed *Boswellia papyrifera* woodlands of Tigray, Northern Ethiopia. *J. Arid Environments, 75*, 695–702.

Tolera, M., Menger, D., Sass-Klaassen, U., Sterck, F. J., Copini, P., & Bongers, F., (2013). Resin secretory structures of *Boswellia papyrifera* and implications for frankincense yield. *Ann. Bot., 111*(1), 61–68.

CHAPTER 5

PHYTOCHEMISTRY OF *BOSWELLIA* SPECIES

LEPAKSHI MD. BHAKSHU[1*] and K. VENKATA RATNAM[2]

[1]*Department of Botany, PVKN Government College, Chittoor – 517002, Andhra Pradesh, India*

[2]*Department of Botany, Rayalaseema University, Kurnool – 518007, Andhra Pradesh, India*

[*]*Corresponding author. E-mail: lmbakshu@gmail.com*

ABSTRACT

The species of the genus *Boswellia* (Family: Burseraceae) are well known to humankind as Frankincense or Olibanum, which is used in sacred ceremonies and rituals in African and Asian countries. It has become a commercially traded related value-addition product and is used as an anti-inflammatory and analgesic drug in rheumatoid arthritis. The oleo-gum resin of the species, such as *B. sacra* (global source for frankincense), *B. serrata, B. dalzielii*, etc., became a source of local people's livelihood as they collected it from their natural locations. The essential oils and gum resins were widely investigated, and explored the occurrence of wide-ranging secondary metabolites that are also being developed as new medicaments for human ailments, including arthritis and analgesics. Moreover, *Boswellia* is an interesting taxon due to its habit, endemism, and unbelievable medicinal effects. Several researchers have explored the phytoconstituents of *Boswellia* species from different parts of the world, and more than 300 phytochemicals are being isolated and characterized. The major phytochemicals include monoterpenes, diterpenes, complex tri/tetra terpenoids, flavonoids/phenolics, etc.

Frankincense – Gum Olibanum: Botany, Chemistry, Extraction, Utilization, Propagation, Biotechnology, and Conservation, T. Pullaiah, K. Venkata Ratnam, Mallappa Kumara Swamy, Lepakshi Md. Bhakshu (Eds.)

5.1 INTRODUCTION

The descent of human civilization has endowed with ever improvement in his lifestyles and found the natural aromatics and felt sacred and was gifted many miraculous articles to the descendants. Frankincense/Olibanum is one among the ancient endeavors which came into light after the birth of Christ and spread throughout the world. It is also well known to Asian countries, especially Arabia, Ethiopia, China, and India, which is being used to worship the Gods, feel to make the Human soul pure. Frankincense is a solidified oozing liquid of the aromatic gum resin, obtained from the incisions of the barks of the Genus *Boswellia*, which belongs to the flowering plant family Burseraceae, used as incense material for 5,000 years. The olibanum-producing trees grow in certain countries, such as Ethiopia, Eritrea, Oman, Somalia, South Arabian region, and India, and has become a source of income and developed as a trade material and exporting to the western countries. In addition, it is also used in several traditional medicinal systems as a pain reliever and immunostimulant (Al-Harrasi et al., 2018). The genus *Boswellia* is represented by 24 species (Thulin, 2020) showing restricted distribution and globally endangered (EN), which may have the threat of extinction.

The important species of *Boswellia*, which are commercially explored and well investigated for their phytoconstituents and pharmacological benefits, include *B. sacra* Flueck. (Syn. *B. carterii* Birdwood), *B. dalzielii*, *B. serrata* Roxb. ex Colebr. and *B. papyrifera*. However, many other related species are still under the investigation by many scientists to explore their pharmacological effects. Literature survey suggested that *B. globosa* Thulin, *B. microphylla* Chiov. and *B. ogadensis* Vollesen are yet to be explored fully for their bioactive phytochemicals and biological effects. The present chapter gives an overview of the phytoconstituents of the aromatic gum resins and extracts obtained from different parts of plants of *Boswellia* species. The chemical structures of all the characterized phytoconstituents of *Boswellia* species are represented in Figure 5.1.

5.2 *BOSWELLIA AMEERO* BALF. F.

Gas chromatographic and mass spectral (GC-MS) analysis of the methanol extract of *B. ameero* revealed the presence of α-campholenal

(13.4%), (3E,5E)-2,6-dimethyl-1,3,5,7-octatetraene (34.9%), 3,4-dimethylstyrene (17.3%), α-terpineol (12.4%) and 1-(2,4-dimethyl-phenyl) ethanol (20.3%) (Ali et al., 2008). Madĕra et al. (2017) investigated the chemical composition of methanolic extraçt of oleo gum resin of *B. ameero*, collected from Socotra Island. They identified components such as α-thujene, α-pinene, sabinene, β-pinene, β-myrcene, α-fenchene, camphene, p-cymene, limonene, cyclodecyne, cis-sabinene hydrate, terpinolene, trans-sabinene hydrate, fenchol, 4-terpineol, α-terpineol, camphor, β-caryophyllene, trans-β-farnesene, α-humulene, caryophyllene oxide, kaurene, and cembrene.

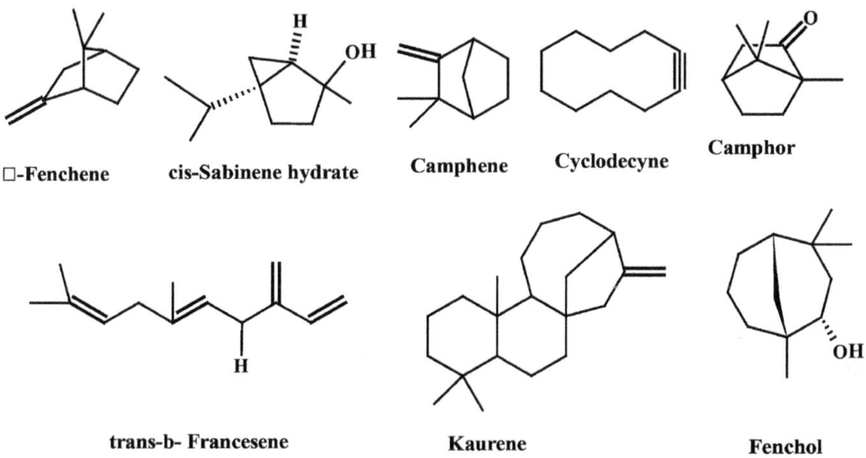

FIGURE 5.1 Chemical structures of some of the phytochemicals of *Boswellia* species.

5.3 *B. BULLATA* THULIN

Chemical composition of methanol extract of *B. bullata* oleo gum resin was analyzed by Madera et al. (2017). The compounds identified were α-thujene, sabinene, p-cymene, unknown, δ-elemene, α-cubenene, α-ylangene, α-copaene, α-gurjunene, calarene, β-caryophyllene, isocaryophyllene, trans-β-farnesene, α-humulene, aromadendrene, α-amorphene, germacrene d, α-elemene, cembrene γ-cadinene, α-cadinol, δ-cadinene, elemol, α-eudesmol, guaiol, guaiylacetate, and neocembrene.

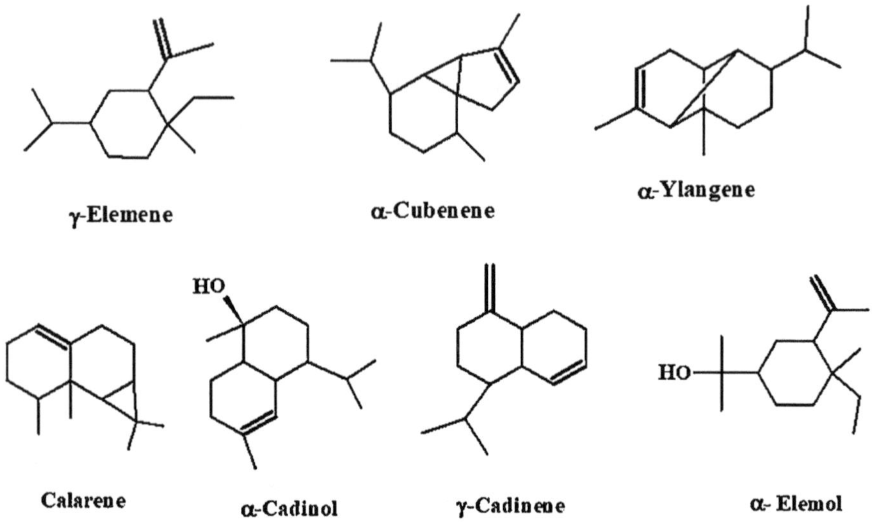

FIGURE 5.1(A).

5.4 *B. DALZIELII* HUTCH.

Hydro-distillation of the 20 *B. dalzielii* oleo-gum resin samples yielded pale-yellow essential oils. About 97 chemical constituents were identified in the oil samples, and rich in monoterpene hydrocarbons. The components observed were santolina triene C, 5,5-dimethyl-1-vinylbicyclo [2.1.1] hexane 1.1, tricyclene, α-thujene, α-pinene, α-fenchene, camphene, thuja-2,4(10)-diene, β-fenchene, sabinene β-pinene, trans-p-mentha-2, 8-diene, myrcene, 3, 3, 7-trimethylcyclohepta-1, 3, 5-triene, (E)-2,6-dimethyl-2, 6-octadiene, δ-2-carene, cis-p-menth-8-ene, p-mentha-1(7), 8-diene, α-phellandrene, 1,5,8-p-menthatriene, o-cresol methyl ether, δ-3-carene, α-terpinene, m-cymene, p-menth-1-ene, p-cymene, 2-acetyl-5-methylfuran, limonene, β-phellandrene,1,8-cineole, (Z)-β-ocimene, o-cymene, (E)-α-ocimene, γ-terpinene, cis-sabinene hydrate, terpinolene, p-cymenene, 6,7-epoxymyrcene, camphenone, rosefuran, perillene, linalool, α-pinene oxide, trans-sabinene hydrate, cis-thujone, (E)-2, 4-dimethylhepta-2, 4-dienal, trans-thujone, myrcenol, exo-fenchol, trans-p-mentha-2, 8-dien-1-ol, chrysanthenone, cis-p-menth-2-en-1-ol, α-campholenal, cis-limonene oxide, trans-limonene oxide, trans-sabinol, trans-pinocarveol, cis-verbenol, trans-verbenol, camphor, α-phellandren-8-ol, trans-pinocamphone, pinocarvone, trans-phellandrene epoxide, p-mentha-1, 5-dien-8-ol, borneol, cis-pinocamphone, terpinen-4-ol, p-cymen-8-ol, α-terpineol, myrtenal, verbenone, trans-carveol, carvone, linalyl acetate, 3,5-dimethoxytoluene, bornyl acetate, thymol, carvacrol, α-terpinyl acetate, α-cubebene, α-copaene, cis-α-bergamotene, β-caryophyllene, trans-β-bergamotene, (Z)-β-farnesene, α-himachalene, α-humulene, β-selinene, α-selinene, caryophyllene oxide, m-camphorene, (3E)-cembrene A, p-camphorene, α-pinacene, cembrenol, and incensole + serratol (DeCarlo et al., 2019).

Kohoude et al. (2017) investigated the chemical composition of *B. dalzielii* leaves by GC-MS (gas chromatography-mass spectrometry) and GC with flame-ionization detection (GC-FID). The oil yielded (1.12%) yellow color oil and identified 50 compounds, representing to 98.87% of all components. Among all the compounds identified, the dominant compounds included 3-carene (27.72%), α-pinene (15.18%), *p*-cymene (9.54%), β-phellandrene (8.48%), isolongifolene (6.15%) and myrcene (5.72%). The minor components were camphene (0.26), E-isolimonene (0.37), α-terpinene (0.45), Z-β-ocimene (0.28), c-terpinene (0.20), isoterpinolene (0.40), E-sabinene hydrate (0.67), 1S-β-fenchol (0.19), 3-terpinenol (0.17), E-pinocamphone (0.69), Z-chrysanthemol (0.64), lavandulol (0.42), terpinen-4-ol (1.41), methyl salicylate (0.52), β-cyclocitral (0.38), z-carveol (0.33), carvenone (0.24), undecan-2-one (0.27), α-cubebene (0.70), α-longipinene (0.23), β-patchoulene (0.25), cyperene (0.20), italicene (0.17), β-caryophyllene (0.75), E-α-bergamotene (0.20), aromadendrene (0.79), Z-β-farnesene (4.53), Z-4,5-muuroladiene (0.24), α-patchoulene (0.47), β-santalene (0.91), 9-epicaryophyllene (0.23), β-chamigrene (0.43), β-selinene (2.13), Z-β-guaiene (0.22), viridiflorene (0.16), β-himachalene (0.20), lilial (0.11), E-α-bisabolene (0.12), caryophyllene alcohol (0.35), viridiflorol (0.41), widdrol (0.38), 1,6-humulanedien-3-ol (1.61), c-eudesmol (0.53), torreyol-a-cadinol (1.93). Kubmarawa et al. (2006) evaluated the volatile components composition of essential oil obtained by hydrodistillation of the air-dried leaves of *B. dalzielii*. The oil yielded (1.25%) with 29 identified components. The major chemical components of the essential were α-pinene (45.7%) and α-terpinene (11.5%).

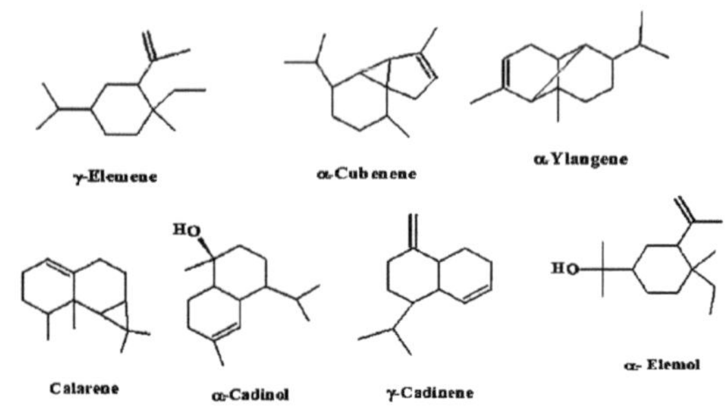

FIGURE 5.1(B).

The essential oil isolated from *B. dalzielii* leaves was dominated with sesquiterpernes (61.09%), zerumbone (33.69%), germacrene D (9.08%), and monoterpenes α-phellandrene (11.8%), α-pinene (9.71%), β-phellandrene (8.19%), and trans-β-caryophyllene (7.45%). The minor components identified in the oil were δ-3-carene (0.45%), γ-terpinene (1.35%), β-ocimene (1.09%), terpinen-4-ol (0.68%), α-terpineol (1.02%), α-cubebene (2.68%), copaene (2.85%), α-bourbonene (0.63%), α-humulene (0.97%), γ-cadinène (1.64%), tau-muurolol (1.32%), and α-cadinol (0.78%) (Bothon et al., 2019). Dalzienoside, α-boswellic acid (αBA), β-boswellic acid (βBA), and acetyl-11-keto βBA were isolated from *B. dalzielii* from the stem bark as a part of bio-assay guided isolation during the characterization of antibacterial effects (Tegasne et al., 2020).

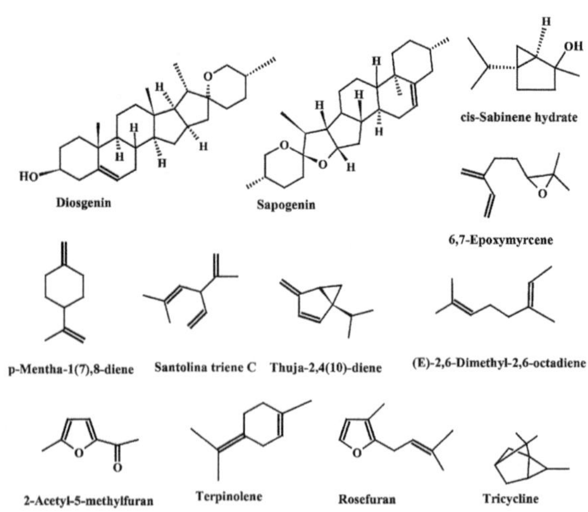

FIGURE 5.1(C).

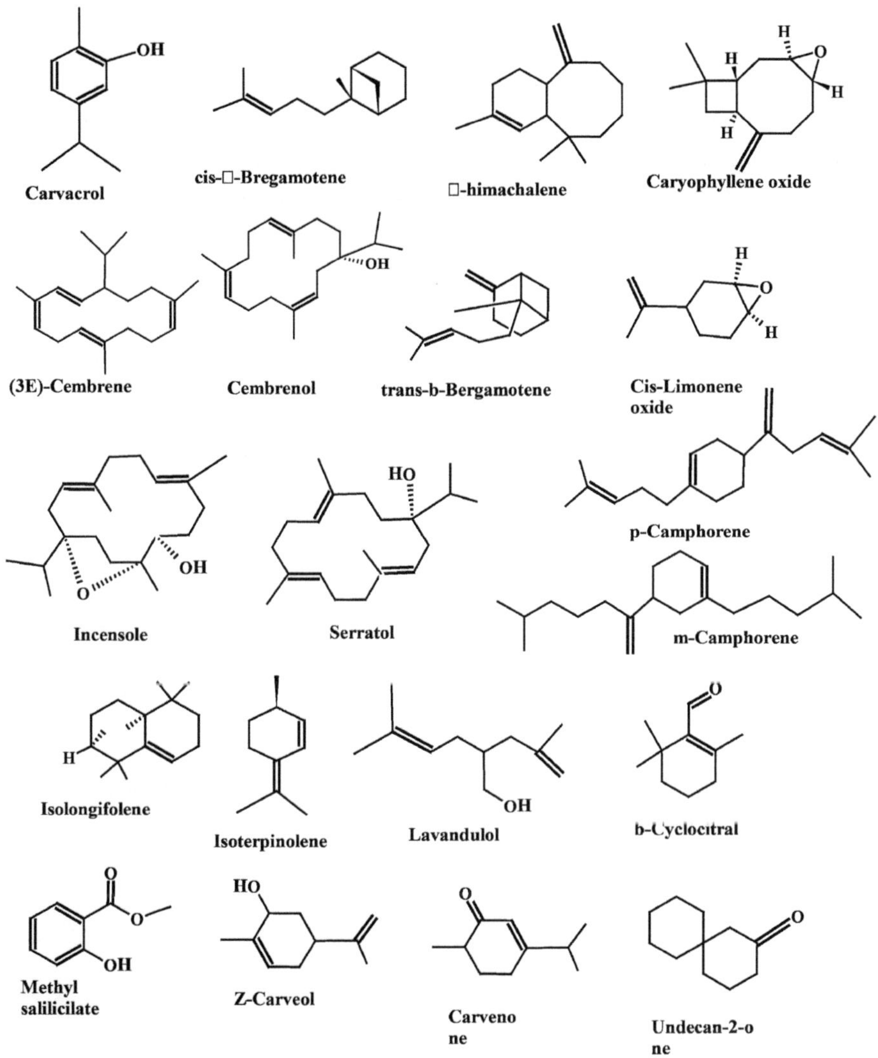

FIGURE 5.1(D).

Phytochemical composition of ethanolic extract of *B. dalzielii* root was analyzed by GC-MS. The results revealed the presence of 15 major compounds, namely glycerin, octanoic acid, 6-pentyl-5,6-dihydro-2*h*-pyran-2-one, decanoic acid, undecanoic acid, *n*-tetradecanoic acid, 2-methy-1-octanol, hexadecanoic acid, *n*-nonyl flouride, Z-11-hexadecenoic acid, octyl ether, 13-octadecenal, (*z*), glycerol 1-palmitate,

2-methyl-*Z,Z*-3,13-octadecadienol, 4-propyiheptadecane (Omoregbeee and Idu, 2017). Incensole, a cembrane diterpenoid, was isolated and characterized from ethyl acetate (EA) fraction of stem bark of *B. dalzielii* by using chromatographic techniques (Alemika et al., 2004). Alemika et al. (2006) isolated phenolic components like gallic acid, protocatechuic acid, 4'-methoxy-(E)-resveratrol-3-orutinoside, phytosterol, β-sitosterol, and chlorocresol from EA fraction of *B. dalzielii* stem bark by bioautography method.

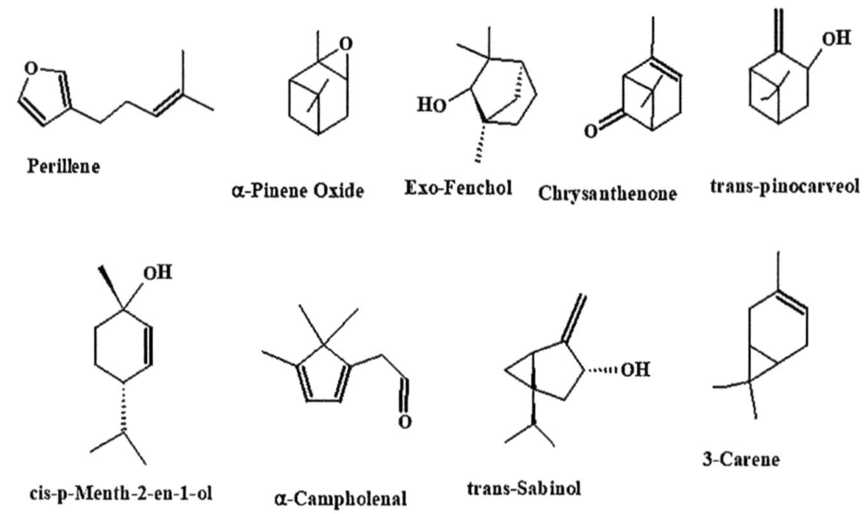

FIGURE 5.1(E).

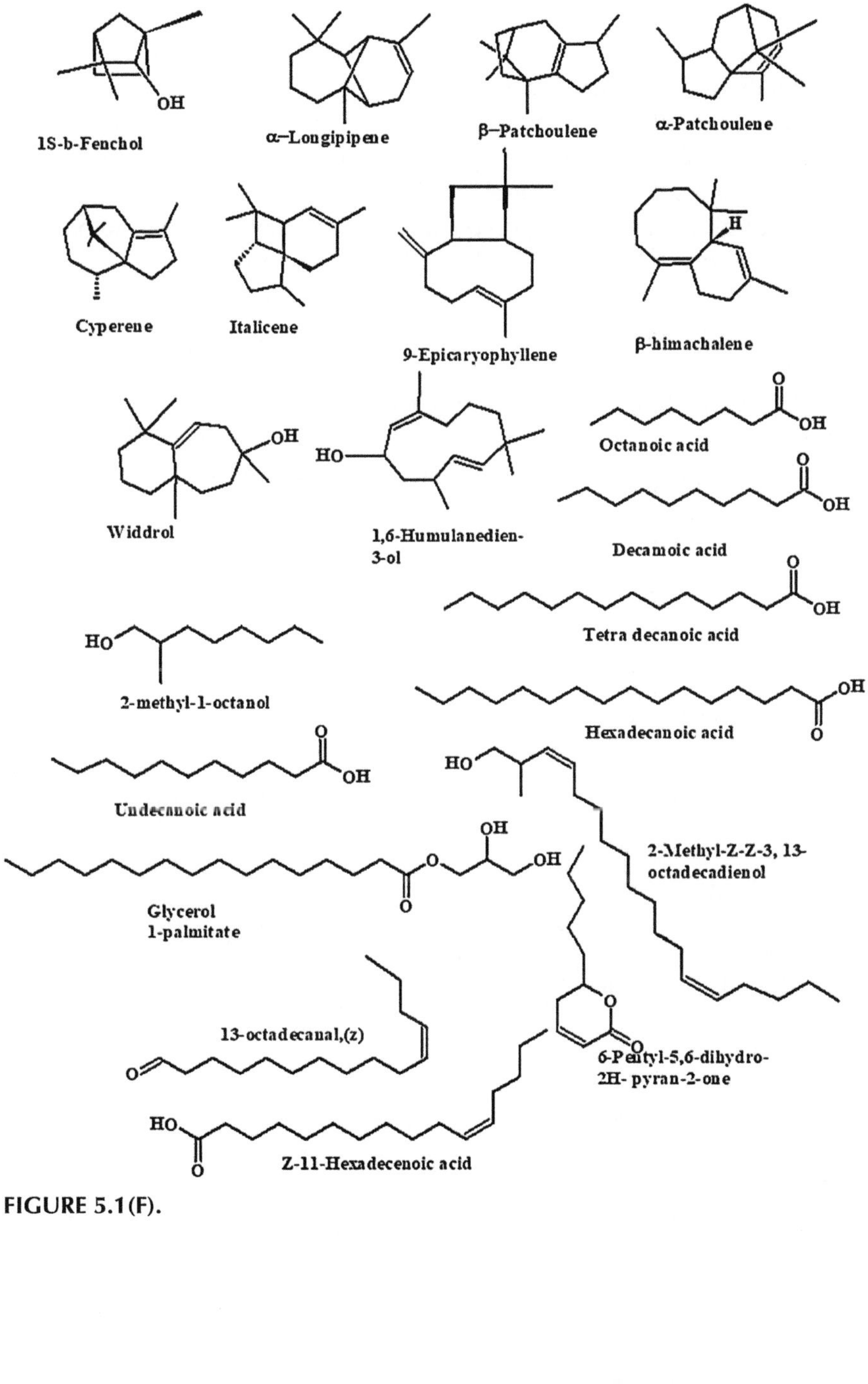
1S-b-Fenchol
α–Longipipene
β–Patchoulene
α-Patchoulene
Cyperene
Italicene
9-Epicaryophyllene
β-himachalene
Widdrol
1,6-Humulanedien-3-ol
Octanoic acid
Decamoic acid
Tetra decanoic acid
2-methyl-1-octanol
Hexadecanoic acid
Undecanoic acid
2-Methyl-Z-Z-3, 13-octadecadienol
Glycerol 1-palmitate
13-octadecanal,(z)
6-Pentyl-5,6-dihydro-2H- pyran-2-one
Z-11-Hexadecenoic acid

FIGURE 5.1(F).

The gas chromatography analysis of the methanolic extracts of air-dried leaves of *B. dalzielii* yielded 1.46 mg/100 g of a total saponin. Around 11 compounds were identified with Sapogenin (89.98%) and saponin (0.09%) being the predominant compound. Other saponins include neochlorogenin, diosgenin, yanogenin, tigogenin, tribuloin, conyzorgin, hispogenin, hecogenin, and salogenin (Onobrudu, 2017).

Quantitative analysis of polyphenolic constituents of the methanolic extract of airdried leaves of *B. dalzielii* yielded a total of 13.54 mg/100 g. Around 24 compounds were identified with protocatechuic acid (32.08%), catechin (23.77%), quercetin (8.35%), luteolin (20.15%), and kaempferol (12.79%) being the dominant compounds of the extract. The phenolic components present in minor quantity are myricetin, daidzein, genistein, butein, epicatechin, biochanin, resveratrol, gallocatechin, robinetin, epigallocatechin, apigenin, epicatechin-3-gallate, isorhamnetin, (–)-epigallocatechin-3-gallate, eupalestin, sinensetin, and biacalin (Onobrudu, 2017).

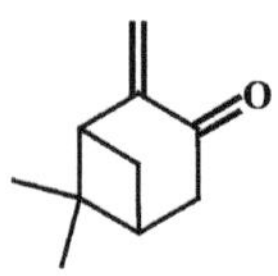

Pinocarvone

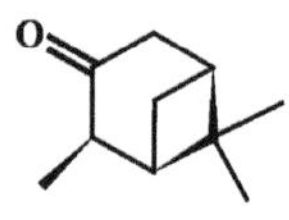

cis-Pinocamphone

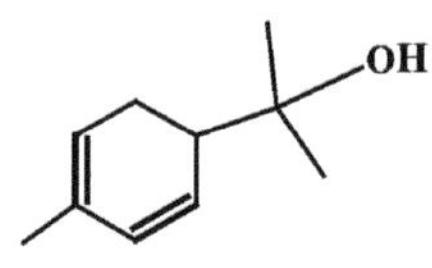

p-Mentha-1,5-dien-8-ol

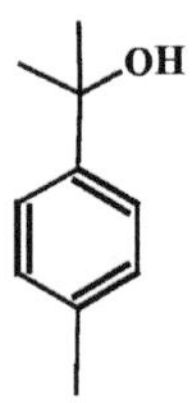

p-Cymen-8-ol

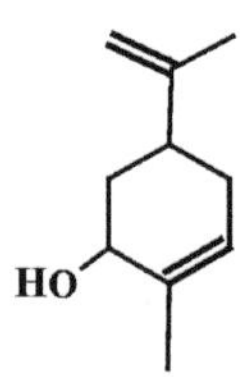

trans-Carveol

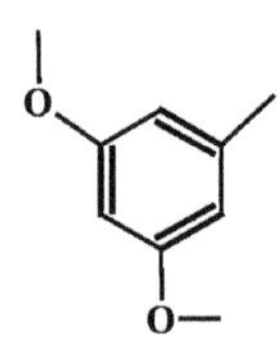

3,5-Dimethoxytoluene

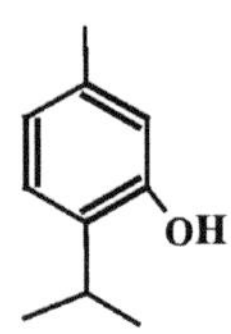

Thymol

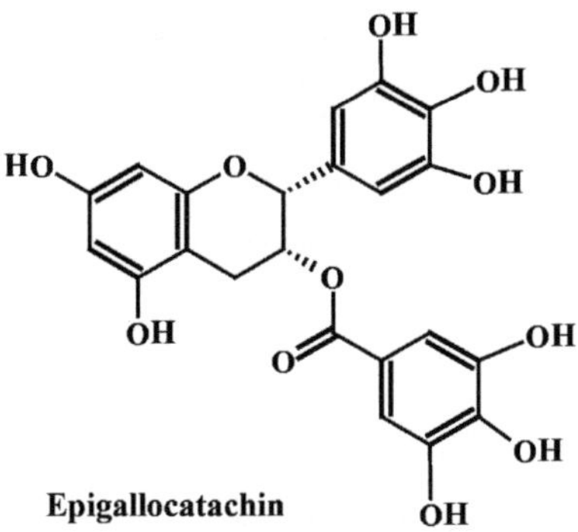

Epigallocatachin

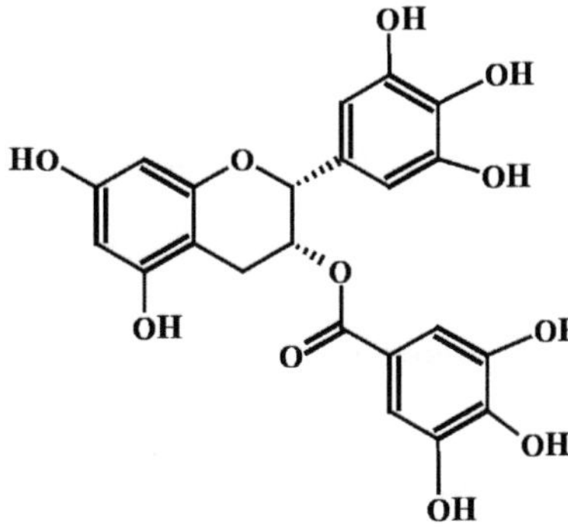

Epicatachin-3-gallate

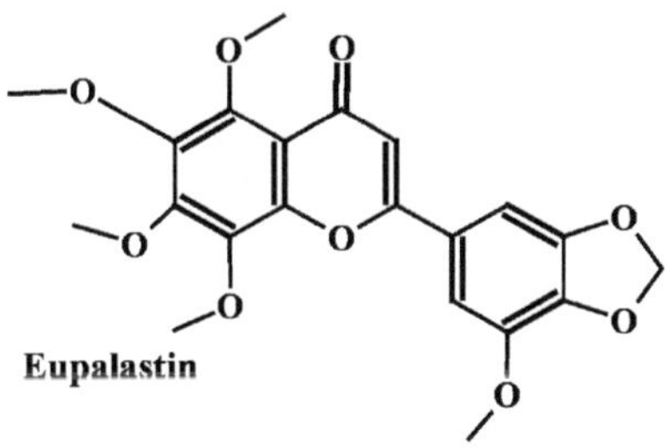

Eupalastin

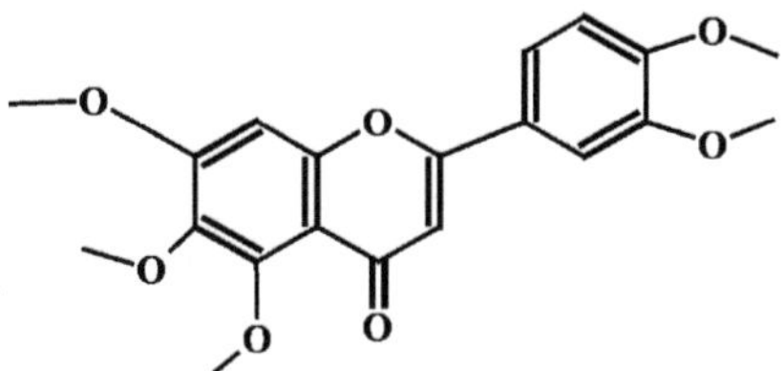

Sinensetin

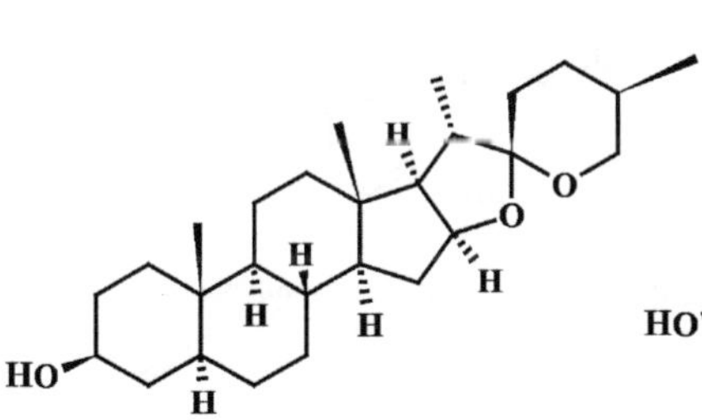

Tigogenin

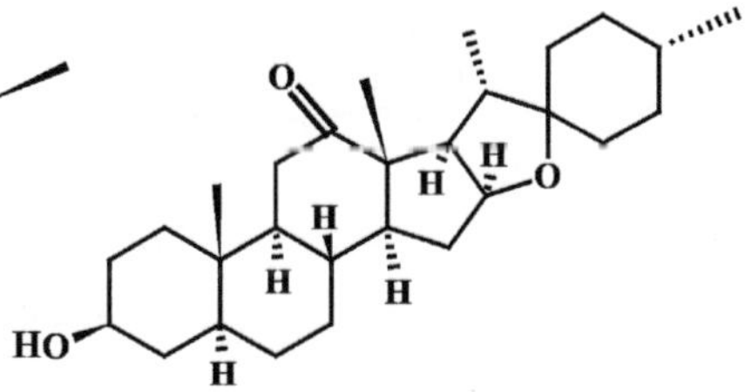

Hecogenin

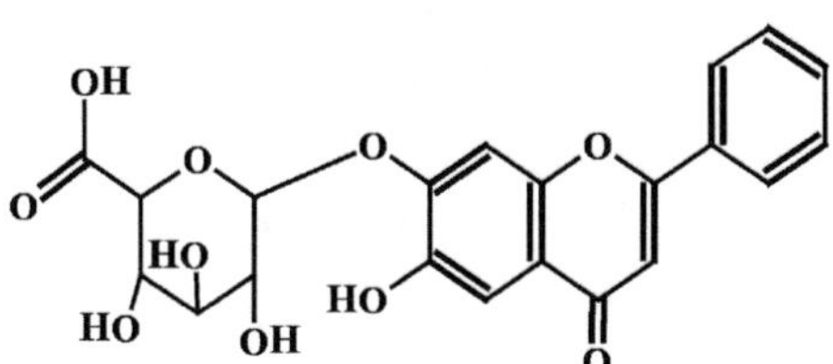

Biacalin

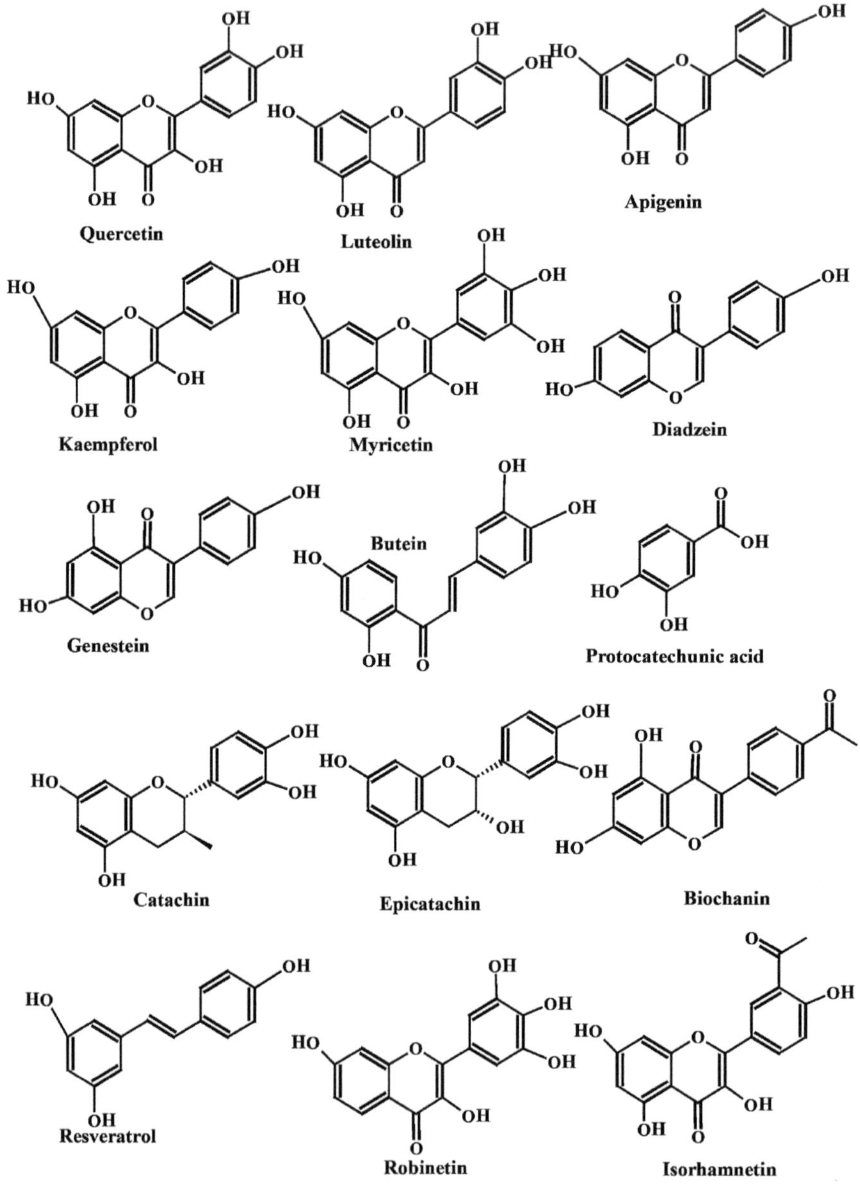
Quercetin
Luteolin
Apigenin
Kaempferol
Myricetin
Diadzein
Genestein
Butein
Protocatechunic acid
Catachin
Epicatachin
Biochanin
Resveratrol
Robinetin
Isorhamnetin

FIGURE 5.1(G).

5.5 *B. DIOSCORIDIS* THULIN

Chemical components, such as α-thujene, α-pinene, sabinene, β-pinene, β-myrcene, p-cymene, limonene, cyclodecyne, β-phellandrene, cis-sabinene hydrate, terpinolene, trans-sabinene hydrate, 4-terpineol, α-terpineol, bornyl acetate, β-caryophyllene, α-humulene were identified in the methanol extract of *B. dioscoridis* oleo gum resin collected from Socotra Island (Madera et al., 2017). The phytochemical investigation of essential oil isolated from the stem bark of *B. dioscoridis* led to the identification of 72 components. The oil was found to have a high content of monoterpene hydrocarbons (32.8%), with α-thujene (9.3%) and α-pinene (8.3%) as the main components (Mothana et al., 2011).

5.6 *B. ELONGATA* BALF. F.

The essential oil isolated from *B. elonagata* oleo gum resin was poor in monoterpenes, a diterpene, verticiol (52.4%) predominated over the sesquiterpene caryophellene (39.1%). Verticiol, was detected for the first time in the *B. elongata* oleo gum resin essential oil. Methyl cycloundecane carboxylate (7.9%) was present in a minor quantity (Ali et al., 2008). Chemical composition of volatile oil isolated from *B. elongata* stem was evaluated by Mothana et al. (2011). The oil was characterized by high diterpene content (31.7%), in which incensol (14.8%) was found to be the major constituent. Madĕra et al. (2017) identified volatile components, i.e., α-thujene, α-pinene, sabinene, β-pinene, β-myrcene, α-fenchene, cyclodecyne, α-terpinene, isobutyl benzene, *p*-cymene, limonene, β-phellandrene, γ-terpinene, cis-sabinene hydrate, terpinolene, 1,2,3,4-tetrahydronaphthalene, unknown 1, trans-sabinene hydrate, cis-verbenol, trans verbenol, 4-terpineol, α-terpineol, bornyl acetate, camphor, calarene, β-caryophyllene, trans-β-farnesene, α-humulene, isocaryophyllene, caryophyllene oxide, guaiyl acetate, neocembrene, and cembrene in methanolic extract obtained from *B. elongata* oleo gum resin. Around 14 triterpene acids, (tirucallane-type (α-tirucalla-8,24-dien-3α-Acetate; 5α-tirucalla-8,24-dien-3α-ol; 3α-acetoxytirucall-8,24-dien-21-oic acid), eight ursane-type (8), oleanane-type (2) and one lupane type (1) along with boswellic aldehyde, α-amyrine, epi-amyrine, acidic long strait-chain, sesquiterpene, and two diterpenes (2 cembrane-type) were characterized from the methanol extract of resin of *B. elongata* (Rehman et al., 2020).

5.7 *B. FREREANA* BIRDW. (ELEMI FRANKINCENSE)

Hydrodistillation of *B. frereana* oleo gum resin yielded light yellow color essential oil, dominated with α-pinene (38%), *p*-cymene (11), α-thujene (8.1%), trans-verbenol (4.2%), bornyl acetate (2.8%), sabinene (2.6%), and limonene (2.4%). The components identified in minor quantity were, β-bourbonene (trace), camphene (trace), trans-carveol (trace), carvone (trace), chrysantenone (trace), 1,8 cineole (trace), α-copaene (trace), m-cymene (trace), β-elemene (trace), isopropyl benzaldehyde (trace), linalool (trace), cis-linalool oxide (trace), trans-linalool oxide (trace), 2,4(8)-p-menthadiene (trace), myrtenal (trace), α-phellandrene (trace), α-phellandrene-dimer (trace), β-pinene (trace), α-pinene-epoxide (trace), pinocarvone (trace), trans-pinocarveol (trace), isopiperitenone (trace), trans-sabinenehydrate (trace), terpinen-4-ol (trace), α-terpineol (trace), thuja-2,4(10)-diene (trace), thujanol (trace), and verbenone (trace) (Basar, 2005).

Hamm et al. (2005) identified α-phellandrene-dimer (20.2%), α-pinene (12.4%), α-thujene (9.8%), cis-bornyl acetate (3.8%), verbenol (3.5%), verbenone (3.1%) and 4-terpineol (2.8%) as major volatiles in *B. frereana* oleo gum resin essential oil. The following chemical components were identified as minor components, i.e., β-bourbonene (0.2%), δ-cadinene (0.1%), camphene (0.6%), α-campholene aldehyde (0.7%), carvacrol (1.9%), trans-carveol (0.9%), carvone (0.1%), caryophylleneoxide (0.1%), α-copaene (0.9%), cumaldehyde (0.5%), cumin alcohol (0.3%), p-cymen-8-ol (0.3%), 1,5,5,8-tetramethyl-12-oxabicyclo-[9.1.0] dodeca-3,7-diene (0.2%), β-elemene (2.7%), eucalyptol (1.7%), β-eudesmene (0.1%), α-cyclo geranylacetate (2.2%), 4-methylene-1-(1-methylethyl)-bicyclo-[3.1.0] hex-2-ene (0.4%), limonene (1.7%), linalool (1.5%), cis-linalool oxide (traces), p-methyl acetophenone (0.2%), 2-methyl-5-(1-methylethyl)-1,3-cyclohexadiene monoepoxide (1.0%), β-myrcene (0.3%), myrtenal (0.7%), myrtenol (0.3%), α-phellandrene (1.1%), β-pinene (1.4%), trans-pinocarveol (1.7%), sabinene (3.2%), cis-sabinol (1.5%), 3,4-dimethoxystyrene (0.5%), γ-terpinene (0.1%), α-terpineol (0.9%), α-thujone (0.8%), and thujol (0.6%).

Chiavari et al. (1991) examined the chemical composition of the methanolic extract of resins of *B. frereana* and *B. sacra* (Syn.: *B. carteri*) by GC and GC/MS. They opined that the data is useful to differentiate between resins of these two species.

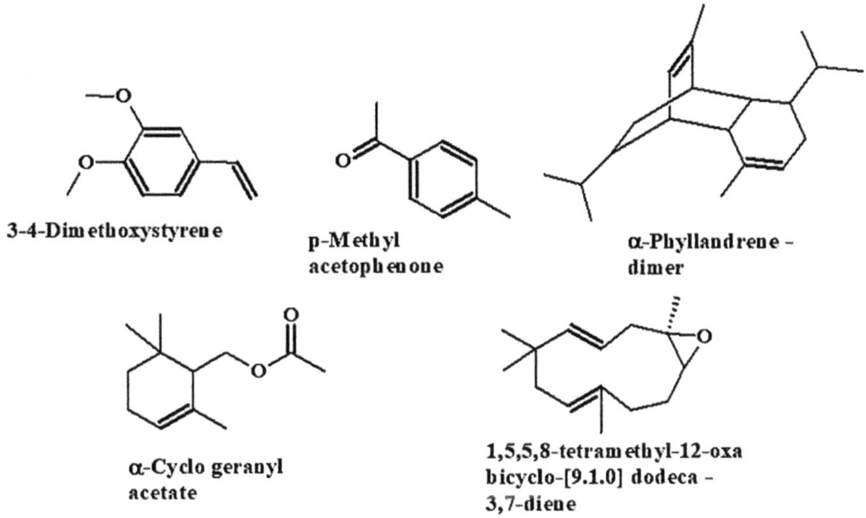

FIGURE 5.1(H).

5.8 *B. NANA* HEPPER

Terpenoids isolated from gum resin of *B. nana* were α-thujene, α-pinene, sabinene, β-pinene, *p*-cymene, limonene, β-phellandrene, bornyl acetate, β-bourbonene, β-caryophyllene, isocaryophyllene, trans-β-farnesene, neocembrene, and cembrene were identified in methanolic extract of *B. nana* gum resin (Madera et al., 2017).

5.9 *B. NEGLECTA* S. MOORE (BORENA TYPE OLIBANUM)

Hydro-distillation of *B. neglecta* gum resin yielded dark yellow color essential oil. GC-MS analysis of the oil indicates that oil was dominated with monoterpenes, i.e., α-thujene (21.3%), α-pinene (21.3%), sabinene (1.3%), δ-3-carene (1.9%), p-cymene (11.8%), terpinene-4-ol (5.3%), verbenone (2.1%). The minor components identified in the oil samples were bornyl acetate, camphene, α-campholene aldehyde, camphor, cara-2,4-diene, carvacrol, trans-carveol, carvone, carvotan acetone, chrysantenone, α-copaene, m-cymene-8-ol, m-cymene, o-cymene, p-cymen-8-ol, p-cymene-9-ol, p-cymene, p-cymenene, β-elemene, isopropyl benzaldehyde, isopropyl benzalcohol, limonene, cis-1,2-limonene epoxide, linalool, linalyl acetate, p-mentha-1,5-diene-7-ol, trans-p-menth-2-en-1-ol, o-methylanisol, myrtenal, myrtenol, (E)-β-ocimene, α-phellandrene, β-pinene, α-pinene oxide, pinocarvone, trans-pinocarveol, dehydro sabina ketone, cis-sabinene hydrate, cis-sabinol, trans-sabinenehydrate, trans-sabinol, cis-1,2-epoxy-terpin-4-ol, g-terpinene, terpinen-4-ol, α-terpineol, terpinolene, a-terpinyl acetate, thuja-2,4(10)-diene, α-thujone, β-thujone, thymol, umbellulone, cis-verbenol, and trans-verbenol (Basar et al., 2001; Basar, 2005).

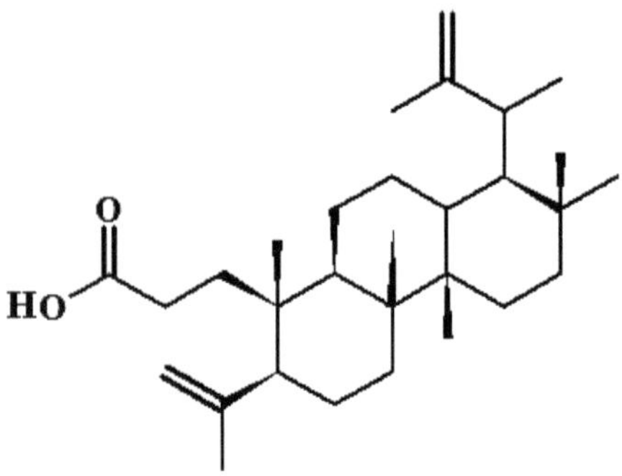

Canaric acid

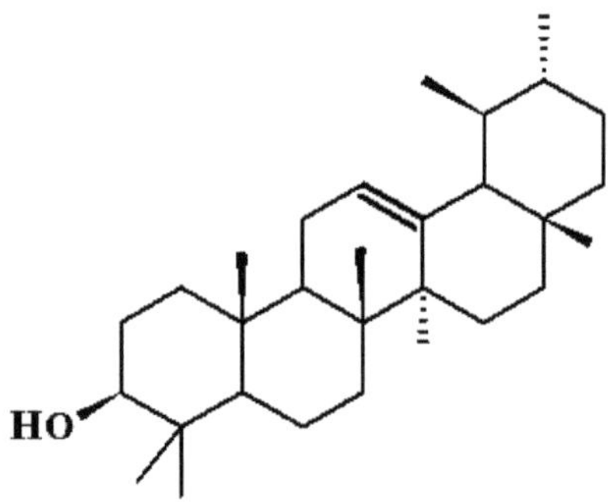

α–Amyrin

FIGURE 5.1(I).

Column chromatographic analysis of petro extract of *B. neglecta* gum resin yielded four known triterpenes, canaric acid, α-amyrin, α-amyrone, and *epi*-α-amyrin. A new ursane-type triterpenes characterized as 3α-acetoxy-28-hydroxy-11-oxours-12-en-24-oic acid, together with 12 known compounds, i.e., 3α,11α-dihydroxyurs-12-en-24-oic acid, 11-oxo-β-boswellic acid, αBA, β-boswellic, ursolic acid, lupeol, β-amyrin, α-amyrin, α-amyrone, quercetin, olean-12-en-3-O-β-glucoside, and catechin-3-O-glucoside were isolated from oleo gum resin of *B. neglecta* (Manguro et al., 2016; Madera et al., 2017).

5.10 *B. OCCULTA* THULIN, DECARLO, & S.P. JOHNSON

B. occulta, recently discovered as new species from Somalia (Thulin et al., 2019). Analytical investigations on chemical composition of commercial oleo gum resins identified that *B. occulta* gum resin oil is a rich source of a methoxy-alkane type of aromatic components (Ayubova et al., 2016; Satyal and Pappas, 2016).

Johnson et al. (2019) investigated the chemical composition of hydrodistilled 12 essential oil samples by GC-MS, GC-FID, collected from individual species from Somalia. The results indicate that the oils had significant level of methyl ethers in all samples (34.5%–62.6%), especially 1-methoxydecane (26.6%–47.9%) and 1-methoxyoctane (3.6%–8.6%). Other components identified in oil samples were 1-methoxyhexane, 1-methoxyheptane, α-thujene, α-pinene, camphene, sabinene, β-pinene, myrcene, α-phellandrene, 1,4-cineole, α-terpinene, p-cymene, 1-methoxyoctane, limonene, β-phellandrene, γ-terpinene, cis-sabinene hydrate, 1-octanol, dihydro myrcenol, terpinolene, (Z)-4,8-dimethylnona-1,3,7-triene, linalool, trans-sabinene hydrate, β-bourbonene, hexyl hexanoate, β-cubebene, α-bourbonene, octyl butanoate, β-elemene, sativene, (*Z*)-caryophyllene, α-gurjunene, dodecanal, decyl acetate, 1-methoxydodecene, β-ylangene, β-caryophyllene, 1-methoxydodecane, β-copaene, guaia-6,9-diene, γ-elemene, *iso*-germacrene d, dihydro curcumene, *cis*-muurola-3,5-diene, (*E*)-β-farnesene, α-humulene, *allo*-aromadendrene, *cis*-cadina-1(6),4-diene, dauca-5,8-diene, 9-*epi*-(*E*)-caryophyllene, *trans*-cadina-1(6),4-diene, γ-muurolene, germacrene d, β-selinene, *trans*-muurola-4(14),5-diene, *epi*-cubebol, α-muurolene, epizonarene, β-dihydro agarofuran, 1-pentadecene, tridecanal, γ-cadinene, cubebol, δ-cadinene, 1-methoxytridecane, *trans*-calamenene, methyl dodecanoate, kessane, *trans*-cadina-1,4-diene, *iso*-kessane, α-cadinene, α-calacorene, elemol, (*E*)-nerolidol, 10-*epi*-liguloxide, dodecanoic acid, palustrol, 1,10-decanediol, longipinanol, germacrene d-4-ol, octyl hexanoate, caryophyllene oxide, decyl butanoate, allo-hedycaryol, guaiol, ledol, dodecyl acetate, copaborneol, 4,10-di-epi-guaiol, 1-epi-cubenol, eremoligenol, γ-eudesmol, τ-cadinol, cubenol, hedycaryol, τ-muurolol, α-muurolol, agarospirol, hinesol, α-cadinol, α-eudesmol, bulnesol, tridecenyl acetate, decyl hexanoate, (3E)-cembrene A, (3Z)-cembrene A, α-pinacene, verticilla-4(20),7,11-triene, manool, neocembrene a, incensole, serratol, incensyl acetate.

5.11 *B. OVALIFOLIOLATA* BAL. & HENRY

Oleo-gum-resin exudates from the stem bark of *B. ovalifoliolata* is yellow color in fresh form and become brownish hard substance after exposing to environment. It has been used as an alternative source of incense of *B. serrata,* by the tribal and rural inhabitants of forest areas of Seshachalam and Nallamala forest regions of Andhra Pradesh. The chief chemical constituents of the gum resin are amyrin type triterpenoids, resin acids, and volatile oils. Stem consists macrocyclic heptanoids, steroids, triterpenoids (Vanimireddy et al., 2003) and volatiles oils (Prasanna et al., 2015). Hydro distillation of *B. ovalifoliolata* gum resin yielded colorless essential oil (5%) with characteristic odor. Around 25 volatile constituents were identified, the dominant components are caryophyllene (28.50%), α-pinene (15.53%), β-myrcene (13.82) and ledol (7.82%). The minor components include α-amarphene (7.73%), viridiflorol (6.84%), β-pinene (3.29%), α-humulene (3.2%), δ-cadinene (2.97%), etc. Leaf yielded pale yellow color oil (0.17%) with 30 identified components.

The oil is dominant with α-pinene (21.55%), β-pinene (16.85%), α-terpineol (14.27%) and caryophyllene (11.78), while, stem bark yielded greenish pale yellow color oil (0.2%) with 25 identified components. The oil is rich in, β-myrcene (17.46), caryophyllene (16.17%), cyclofenchene (8.99%), α-amarphene (10.32%) and cedr-8-ene (7.99%). The minor components present in *B. ovalifoliolata* gum, stem bark and leaf oils are (–)-zingiberene, γ-eudesmol, decalin, cineol, α-bergamotene, β-bisabolene, cedr-8-ene, α-santatol, guaiol, eucalyptol, myrcenol, terpinene-4-ol and cis-β-ocimene (Prasanna et al., 2015). Geetha and Chakravarthula (2018) identified β-farnesene (13.44%), caryophyllene oxide (10.45%) and spathulenol (2Z, 6E) (6.6%) as major components in essential oil obtained from *B. ovalifoliolata* stem bark. The leaf oil consists spathulenol (2Z, 6E) (11.1%), caryophyllene oxide (9.0%), and decyl acetate (6.3%) as major components identified by GC-MS analysis. Benelli et al. (2018) investigated the chemical composition of *B. ovalifoliolata* leaf essential oil, which consists of 20 components. The oil is rich in β-pinene, α-pinene, α-terpineol, and caryophyllene.

Phytoconstituents isolated from *B. ovalifoliolata* stem and oleo gum resin revealed that two new macrocyclic diaryl ether heptanoids, ovalifoliolatin A and B, along with three known compounds; acerogenin C, 3a-hydroxyurs-12-ene, and sitost-4-en-3-one were isolated from stem part

(Vanimireddy et al., 2003). Acetogenin C,52 3β-urs-12-en-3-ol (α-amyrin) and sitost-4-en-3-one were isolated (Gonzalez & Zhu, 1997; Ahmad & Rahman, 1994).

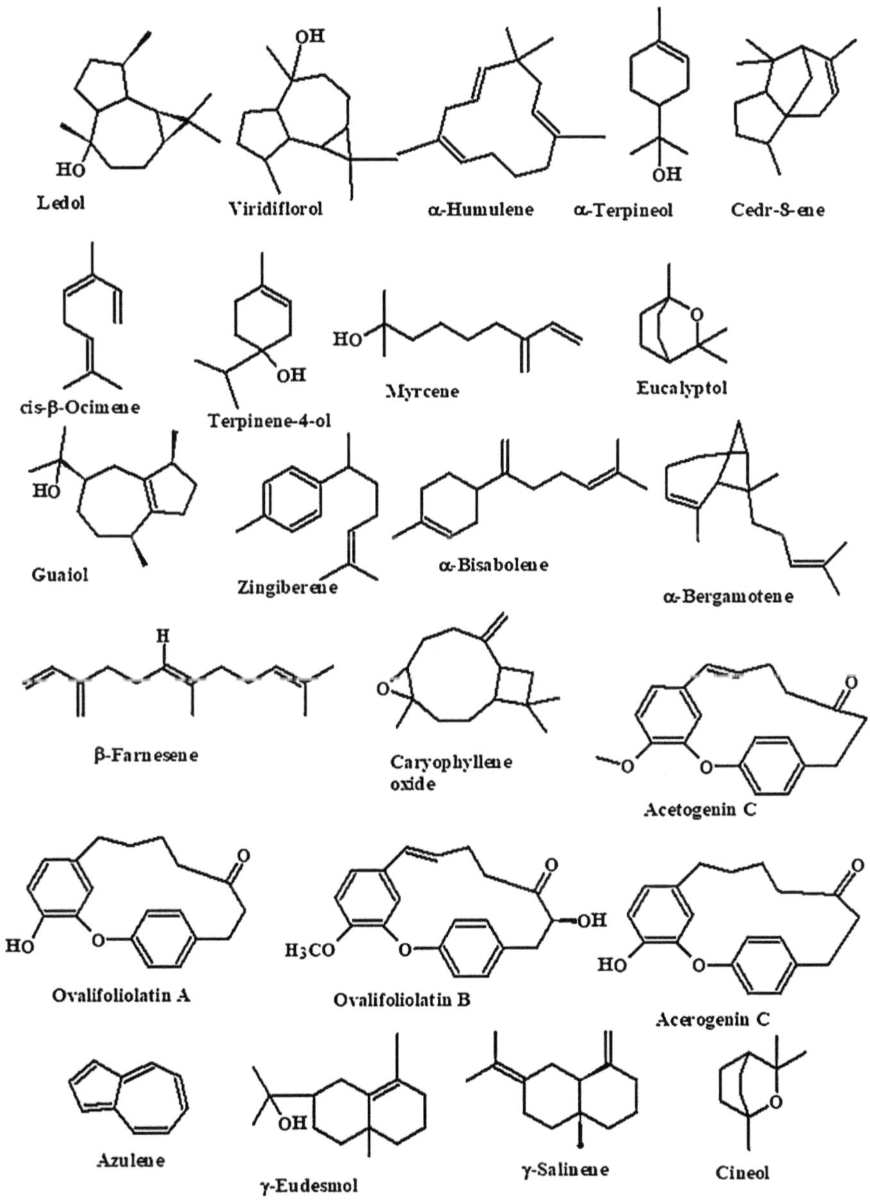

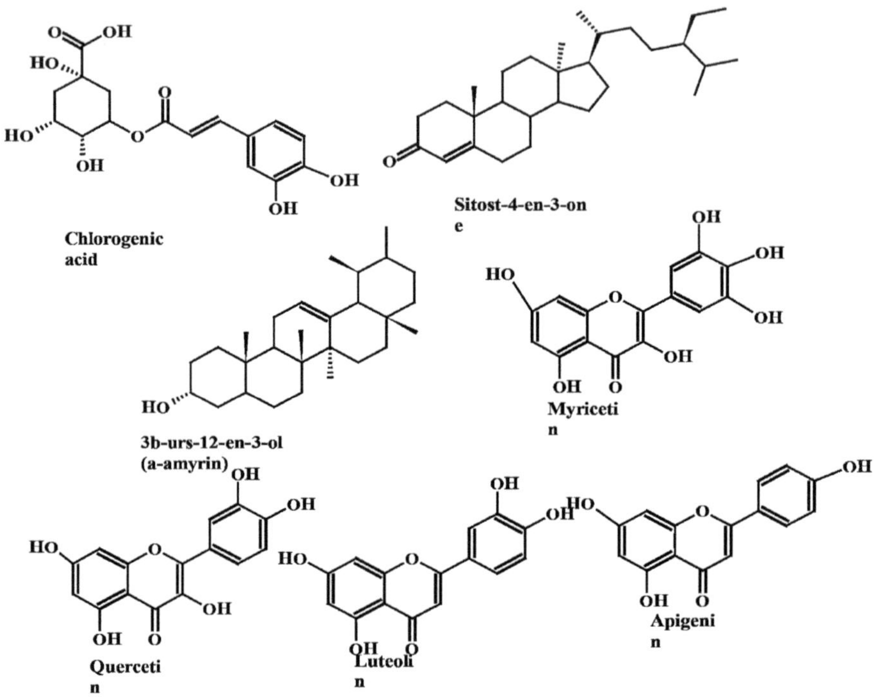

FIGURE 5.1(J).

The phenolic components isolated from *B. ovalifoliolata* gum and stem bark are chlorogenic acid, gallic acid, gentisic acid, phloretic acid, vanillic acid, melilotic acid, coumarin, salycylic acid, and cinnamic acid. Flavonoids identified are quercetin, rutin, myricetin, luteolin, apigenin, orientin, and vitexin (Savithramma et al., 2010, 2014).

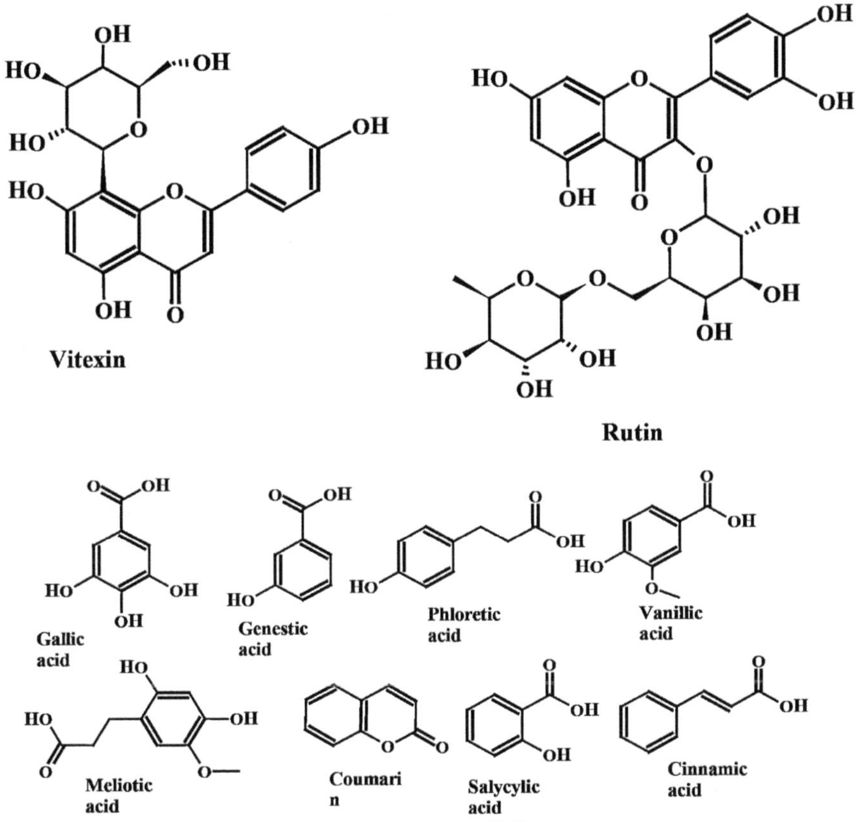

FIGURE 5.1(K).

5.12 *B. PAPYRIFERA* (CAILL. EX DEL.) HOCHST (SUDANESE FRANKINCENSE)

The oleo gum resin of *B. papyrifera* is claimed to be the principal source of frankincense from ancient times, coming from Ethiopia, Sudan, or East Africa, and was considered an inferior form of frankincense (Tucker, 1986). The natural oleo gum resin samples of *Boswellia* species are composed of about 5–9% aromatic essential oils, 65–85%

alcohol-soluble resins, and the remaining water-soluble gums consist of polysaccharides (Tucker, 1986; Khan and Farooqi, 1991). The chemical composition of *B. papyrifera* gum resin is markedly different from that of other *Boswellia* species, with the predominance of octyl acetate, a high content of octanol and presence of monoterpenes, namely α-pinene, limonene, linalool, and only traces of sesquiterpenes (Basar et al., 2001).

Chemical composition of colorless essential oils obtained from three authenticated oleo gum resin samples of *B. papyrifera*, namely, Kordofan, Nagawa, and Damazine was evaluated by Mustafa (2012). The major compounds of Kordofan essential oil were octyl acetate (46.50%) accompanied by 1-octanol (17.52%), verticiol (9.54%) and androstan-6-one, 3-(acetyloxy)-5-hydroxy-, (3.β, 5.α) (7.61%). In Damazine frankincense, the main constituents were octyl acetate (39.31%), 1,2-benzene dicarboxylic acid, mono (2-ethylhexyl) ester (21.55%), advastab (14.17%) and phenol, 4,4-methylene bis [2,6-bis (1,1-dimethylethyl)] (7.35%). The essential oil of Nagawa olibanum sample composed mainly of octyl acetate (35.31%) and 1-octanol (25.77%), along with considerable quantities of d-limonene (10.19%) and androstan-6-one, 3-(acetyloxy)-5-hydroxy-, (3.β, 5.α) (5.22%). The minor components identified in three oil samples as acetic acid, decyl ester, borneol, citronellol acetate, 8-isoproponyl-1, 5-dimethyl-1, 5-cyclodecadiene, β-linalool, α-pinene, anisol, o-methyl, α-phellandrene, β-pinene, p-undecycloxybenzoic acid methyl ester, cycloheptane, 4-methylene-1-methyl-2-(2-methyl-1-propen-1-yl)-1-vinyl, 1-ethyl-2-pyrrolidone-4-carboxamide, 1,3,6-octatriene, 3,7-dimethyl-, (Z), bicyclo [4.1.0] hept-3-ene, 3,7,7-trimethyl-, (1S)-, benzo [f] chromen-3-one, perhydro-2-acetyl-4a, 7,7,10-b-tetramethyl, methanone, (3,5-dimethyl-1-piperidin) (morpholyl), β-myrcene, phenol, 4,4-methylene bis [2,6-bis(1,1-dimethylethyl)], carvotan, acetone, cyclohexene, 1-ethenyl-1-methyl-2,4-bis(-methylethenyl)-, [1s-(1.α., 2.β., 4.β)].

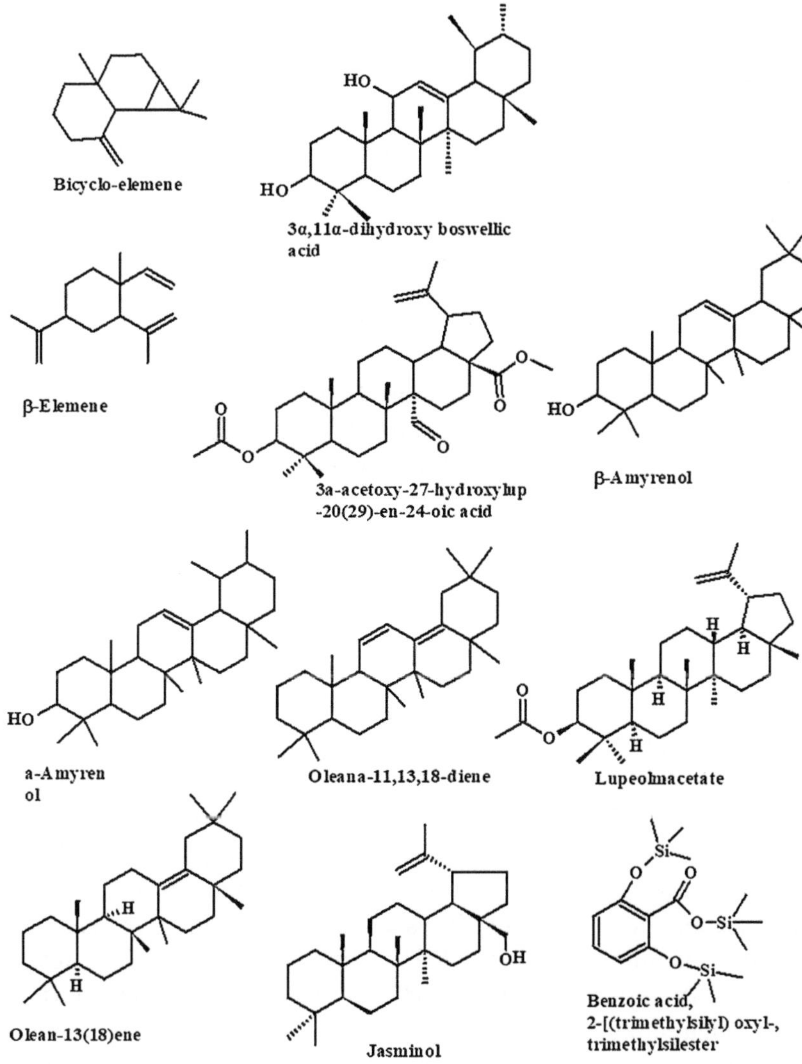

FIGURE 5.1(L).

Hydro-distillation of *B. papyrifera* oleo gum resin yielded (3%, w/w) pale yellow-colored volatile oil with an agreeable perfumery odor. A total of 16 compounds were identified, comprising 99.30% of the essential oil and dominated by the presence of alcohol and ester derivatives (46.82%) followed by monoterpenes (16.30%), sesquiterpenes (27.70%) diterpenes

(5.08%) and fatty acids which represent 5.51% of the total oil. The identified components were α-pinene, β-myrcene, i-limonene, β-ocimene (Z)-, 1-octanol, linalool, octyl acetate, bicycloelemene, β-elemene, elemol, (4E)-1,5,9-trimethyl-1-vinyl-4,8-decadienyl formate, nerolidol-epoxyacetate, n-hexadecanoic acid, heptadecene-(8)-carbonic acid-(1), elaidinsaeure, and verticiol (Yagi et al., 2016).

Hamm et al. (2005) reported *o*-campholene aldehyde (traces), trans-carveol (0.3%), carvone (0.3%), cembrene a (1.7%), *p*-cymene (0.2%), n-decanoic acid (0.1%), n-decylacetate (0.9%), eucalyptol (0.1%), n-hexanoic acid (0.2%), hexylcaprylate (1.2%), n-hexyl hexanoate (0.9%), incensol (1.0%), incensolacetate (10.8%), incensol oxide (0.4%), incensol oxide acetate (0.1%), linalool (0.2%), cis-linalool oxide (0.2%), nerylacetate (0.1%), n-nonanoic acid (0.5%), octanol (13.9%), n-octanoic acid (0.8%), n-octylacetate (64.6%), *α*-pinene (0.5%), trans-pinocarveol (0.1%) in gum resin oil of *B. papyrifera*. Camarda (2007) determined the chemical composition of *B. papyrifera* gum resin essential oil. The identified components were benzene, 1 methoxy-2-methyl (0.3%), canvone (0.1%), *p*-cymene (0.4%), decylacetate (0.3%), eucalyptol (0.5%), geranyl acetate (0.1%), incensol acetate (1.7%), incensol (0.7%), *α*-terpinene (0.4%), terpinene (0.2%), terpinen-4-ol (0.1%), *α*-thujene (0.1%), thuribergene (0.1%), verticilla-4(20),7,11-triene (2.3%).

Dekebo et al. (1999) evaluated the phytochemical composition of essential oil obtained from *B. papyrifera* gum resin. The components identified were, endo-borneol (1.5%), bornyl acetate (0.7%), 1,8 cineole (2.2%), β-elemene (1.1%), hexylacetate (1.3%), *p*-mentha-6,8-dien-2-one (0.7%), octanol (8.0%), terpinolene (0.4%) and verticellol (0.5%). Common components reported in essential oil samples were camphene (0.1%, 0.3%), limonene (4.7%, 6.5%), linalool (0.7%, 3.2%), mycene (0.3%, 0.7%), n-octylacetate (63.5%, 56.0%) and *α*-terpineol (0.2%, 0.5%) (Camarda, 2007; Dekebo et al., 1999). Halim et al. (2020) studied the chemical composition of *B. papyrifera* olio-gum resin collected from two different locations, i.e., Abergelle and Metema of Ethiopia, revealed that n-octyl acetate (51.2–58.56%), ethylpent-3en-1yl (11.72–12.91%) and 1-octanol (8.83–17.66%) observed as major constituents. Essential oil isolated by direct steam distillation method revealed the presence of 25 components. The major constituent was found to be 2-ethyl hexyl acetate. Glucuronic acid, galactose, and arabinose were isolated from water extract by using TLC technique (Osman, 2007).

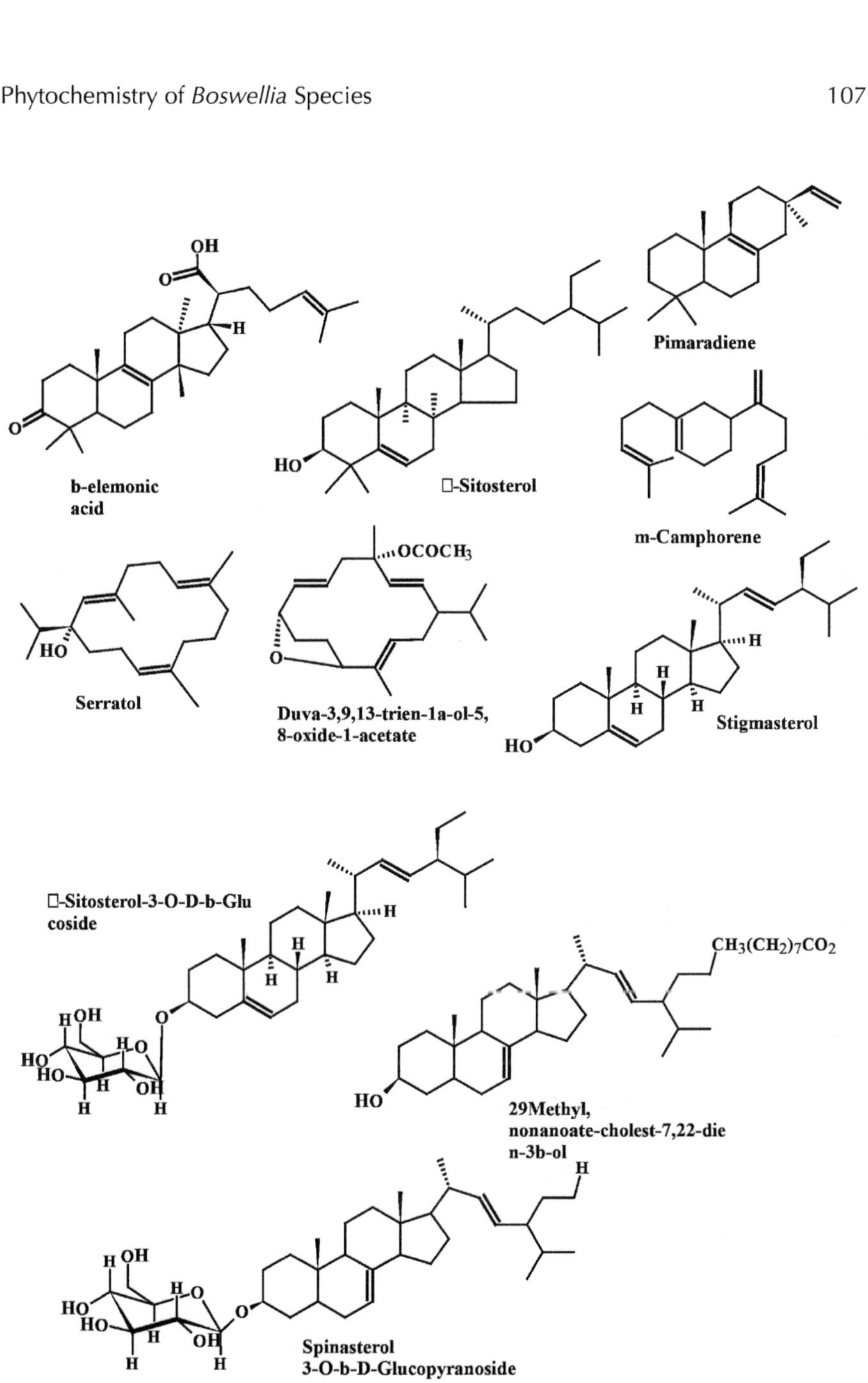
b-elemonic acid
□-Sitosterol
Pimaradiene
m-Camphorene
Serratol
Duva-3,9,13-trien-1a-ol-5, 8-oxide-1-acetate
Stigmasterol
□-Sitosterol-3-O-D-b-Glu coside
29Methyl, nonanoate-cholest-7,22-die n-3b-ol
Spinasterol 3-O-b-D-Glucopyranoside

FIGURE 5.1(M).

The major compounds identified in petroleum ether, methanol, and acid fraction extracts obtained from *B. papyrifera* gum resin collected from Kordofan were Nophtho [1,2b] furan-2-one, 2, 3, 3a, 4, 5, 5a, 6, 7, 9a, ab-decahydro-3, 5a, 9-trimethyl-7, 9a-peroxy (24.23%), ergost-25-ene-6, 12-dione, 3,5-dihydro-, (3-β, 5-α.) and hexadecanoic acid, trimethylester, lycopene, 1,1,2,2,-tetrahydro-1, 1-dimethoxy-, all-trans (13.78%) and naphthalene, decahydro-1, 4-dimethoxy-, (1-α., 4.α. 4a.α., 8a.β.) (10.16%), tricosane-1, 15-diol, bis (O-trimethylsilyly). In addition to these components, the following pentacyclic triterpenes were found; ursa-9 (11), 12-dien-28-oic acid, 3-(-acetyl (oxy), methyl ester, (3-β), β-amyrin methyl ester, β-amyrenone, β-amyrene, β-amyrenol (0.31%) and urs-11-en-28-oic acid, 2, 13-dihydroxy-3, 23-[(-methyl ethylidene) bis (oxy)]-, urs-12-ene, 1,2,11-triol,3-(acetyloxy)-, (1.β., 2.α., 3.β., 11. α) (1.43%), γ-lactone, (2-α-, 3-β,4-α) (Mustafa, 2012).

The chemical components identified in petroleum ether, methanol, and acid fraction extracts of *B. papyrifera* gum resin collected from damazine were found to be oleic acid, trimethylsilylester, higher concentrations of 3-oxatricyclo [20, 8-O-O-(7,16) triaconta-1(22), 7(16), 9, 13, 23, 29, hexaene, hexadecanoic acid, trimethylsilylester (7.90%) and pentacyclic triterpenes such as β-amyrene, β-amyrenol, β-amyrenone, lupeol acetate, olean-13 (18) ene, urs-11-en-28 oic acid, 2, 13-dihydroxy-3, 23-[(1-methylethylidene) bis (oxy)]- and γ-lactone, (2.α., 3.β.,4.α) (Mustafa, 2012). GC-MS analysis of petroleum ether, methanol, and acid fraction extracts of *B. papyrifera* gum resin collected from Nagawa yielded major compounds as octadecanoic acid, trimethylsilylester (20.45%), followed by hexadecanoic acid, lupeol, jasminol, α-amyrin, β-amyrene, β-amyrin methylether, α-amyrin acetate, α-amyrenol, β-amyrenone, urs-11-en-28 oic acid, 2, 13-dihydroxy-3, 23-[(1-methylethylidene) bis (oxy)]-, γ-lactone, (2.α., 3. β., 4.α), benzoic acid, 2-[(trimethylsilyl) oxy]-, trimethylsilylester; oleanane 3,16-diol, 13, 28-epoxy, diacetate (3.β., 16.α) and oleana-11, 13, 18-diene (Mustafa, 2012).

Atta-Ur-Rahman et al. (2005) isolated a new triterpene, 3α-acetoxy-27-hydroxylup-20(29)-en-24-oic acid, two new stilbene glycosides, trans-4,'5-dihydroxy-3-methoxystilbene-5-O-[α-L-rhamnopyranosyl-(1-2)-(α-L-rhamnopyranosyl-(1-6)-β-D-glucopyranoside, trans-4,'5-dihydroxy-3-methoxystilbene-5-O-(α-L-rhamnopyranosyl-(1-6)-β-D-glucopyranoside along with five known terpenoids such as 11-keto-β-boswellic acid, β-elemonic acid, 3α-acetoxy-11-keto-β-boswellic acid, βBA, and β-sitosterol from stem bark of *B. papyrifera*.

These stilbene glycosides have demonstrated phosphodiesterase I and xanthine oxidase enzyme inhibition activities and triterpenes demonstrated prolyl endopeptidase inhibitory activities. Whereas, Abdurrahman et al. (2019) isolated and characterized nine diterpenes, i.e., m-camphorene, pimaradiene, verticilla-4(20),7,11-triene, serratol, incensole, incensyl acetate, duva-3,9,13-trien-1α-ol-5,8-oxide-1-acetate, incensole oxide and incensole oxide acetate from EA extract of *B. papyrifera* stem bark. A new sterol namely, 29-methyl, nonanoate-cholest-7, 22-dien-3β-ol, along with three known related sterol compounds, stigmasterol, β-sitosterol-3-O-β-D-glucoside and spinasterol 3-O-β-D-glucopyranoside were isolated and characterized by Abdurrahman et al. (2020).

5.13 *B. PIROTTAE* CHIOV.

The resin of *B. pirottae* is a rare commodity. Essential oils obtained from pounded resins using hydrodistillation yielded 5%, analyzed and disclosed its components. The oil is highly pleasant and aromatic specialized with rich content of trans-verbenol (15.5%) and terpinen-4-ol (14.6%). Basar et al. (2003) reported, the essential oil components from gum-resin obtained by hydrodistillation, such as bornyl acetate (1.6%), camphene (0.1%), γ-campholene aldehyde (0.2%), α-campholene aldehyde (1.6%), carvacrol (0.3%), trans-carveol (1.6%), chrysantenone (0.2%), cumin alcohol (0.7%), *p*-cymen-8-ol (1.6%), dodecanol (0.2%), limonene (0.4%), linalool (2.4%), cis-linalool oxide (0.2%), trans-linalool oxide (0.2%), linalyl acetate (1.3%), p-mentha-1,5-diene-7-ol (1.1%), cis-p-menth-2-en-1-ol (0.4%), trans-p-menth-2-en-1-ol (0.5%), myrtenal (0.8%), myrtenol (2.1%), α-pinene (5.2%), β-pinene (0.4%), α-pinene oxide (0.2%), pinocarvone (0.6%), trans-pinocarveol (2.4%), sabinaketone (0.5%), sabinene (6.0%), cis-sabinenehydrate (1.9%), cis-sabinol (1.8%), trans-sabinenehydrate (2.4%), trans-sabinol (1.2%), terpinen-4-ol (14.6), α-terpineol (1.1%), α-terpinyl acetate (0.9%), thuj-3-en-10-al (0.5%), α-thujene (1.5%), α-thujone (0.1%), β-thujone (0.6%), thymol (0.4%), umbellulone (0.5%), verbenone (4.9%), cis-verbenol (1.2%), trans-verbenol (15.5%).

5.14 *B. POPOVIANA* HEPP.

Phytochemical composition of methanol extract of *B. popoviana* gum resin was evaluated by using GC-MS analysis (Madera et al., 2017). The components were identified as α-thujene, α-pinene, sabinene, β-pinene, α-fenchene, camphene, β-myrcene, 3-δ-carene, p-cymene, limonene, isobutylbenzene, β-phellandrene, γ-terpinene, terpinolene, fenchol, 4-terpineol, α-terpineol, camphor, β-caryophyllene, trans-β-farnesene, α-humulene, δ-cadinene, caryophyllene oxide, isocaryophyllen, neocembrene, and cembrene (Madera et al., 2017).

5.15 *B. RIVAE* ENGL. (ETHIOPIAN FRANKINCENSE)

The gum resin is woody with a spicy fragrance. Hydro-distillation of gum resin yielded thick yellow oil. The gas chromatographic analysis of oil revealed that the major components were limonene (14.8%, 21.1%, 28.0%), δ-3-carene (9.6%, 17.3%, 15.7%), *p*-cymene (2.9%, 3.2%, 7.1%), α-pinene (5.3%, 16.7%, 13.3%), α-thujene (1.3% 2.9%, 1.7%), and octanol (nd, nd, 17.8%) (Baser et al., 2001; Baser, 2005; Camarda, 2007). Basar (2005) identified 24-norursa-3, 12-diene (18.7%), *α*-amyrin (4.2%), *β*-amyrin (0.9%), *α*-amyrenone (2.8%), *β*-amyrenone (2.3%), and epi-*β*-amyrin (0.9%) components in the pyrolysate of *B. rivae*. The minor components were bornyl acetate, camphene, γ-campholene aldehyde, α-campholene aldehyde, cara-2,4-diene, carvacrol, cis-carveol, trans-carveol, carvone, trans-dihydrocarvone, 1,8 cineole, α-copaene, cumin alcohol, m-cymen-8-ol, m-cymene, o-cymene, p-cymen-8-ol, p-cymen-9-ol, p-cymenene, β-elemene, eucarvone, isopropyl benzaldehyde, isopropyl benzalcohol, cis-1,2-limonene epoxide, 8,9-limonene epoxide ii, 8,9-limonene-epoxide i, trans-1,2-limonene epoxide, linalool, cis-linalool oxide, trans-linalool oxide, p-mentha-1,8-diene-4-ol, cis-p-mentha-1(7),8-diene-2-ol, cis-p-mentha-2,8-diene-1-ol, trans-p-mentha-1(7),8-diene-2-ol, trans-p-mentha-2,8-diene-1-ol, o-methylanisol, p-methylanisol, myrcene, myrtenal, myrtenol, (E)-β-ocimene, (E)-β-ocimene epoxide, α-phellandrene, α-phellandrene-8-ol, β-pinene, pinocamphone, pinocarvone, trans-pinocarveol, piperitenone, dehydro-sabinaketone, sabinaketone, sabinene, cis-sabinene hydrate, trans-sabinenehydrate, trans-sabinol, 2,5 dimethyl styrene, α-terpinene, γ-terpinene, terpinen-4-ol, α-terpineol, terpinolene, α-terpinyl acetate, thuj-3-en-10-al, thuja-2,4(10)-diene, α-thujone, β-thujone, thymol, umbellulone, verbenone, cis-verbenol, trans-verbenol.

Essential oil composition of *B. rivae* gum resin collected from three geographical locations of Ethiopia, revealed the presence of *α*-thujene, *α*-pinene, o-cymene, δ-3-carene, p-cymene, limonene, *α*-campholene aldehyde and trans-verbenol (Bekana et al., 2014). Gas chromatographic and mass spectra analysis of commercial essential oil samples of *B. rivae* purchased from Ethiopia, consists of the following chemical constituents, i.e., α-pinene, β-pinene, δ-3-carene, p-cymene, limonene, and α-thujene (Tihar & Dagne, 2014).

Methanol extract of *B. rivae* and *B. neglecta* was also found to contain monoterpene (*α*-pinene) and triterpenes, namely, *β*-amyrin, *α*-amyrin, *β*-amyrenone, and *α*-amyrenone (Bekana et al., 2014). Two new ursane-type triterpenes, 3-oxo-24-acetoxy-11α-hydroxy-urs-12-ene and methyl 3α-acetoxy-11α-methoxy-urs-12-en-24-oate, along with known 9 compounds were isolated from *B. rivae* oleo-gum resin (Manguro & Wagai, 2016).

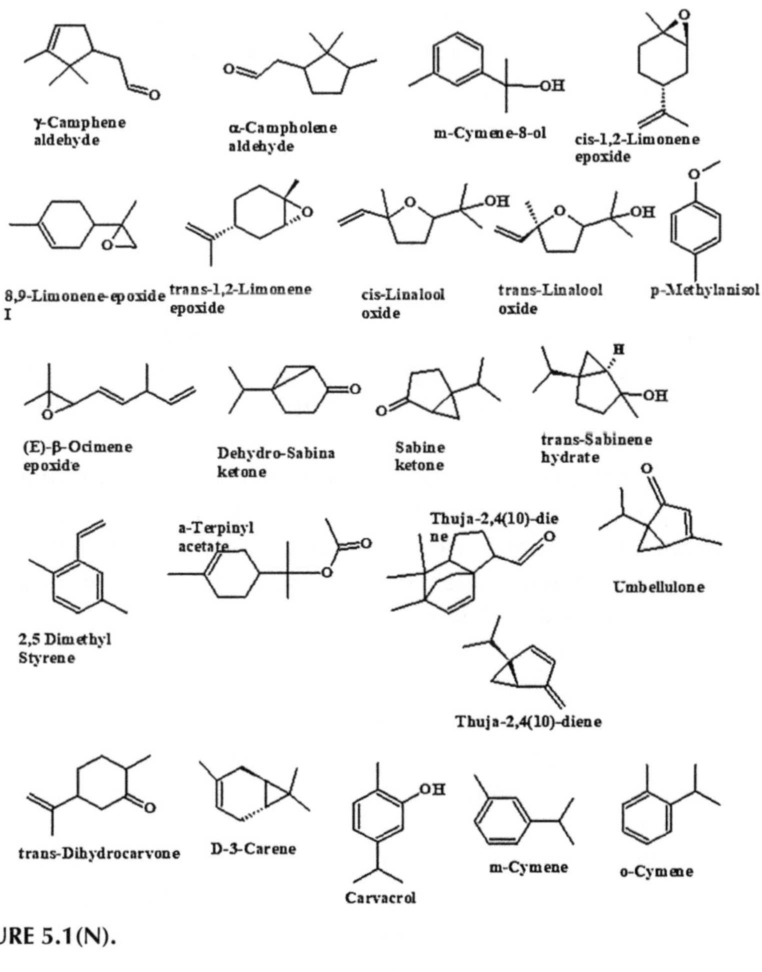

FIGURE 5.1(N).

5.16 *B. SACRA* FLÜCK. (SYN.: *B. CARTERII* BIRDW.) (OMANI LUBAN OR FRANKINCENSE)

B. sacra (Syn.: *B. carteri*) yields a yellowish-brown oleo-gum resin, called the frankincense, which is in high demand for various commercial products such as cosmetic ingredients (soaps, lotions, and ointment formulation), food flavors, and perfumes (Al-Harrasi et al., 2018) and it accounts for about two-thirds of global frankincense production. In 1998, the International Union for Conservation of Nature (IUCN) warned that one of the primary frankincense species, *B. sacra*, is "near threatened." The gum resin is composed of essential oil (5±15%), mucus-like cluster (12±23%) and a lipophilic part (55±66%). The lipophilic part was rich in terpenoids, among which are the medicinally important group of boswellic acids (BAs) (Shen and Lou, 2008).

Hydro-distillation of the finely powdered oleo gum resin of *B. sacra* yielded (5.5%) colorless oil, with a balsamic slightly spicy and perfumery odor (Al-Harrasi and Al-Saidi, 2008). The oil was rich in monoterpenes (34) which constituted 97.3% in which *E*-β-ocimene and limonene were the major constituents, while the sesquiterpenes (16) constituted 2.7%, in which the *E*-caryophyllene was the major constituent. The identified monoterpenes were 2-β-pinene (0.1%), α-thujene (6.6%), *E*-β-ocimene (32.3%), sabinene (5.2%), 1-β-pinene (1.8%), myrcene (6.9%), α-pinene (5.3%) the major component identified was limonene (33.5%), and few of the components showed less than 5%, such as 2-carene, *Z*-β-ocimene, γ-terpinene (1.0%), terpinolene, 2,4(10)-thujadiene, camphene, *p*-cymene, 1,4-cyclohexadiene, perillene, isopentyl-2-methyl butanoate, isoamyl valerate, 1,3,6-trimethylenecycloheptane, β-thujone, α-campholene aldehyde, *allo*-ocimene, *trans*-pinocarveol, *p*-mentha-1,5-dien-8-ol, 4-terpineol, sabinyl acetate, myrtenal, α-terpineol, α-phellandrene epoxide, verbenone, *trans*-(+)-carveol, carvone, and 1-bornyl acetate and the sesquiterpenes, α-cubebene, α-copaene, β-bourbonene, β-elemene, α-gurjunene, E-caryophyllene, α-humulene, allo aromadendrene, α-amorphene, germacrene D, β-selinene, α-selinene, α-muurolene, γ-cadinene, caryophyllene oxide and γ-muurolene.

Ni et al. (2012) evaluated the phytochemical composition of hydro distilled four fractions of Omani luban essential oils. They collected four fractions at different time and temperature, i.e., Fraction I between 0 h and 2 h (at 78°C), Fraction II between 8 h and 10 h (at 78°C), Fraction III between 11 h and 12 h (at 78°C), and Fraction IV between 11 h and 12 h (at 100°C).

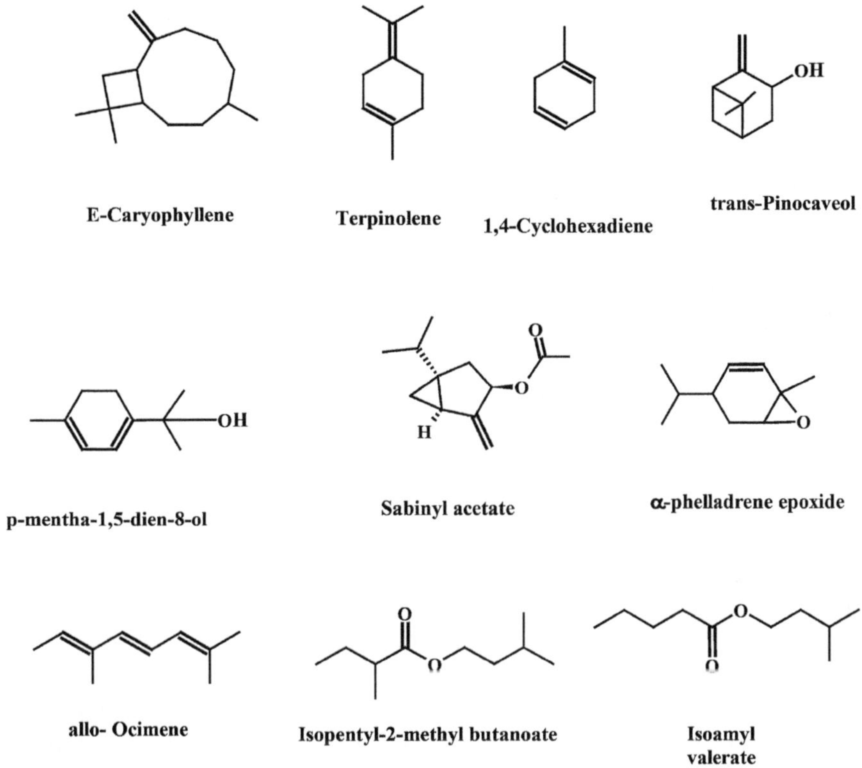

FIGURE 5.1(O).

The major chemical constituents present in four fractions were α-pinene (78.45%, 76.29%, 65.49%, 59.40%), camphene + verbenene (4.09%, 4.39%, 3.42%, 3.46%), myrcene (2.56%, 2.82%, 7.46%, 5.36%), limonene (5.55%, 6.92%, 8.43%, 8.99%), β-pinene (2.16%, 2.33%, 2.27%, 2.38%) and sabinene (1.41%, 1.11%, 2.12%, 3.63%). The minor components identified in four fractions, which constituted less than 1% each, were toluene, α-thujene, sabinene, ortho-methyl anisole, α-phellandrene, δ-3-carene, para-cymene, cis-β-ocimene, γ-terpinene, para-cymenene, α-terpinolene, myrcenol, α-campholene aldehyde, trans-pinocarveol, cis-verbenol, pinocamphone, α-phellandren-8-ol, para-cymene-8-ol, terpinene-4-ol, α-terpineol + myrtenal, myrtenol, cis-carveol, bornyl acetate, α-terpenyl acetate, α-copaene, β-elemene, trans-β-caryophyllene, α-humulene, β-selinene and α-selinene. The components present in fraction II, III, and IV were verbenone) γ-muurolene and δ-cadinene. The components present in only fraction III and IV were borneol, dimethyl ether orcinol, allo-aromadendrene, germacrene D, γ-cadinene, and caryophyllene oxide.

Van Vuuren et al. (2010) analyzed the chemical composition of 20 frankincense oil samples, purchased at various herbal shops or pharmacies, which include two samples of *B. sacra*. Omani luban oil is dominated with α-pinene (22.5%, 18.3%), limonene (11.2%, 13.1%), β-caryophyllene (7.6%, 7.2%), sabinene (6.9%, 4.1%), *p*-cymene (5.9%, 4.7%), myrcene (5.5%, 3.5%), α-thujene (3.9%, 11.2%), α-humulene (2.5%, 2.7%), δ-cadinene (1.9%, 2.1%) α-copaene (1.5%, 1.3%), β-pinene (1.1%, 0.9%), β-caryophyllene oxide (0.0%, 3.0%). Woolley et al. (2012) evaluated the phytochemical composition of the Somalian (*B. carterii*) and Omani/Yemeni (*B. sacra*) frankincense species. The components identified with their average percent for *B. sacra* are α-pinene (68.2%), limonene + β-phellandrene (6.2%), sabinene (2.9%), camphene (2.1%), β-pinene (2.0%), myrcene (0.7%), and α-thujene (0.6%), respectively.

Al-Saidi et al. (2012) evaluated the chemical composition of four commercial *B. sacra* oleo gum resins. Alcoholic extract of oleo gums resins of *B. sacra* and *B. carterii* possess rich amount of lupeolic acids and BAs, i.e., 11-keto-β boswellic acid (KBA), acetyl-11-keto-β-boswellic acid (AKBA), αBA, 3-O-acetyl α boswellic acid, acetyl β boswellic acid (Rashan et al., 2019). A new boswellic acid derivative, 11α-ethoxy-β-boswellic acid, and a new ursane-type triterpene, nizwanone, together with two known compounds papyriogenin B and rigidenol were isolated from Omani frankincense *B. sacra* (Al-Harrasi et al., 2013a). Further, Wang et al. (2019) isolated four cembrane-type diterpenoids, boscartins L to O and five known compounds (1S, 3R, 11S, 12R, 7E)-1,12-epoxy-4-methylenecembr-7-ene-3,11-diol, isoincensole oxide, incensole oxide, incensole acetate, and incensole oxide acetate were from the gum resin of *B. sacra*.

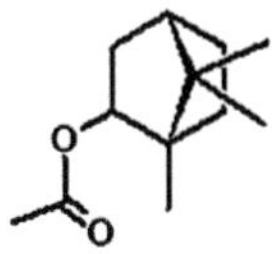

1-bornyl acetate

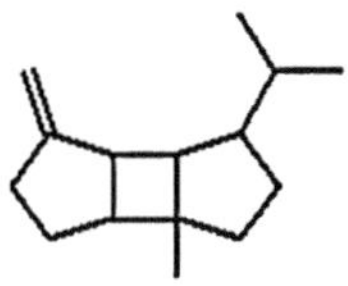

β-Bourbonene

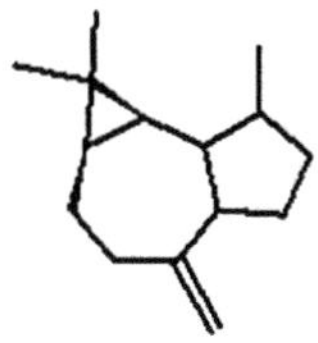

allo aromadendrene

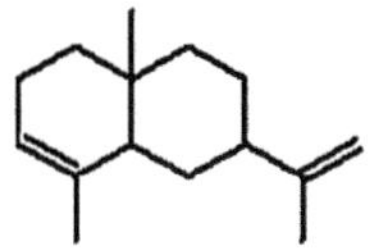

α-selinene

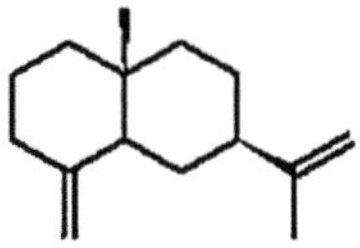

β-selinene

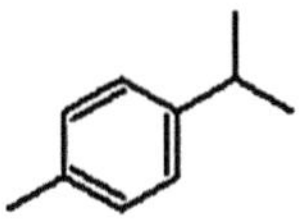

para Cymene

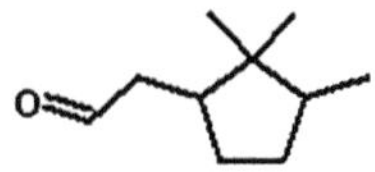

α-campholene aldehyde

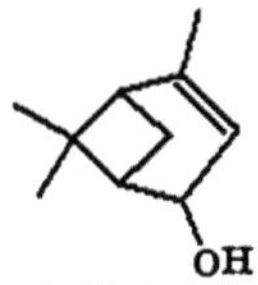

cis-Verbenol

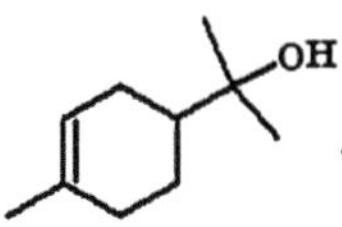

α–Phellandren-8-ol

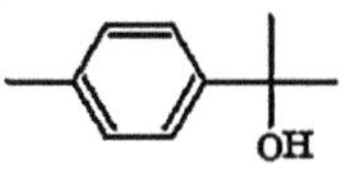

para-Cymene-8-ol

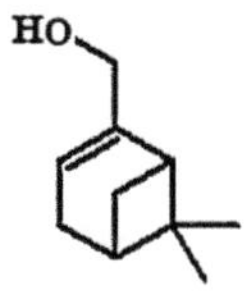

Myrtenol

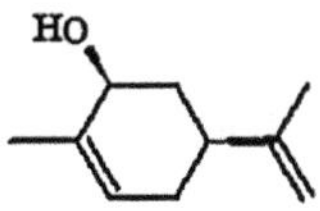

cis- Carveol

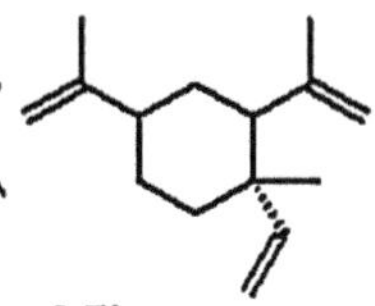

β-Elemene

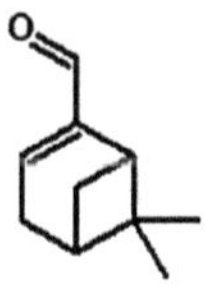

Myrtenal

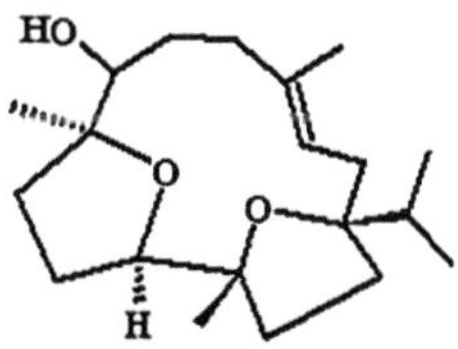

Isoincensole oxide

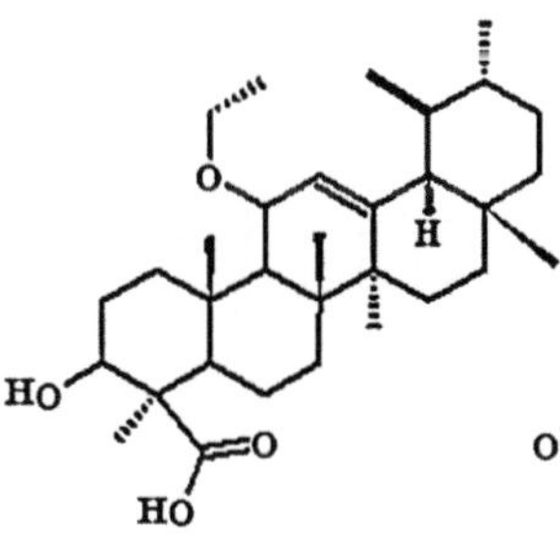

11-a-ethoxy-b-boswellic acid

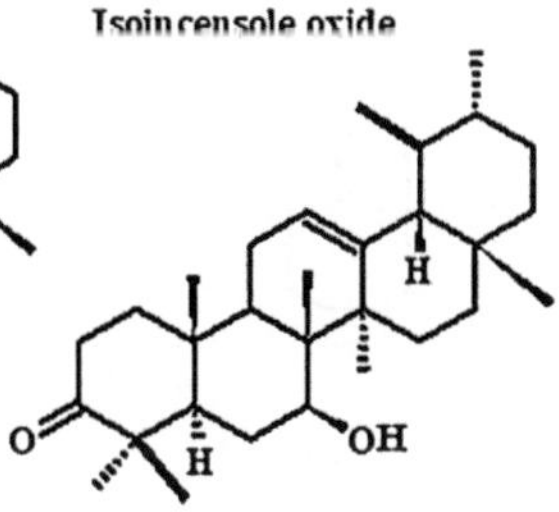

Nizwanone

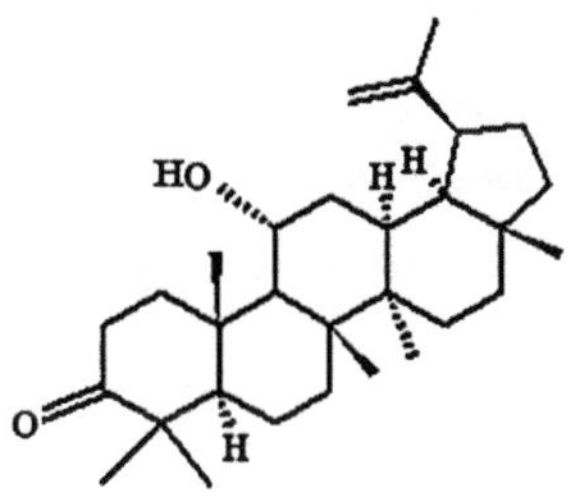

Rigidenol

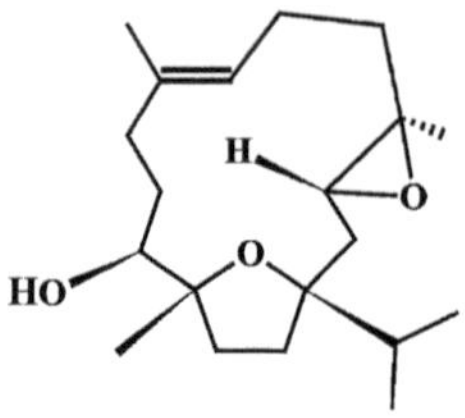

Incensole oxide

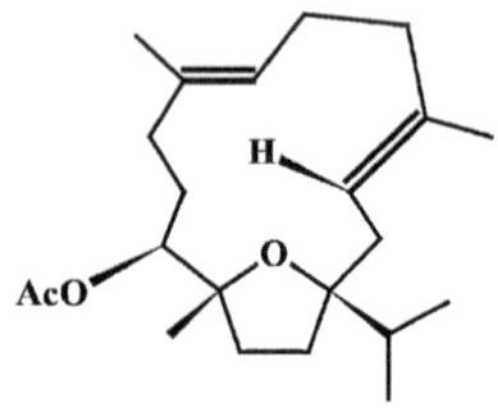

Incensole acetate

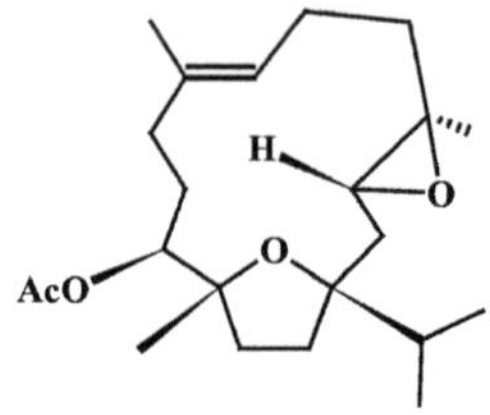

Incensole oxide acetate

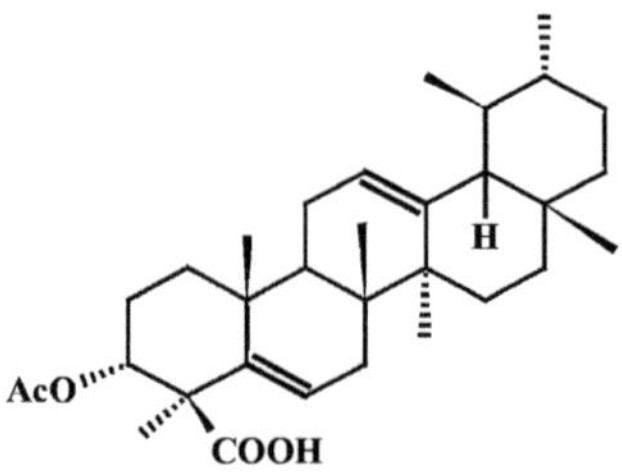

3a-acetoxyurs-5:12-dien-24-oic acid

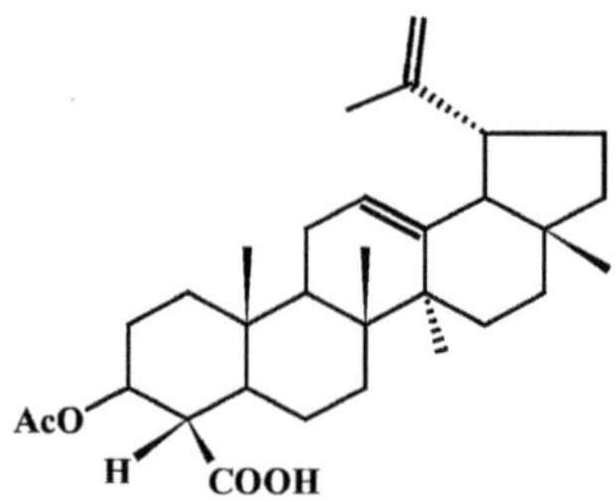

3-a-acetoxylup-12:20(29)-dien-2 4-oic acid

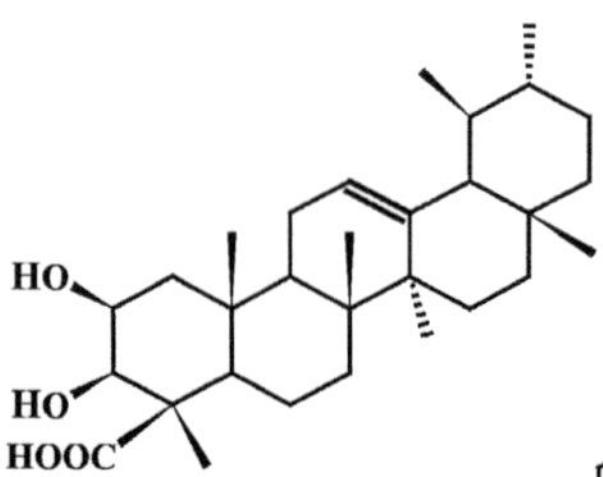

Commic acid-D

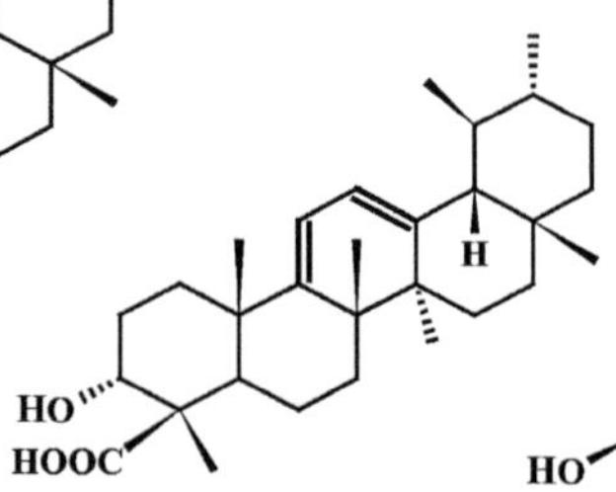

9,11-dehydro-b-boswellic acid

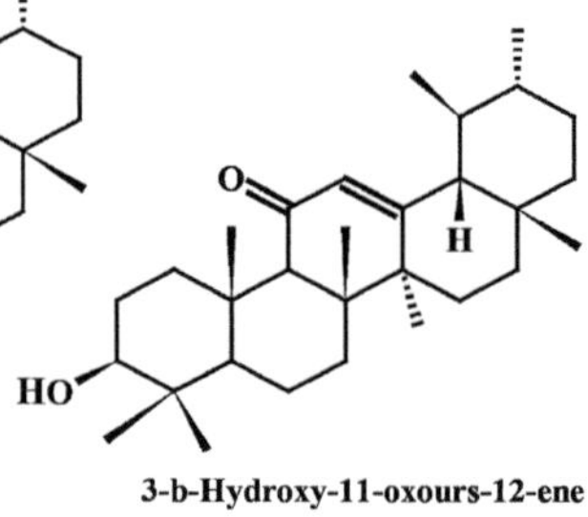

3-b-Hydroxy-11-oxours-12-ene

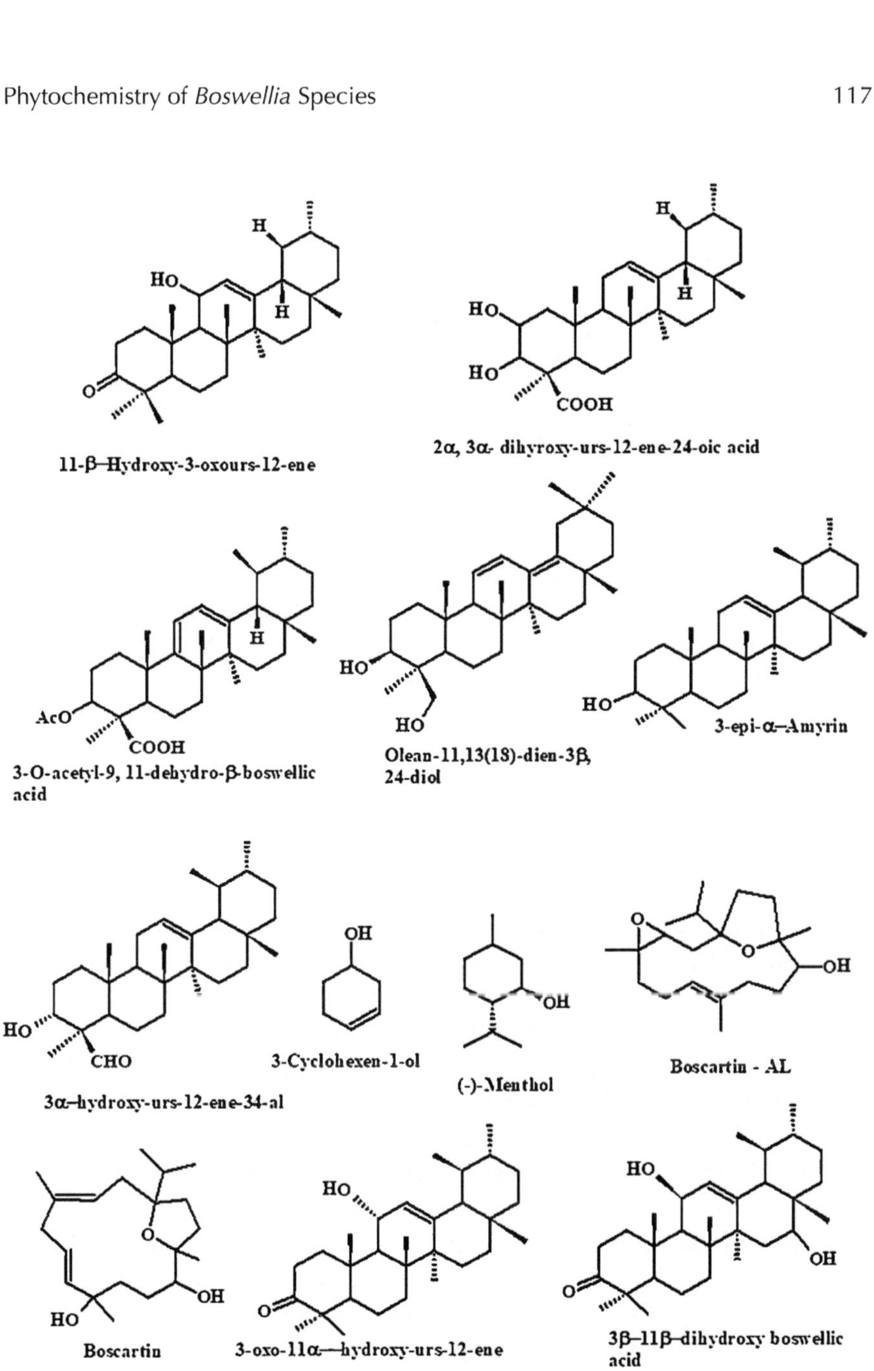
11-β–Hydroxy-3-oxours-12-ene
2α, 3α- dihyroxy-urs-12-ene-24-oic acid
3-O-acetyl-9, 11-dehydro-β-boswellic acid
Olean-11,13(18)-dien-3β, 24-diol
3-epi-α–Amyrin
3α–hydroxy-urs-12-ene-34-al
3-Cyclohexen-1-ol
(-)-Menthol
Boscartin - AL
Boscartin
3-oxo-11α—hydroxy-urs-12-ene
3β–11β–dihydroxy boswellic acid

FIGURE 5.1(P).

Ali et al. (2014) subjected 2% methanol/dichloromethane obtained from methanol extracts on HPLC yielded two new O-acetyl derivatives of pentacyclic triterpenic acids, 3α-acetoxyurs-5:12-dien-24-oic acid and 3-α-acetoxylup-12:20 (29)-dien-24-oic acid along with four known compounds: commic acid-D, 9,11-dehydro-β-boswellic acid, 3-β-hydroxy-11-oxours-12-ene, and 11-β-hydroxy-3-oxours-12-ene from *B. sacra* gum resin. GC-MS analysis of water extract of *B. sacra* gum resin revealed the presence of 1,8-cineole (4.16%), 1-octanol (6.91%), (–)-menthol (10.81%), 3-cyclohexen-1-ol (21.33%), octanoic acid (18.91%), thymol (6.14), carvacrol (5.67%). Nine terpenoid components were isolated from *B. sacra* gum resin βBA, lupeolic acid, 3α-O-acetyl-11-keto-β-boswellic acid, 11-keto-β-boswellic acid and lupeol, 2α,3α-dihydroxy-urs-12-ene-24-oic acid, 3-O-acetyl-9,11-dehydro-β-boswellic acid, olean-11,13(18)-dien-3β,24-diol and 3-oxo-11α-hydroxy-urs-12-ene (Al-Harrasi et al., 2013b). Around 10 undescribed cembrane-type diterpenes boscartins AL-AU, along with five known analogs were isolated from *B. sacra* gum resin (Wang et al., 2019). Al-Harrasi et al. (2018) characterized two new chemical constituents, i.e., β-boswellic aldehyde and 3β, 11β-dihydroxy boswellic acid) from the *B. sacra* resins along with known α-amyrin (3-epi-α-amyrin, β-amyrin, and α-amyrin).

Hanus et al. (2007) evaluated the chemical composition of n-hexane extract of *B. sacra* (=*B. carterii*). The components identified namely, incensole (21.5%), incensole acetate (19.9%), octyl acetate (8.9%), verticilla-4(20),7,11-triene (4.2%), incensole oxide (3.15%), incensole oxide acetate (2.5%), cembrene (0.1%), cembrene A (0.9%), cembrene C (0.6%), 24-noroleana-3,9(11),12-triene (0.1%), 24-norursa-3,9(11),12-triene (0.6%), 24-noroleana-3,12-diene (2.4%), 24-norursa-3,12-diene (4.3%) and 24-norursa-3,12-diene-11-one (2.2%). Compounds identified in n-hexane extract of resin after silylation with BSTFA such as trimethylsilyl 3-O-trimethylsilyl-α-boswellate (0.2%), trimethylsilyl 3-O-trimethylsilyl-β-boswellate (0.5%), trimethylsilyl 3-oxo-8,24-diene-tirucallate (1.0%), trimethylsilyl 3-O-acetyl-11-methoxy-β-boswellate (0.5%), trimethylsilyl 3-O-acetyl-α-boswellate (0.5%) and trimethylsilyl 3-O-acetyl-β-boswellate (0.7%) (Hanus et al., 2007).

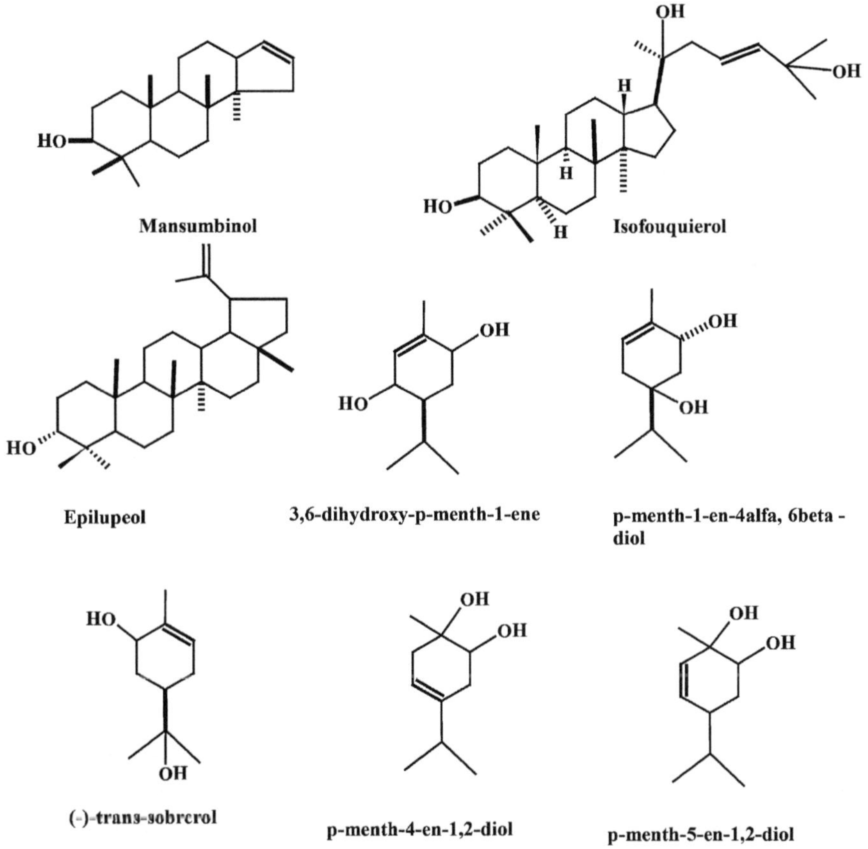

FIGURE 5.1(Q).

Hydro-distilled *B. sacra (=B. carterii*) oleo gum resin yielded yellow color essential oil, which constituted with β-bourbonene, bornylacetate, 3-carene, camphene, cembra-1,3,7,11-tetraene, Cembra-3,7,11,15-tetraene, campholenealdehyd, cembrene, cembrene a, cembrene c, 1,8-cineole, citronellylacetate, chrysanthenone, o-cymene, *p*-cymene, decylacetate, 3,5-dimethoxytoluene, geranylacetate, 1-hexanol, hexyl acetate, incensole, incensole acetate, limonene, linalool, myrcene, myrtenal, nerylacetate, (S)-trans-nerolidol, (E)-nerolidol, 1-octanol, octyl acetate, (E)-β-ocimene, (Z)-b-ocimene, α-pinene, α-phellandrene epoxide, β-pinene, trans-pinocarveol, sabinene, α-terpineol, terpinene-4-ol, α-thujene, β-thujone, verticilla-4(20),7,11-triene, verbenone, cis-verbenol, trans-verbenol (Obermann, 1977; Basar et al., 2003; Baser, 2005).

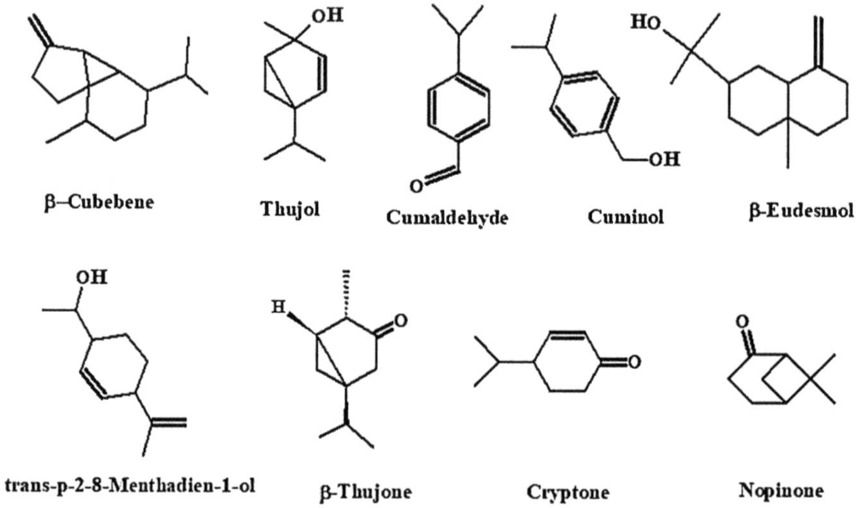

FIGURE 5.1(R).

In addition to these components, other components reported from oleo gum resin essential oil were abieta-8,12-diene, α-amorphene, allo-aromadendrene, benzylbenzoate, beyerene, bisabolene, isopentyl-2-methylbutanoate, cis-calamenene, γ-cadinene, δ-cadinene, t-cadinol, (E)-caryophyllene, caryophyllene oxide, cis-carveol, trans (+) carveol, carvone, α-cedrene, cedrol, β-citronellol, α-copaene, α-cubebene, β-cubebene, 1,4-cyclohexadiene, p-cymen-8-ol, decanol, 2,6-dimethoxytoluene, duva-3,9,13-trien-1,5α-diol, duva-4,8,13-trien-1,3α-diol, duva-3,9,13-trien-1-α-hydroxyl 5,8-oxide-1-acetate, duva-3,9,13-trien-1,5α-diol-1-acetate, duva-3,9,13-triene-1α-ol-5,8-oxide-1-acetate, β-elemene, farnesyl acetate, germacrene d, α-gurjunene, hedycariol, 1,3,6-trimethylencycloheptane, hexyl hexanoate, α-humulene, isocaurene, isocembrene, isophyllocladene (kaur-15-ene), kaurene, ledol, maaliane, p-mentha-1,5-dien-8-ol, o-methyl anisole, α-muurolene, γ-muurolene, α-muurolol, naphthalene decahydro-1,1,4a-trimethyl-6-methylene-5-(3-methyl-2-pentenyl), naphthalene 1,2,3,4,4a,7-hexahydro-1,6-dimethyl-4-(1-methylethyl, neocembrene a, cis-nerolidol, n-octanol, octanolacetate octyl formate, allo-ocimene, cis-ocimene, perillene, phenanthrene-7-ethenyl-1,2,3,4,4a,5,6,7,8,9,10,10a-dodeca-hydro-1-1-4a-7-tetramethyl, α-phellandrene, β-phellandrene, phyllocladene, α-pinene-epoxide,

1-β-pinene, 2-β-pinene, isopinocampheol, piperitone, piramidene, sabinene, cis-sabinol, sabinyl acetate, sandaracopimara-8(14)-15-diene, sclarene, α-selinene, β-selinene, δ-selinene, trans-terpine, γ-terpinene, 4-terpineol, terpinolen, isoterpinolene, 2,4(10)-thujadiene, thujopsene, thunbergol, isomyl-valerate, verticiol, viridiflorol (Mikhaeil et al., 2003; Marongiu et al., 2006; Camarda et al., 2007; Al-Harrasi and Al-Saidi, 2008).

Paul (2012) isolated ß-caryophyllene oxide, serratol, τ-cadinol, iso-serratol, incensole, 3-ß-OH-tirucallol, 3-α-OH-11-keto-12-ursen from yellow-orange color essential oil obtained from *B. carterii* using chromatographic techniques. Further, Hamm et al. (2005) evaluated the volatile constituent's composition from the headspace of *B. sacra* (=*B. carterii*), obtained by solid-phase micro-extraction (SPME) method. The identified components were allo-aromadendrene, bornylacetate, β-bourbonene, δ-cadinene, γ-cadinene, t-cadinol, cis-carveol, trans-carveol, carvone, β-caryophyllene, caryophyllene oxide, camphene, campholene aldehyd, cembrene a, cembrene c, α-copaene, α-cubebene, β-cubebene, cumaldehyde, cuminol, p-cymene, p-cymene-8-ol, 1,hydroxy-1,7-dimethyl-4-isopropyl-2,7-cyclodecadiene, 1,5,5,8-tetramethyl-12-oxabicyclo-[9.1.0] dodeca-3,7-diene, 3,5 dimethoxytoluene, β-elemene, β-eudesmene, α- or β-eudesmol, γ-eudesmol, germacrene d, 4-methylene-1-(1-methylethyl)-bicyclo[3.1.0]hex-2-ene, 1-methyl-4-(1-methylethenyl)-1,2-cyclohexanediol, 2-methyl-5-(1-methylethyl)-1,3-cyclohexadiene monoepoxide, α-humulene, isoincensole + isoincensoleacetate, limonene, linalool, o-methylanisole, trans-p-2,8-menthadien-1-ol, trans-p-mentha-2,8-dienol, α-muurolene, γ-muurolene, myrcene, myrtenal, myrtenol, 1,2,3,4,6,8a-hexahydro-1-isopropyl-4,7-dimethyl-naphthalene, 2-isopropenyl-4a,8-dimethyl-1,2,3,4,4a,5,6,8a-octahydronaphthalene, perillol, α-phellandrene, α-phellandrene-dimer, α-pinene, β-pinene, trans-pinocarveol, sabinene, cis-sabinol, 3,4 dimethoxystyrene, γ-terpinene, α-terpineol, 4-terpineol, terpinylacetate, (+)-a-thujene, β-thujone, thujol, m-thymol, verticilla-4(20),7,11-triene, verbenone, cis-verbenol.

Volatile components identified in pyrolyzed *B. sacra* (=*B. carterii*) oleo gum resin were 1-acetyl-4-isopropenylcyclopentene, 2,4-dimethyl-acetophenone, α-amyrenone, β-amyrenone, 10-hydroxy-4-cadinen-3-one, α-campholenaldehyde, γ-campholenaldehyde, carvone, carvonan acetone, 2-hydroxy-1,4-cineole, cryptone, cuminaldehyde, eucarvone, isopropylidencyclohexane, 1,2,4-trihydroxy-p-menthane, d-4-p-menthen-2-one, 5-hydroxy-p-menth-6-en-2-one, myrtenal, myrtenoic acid, nopinone,

3,6,6,-trimethyl-norpinan-2-one, o-methylacetophenone, perillaaldehyde, phellandral, pinocamphone/isopinocamphone, piperitone, thujone, and erbenone (Pailer et al., 1981).

The hydrodistillation of *B. sacra (=B. carterii)* oleo gum resin yielded pale yellow oil (2.44%, v/w) on a dry weight (DW) basis (Mohamed et al., 2015). The GC-MS analysis of the oil led to the identification of 40 constituents, dominated with oxygenated monoterpenes (13.31%). The components identified were verticiol (14.48%), α-phellandrene (0.25%), α-pinene (3.09%), sabinene (0.77%), α-myrcene (0.86%), octanal (0.28%), 1-hexyl acetate (1.11%), 9-oxabicyclo[6.1.0]non-3-yne (9.12%), trans-α-ocimene (0.63%), 1,4-cyclononadiene (2.86%), isobutyl-cyclopentane (12.25%), tetrahydro-5-hydroxymethyl-2-furaldehyde-3-one (5.91%), 1-butanol (4.28%), cis-á-terpineol (1.82%), trans,trans-3,5-heptadien-2-one (2.63%), n-octyl acetate (9.20%), 1-decanol (2.07%), carvone (0.22%), 3,5-dimethoxytoluene (0.25%), (Z)-3-methyl-1,3-nonadiene (0.82%), geranyl acetate (0.89%), lavandulyl acetate (1.44%), 3-methyl-7-oxabicyclo-[4.1.0]-heptane (2.11%), (e)2-decenyl acetate (0.38%), butylated hydroxytoluene (0.21%), dodecanoic acid (0.33%), octanoic acid, hexyl ester (0.57%), Z-4-tridecen-1-yl acetate (0.3%), geranyl-à-terpinene (0.2%), cembrene (2.07%), germacrene-d (0.58%), verticellol (6.05%), cycloheptane,4-methylene-1-methyl-2-(2-methyl-1-propen-1-yl)-1-vinyl (0.44%), cembrene C (0.24%), cembrene A (0.82%), trans-geranylgeraniol (0.51%), azulene,1,2,3,3a,4,5,6,7-octahydro-1,4-dimethyl-7-(1-methylethenyl), [1R(13aá,47á)] (0.69%), à-humulene (0.38%), benzo[f]chromen-3-one, Perhydro-2-acetyl-4-a,-7,7,10b-tetramethyl (2.69%) and 2,5-dioxa-8,12-dimethyl-13-acetoxytetracyclo-[6.11.0.0-(9,17).0(12,16)]-nonadecane (6.12%).

The chemical composition of essential oils of the resin samples of *B. sacra (=B. carterii)* used as food and non-food grade, obtained by different methods of distillation, i.e., water distillation, steam distillation, and hydro-steam distillation, were analyzed by gas chromatography, coupled to mass spectrometry (GC-MS). The major components identified in a non-food grade gum resin by hydro-steam distillation were dihydro citronellol acetate (48%), 2-phenyl ethyl anthranilate (11.5%), α-santonine (7.7%), and with water distillation were dihydro citronellol acetate (48.2%), borneol (8.6%), methyl decanoate (7.9%), and with steam distillation were dihydro citronellol acetate (60.6%), borneol (9.7%), (Z)-β-ocimene (5.3%), respectively. Whereas the major components identified in food grade by hydro-steam distillation were dihydro citronellol acetate

(55.6%), α-santonine (9%), 2-phenyl ethyl anthranilate (7.3%), and with water distillation were dihydro citronellol acetate (63.7%), borneol (9.1%), 2-phenyl ethyl anthranilate (3%), and with steam distillation were dihydro citronellol acetate (55.9%), (E)-phytol acetate (7.5%), borneol (7.3%), respectively (Dounchaly et al., 2016).

Eight new cembranoids, boscartins A–H (1, 2, and 4–9), and the known incensole oxide were isolated from oleo gum resin (Ren et al., 2015). Seven new diterpenoids, boscartins I-K, three cembrane-type diterpenoids and four boscartols K-N, prenyl-aroma-dendrane-type diterpenoids, along with nine known diterpenoids, were isolated from the gum resin (Yu et al., 2018a). Further, three new terpenes, i.e., one cembrane-type diterpenoid (rel)-(1S,5R,7E,11E)-1-isopropyl-8,12-dimethyl-4-methylenecyclotetradeca-7,11-diene-1,5-diol, one dinortirucallane-type triterpene 20,21-dinortirucalla-8,24-diene-3β-ol-7-one and one ursane-type triterpene 21β-hydroxy-3-O-acetyl-11-keto-boswellic acid were isolated from the *B. sacra (=B. carterii*) gum resin (Yu et al., 2018b).

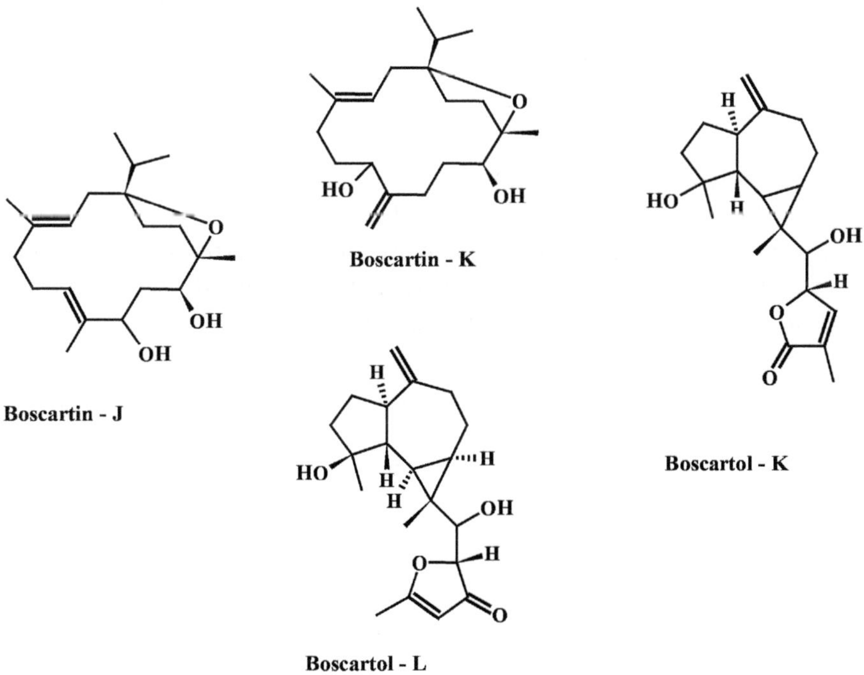

FIGURE 5.1(S).

The structures of three new tirucallane-type triterpenes, i.e., boscartene L, M, and N compounds isolated from the gum resin of *B. sacra* (=*B. carterii*), were elucidated by 1D and 2D NMR spectroscopy and mass spectrometry (Yang et al., 2018).

Three new monoterpenes, olibanumols A, B, C, and three new triterpenes, olibanumols H, I, and J together with seven known terpenoids, i.e., 3,6-dihydroxy-p-menth-1-ene, p-menth-1-en-4α,6β-diol, (–)-trans-sobrerol, p-menth-4-en-1,2-diol, p-menth-5-en-1,2-diol, isofouquierol, and epilupeol were isolated from aqueous acetone extract of the gum-resin from *B. carterii* purchased at Cairo of Egypt (Yoshikawa et al., 2009). Further, intensive chromatographic studies on the aqueous acetone extract of gum-resin from *B. sacra (=B. carterii*) resulted in the isolation of four terpenoids, i.e., olibanumols K, L, M, and N together with seven lupane-type triterpenes, epilupeol acetate, lup-20(30)-ene-3α,29-diol, glochidiol, lupeol, lup-20(29)-ene-2α,3β-diol, 3β-acetoxylup-20(29)-en-11β-ol and lupenone, five ursane-type triterpenes, urs-9(11),12-dien-3β-ol, neoilexonol, neoilexonol acetate, urs-12-ene-3β,11α-diol and urs-12-ene-3α,11α-diol and five dammarane-type and two nordammarane-type triterpenes, dammarenediol II, dammarenediol II acetate, 3-O-acetyl-3β, 20S, 24-trihydroxydammar-25-ene, isofouquierol acetate, ocotillyl acetate, 3β-hydroxymansumbin-13(17)-en-16-one and mansumbinol (Morikawa et al., 2010).

Eight newly described cembrane-type diterpenes Boscartins AH, AI, AJ, AK, AL, AM, AN, and AO and two known cembrane boscartin-Q and boscartin-Z were isolated and characterized from partitioned petroleum ether extract from 20% aqueous ethanol extract obtained from gum resin of *B. sacra* (Syn.: *B. carterii*) (Sun et al., 2020).

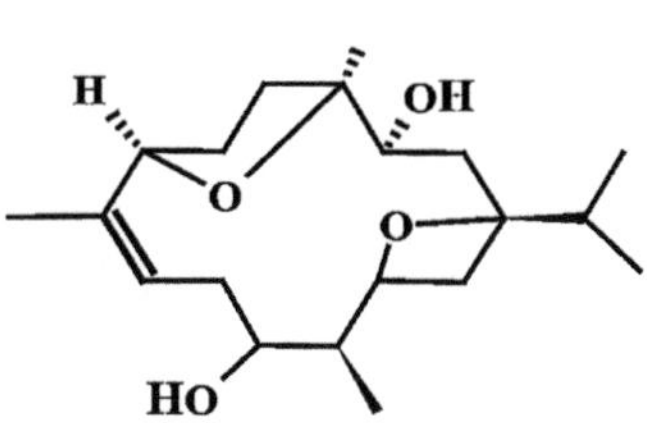

Boscartin A

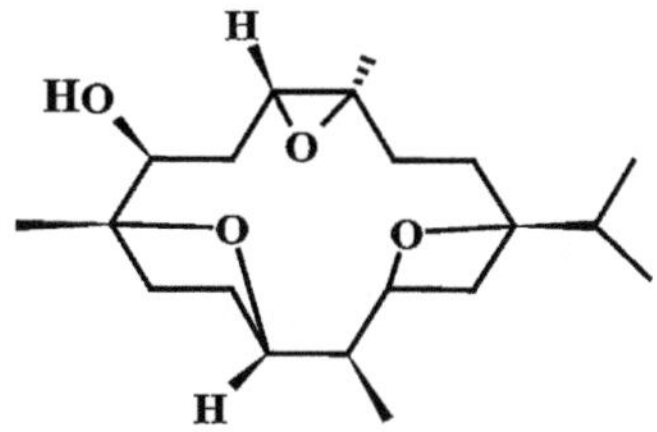

Boscartin B

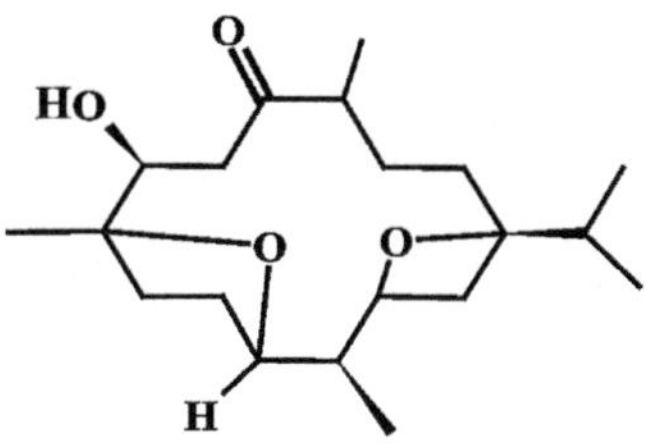

Boscartin C

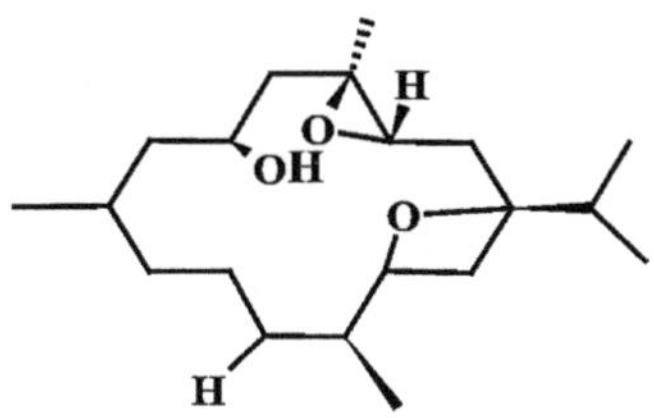

Boscartin D

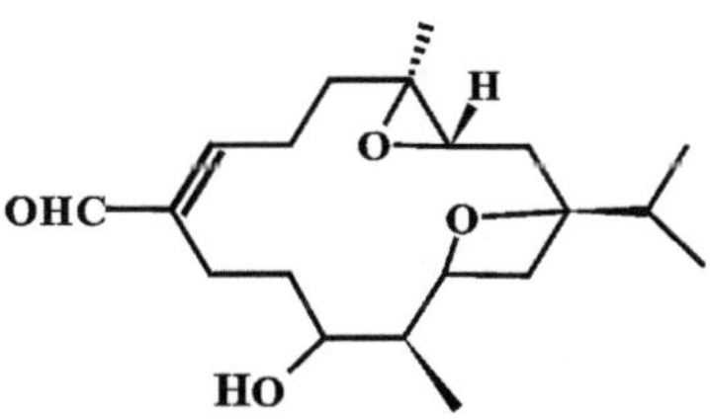

Boscartin E

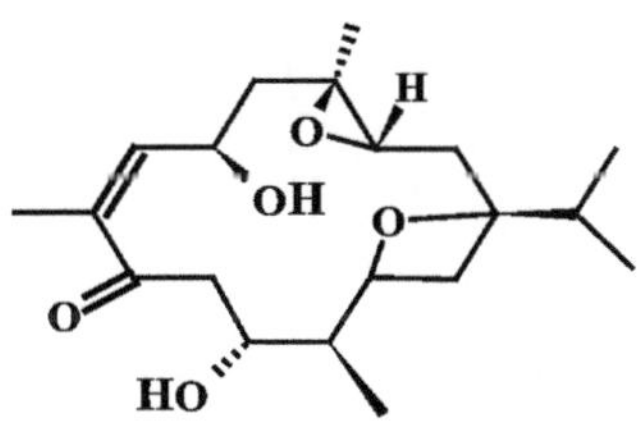

Boscartin F

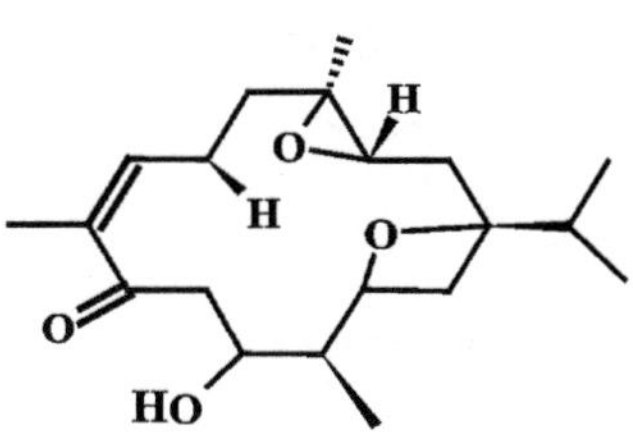

Boscartin G

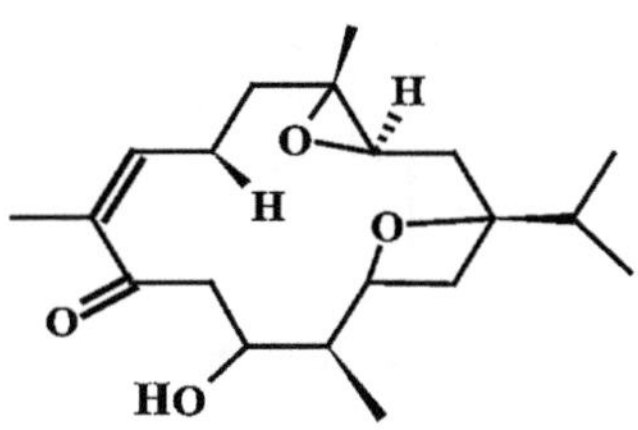

Boscartin H

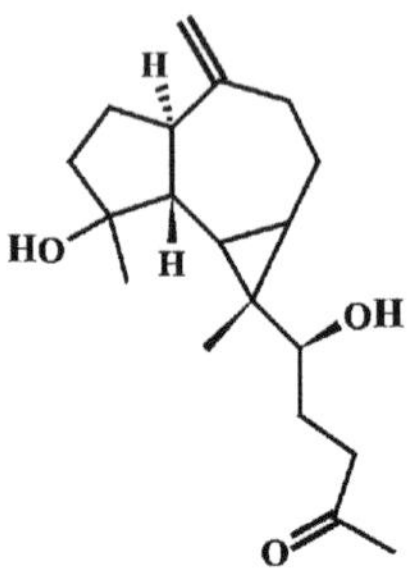

Boscartol - M

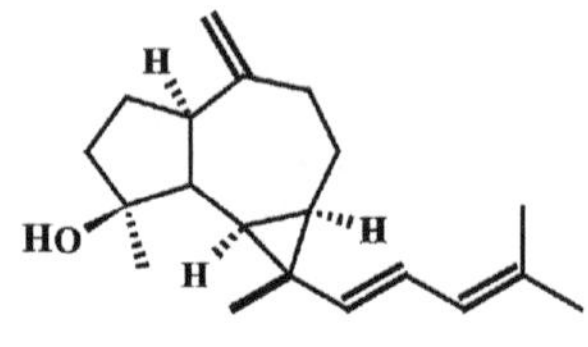

Olibanumol-D

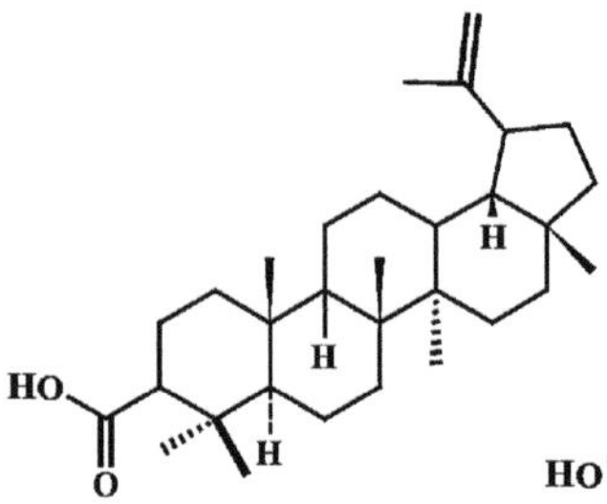

Olibanumol - F

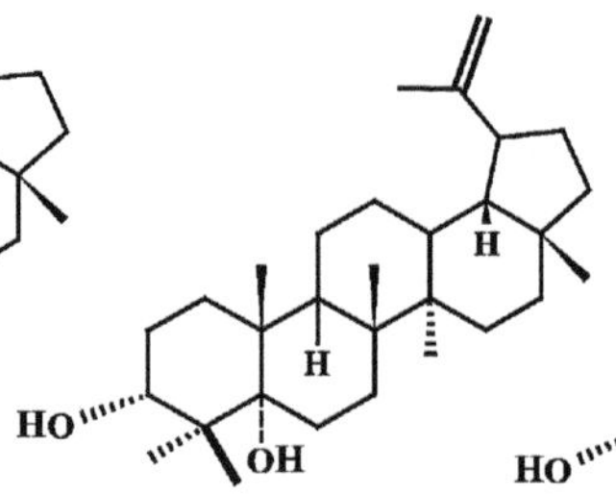

Olibanumol - G

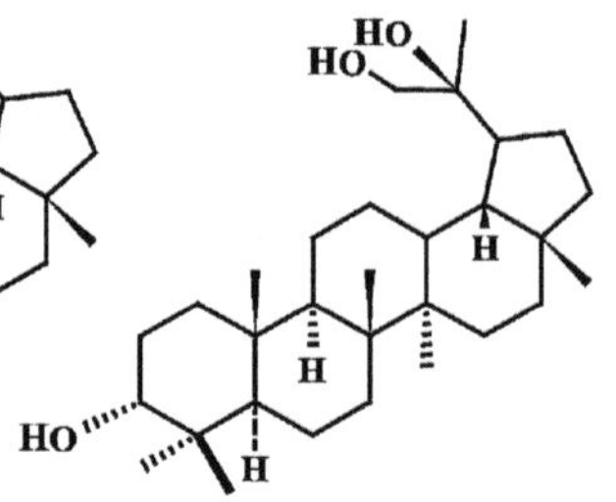

Olibanumol - H

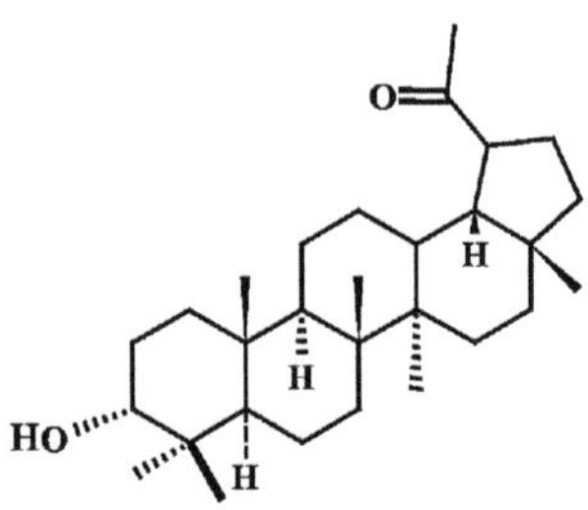

Olibanumol - I

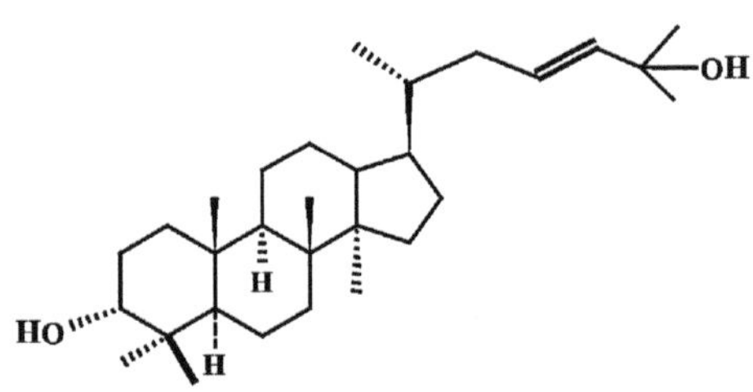

Olibanumol - J

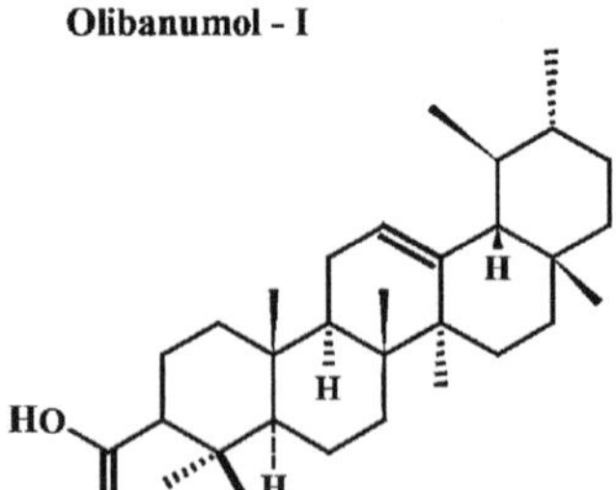

Olibanumol - K

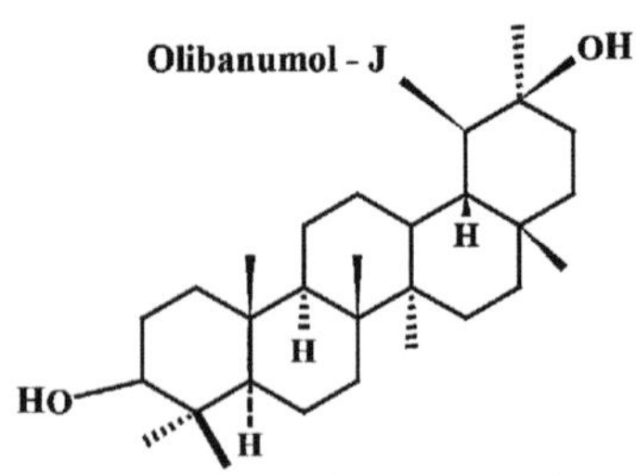

Olibanumol - L

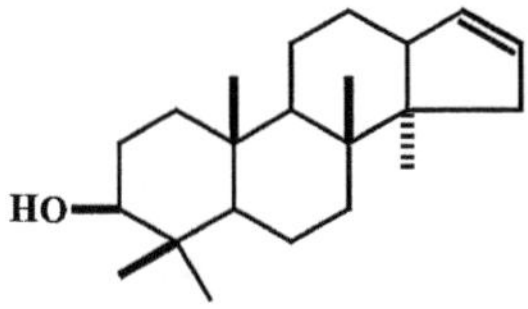

Mansumbinol

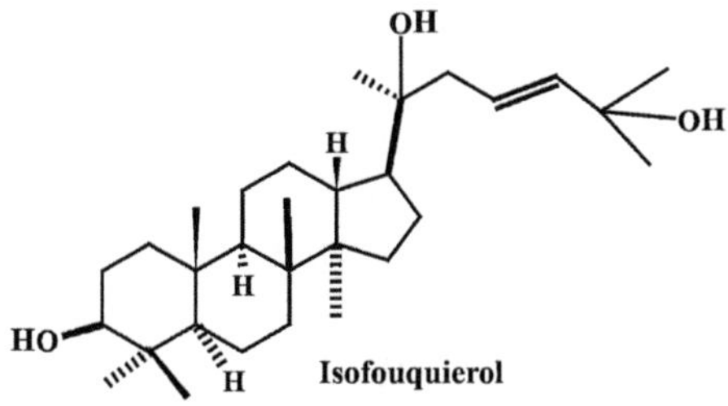

Isofouquierol

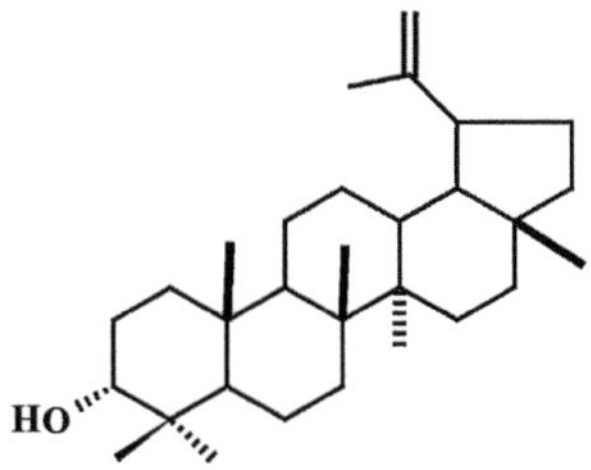

Epilupeol

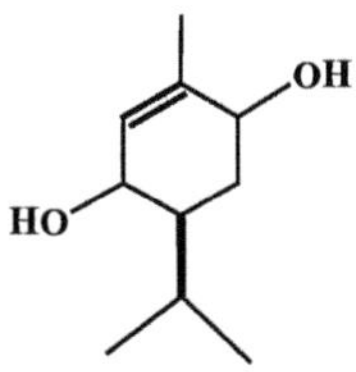

3,6-dihydroxy-p-menth-1-ene

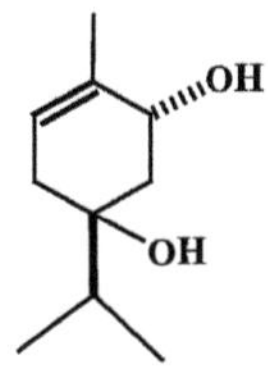

p-menth-1-en-4alfa, 6beta - diol

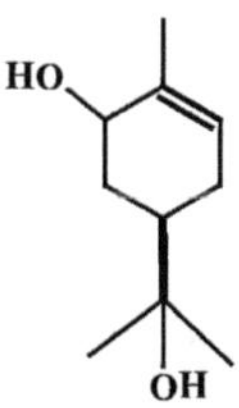

( ) trans sobrerol

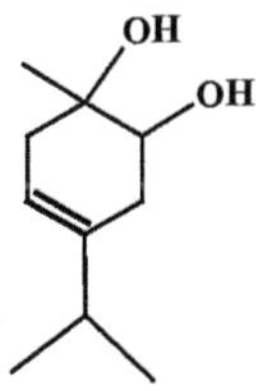

p-menth-4-en-1,2-diol

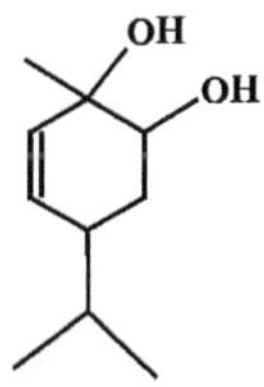

p-menth-5-en-1,2-diol

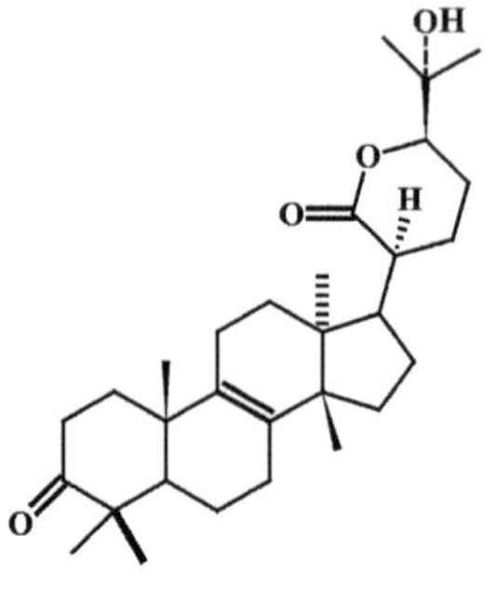

Boscartin - L

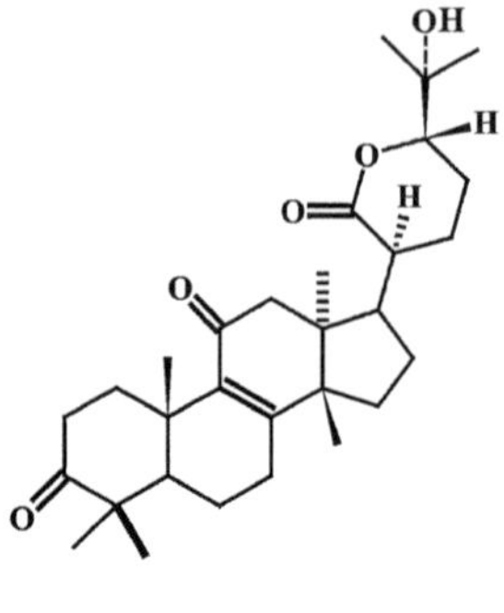

Boscartin- M

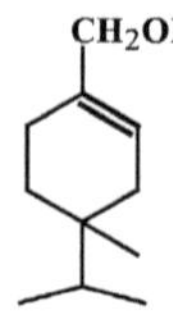

Olibanumol - C

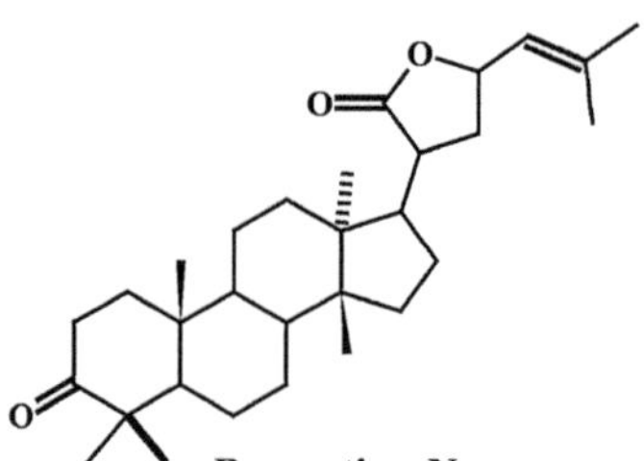

Boscartin - N

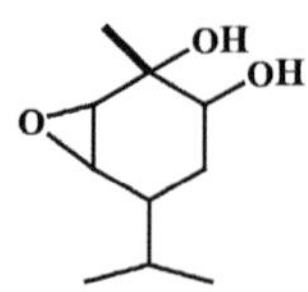

Olibanumol - A

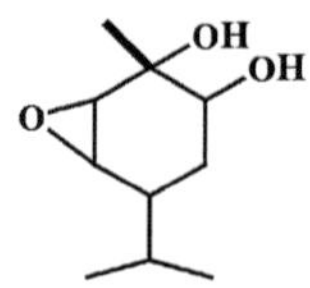

Olibanumol - B

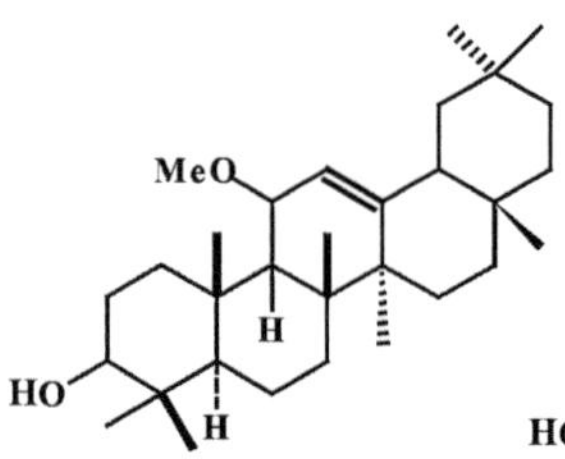

Olibanumol - E

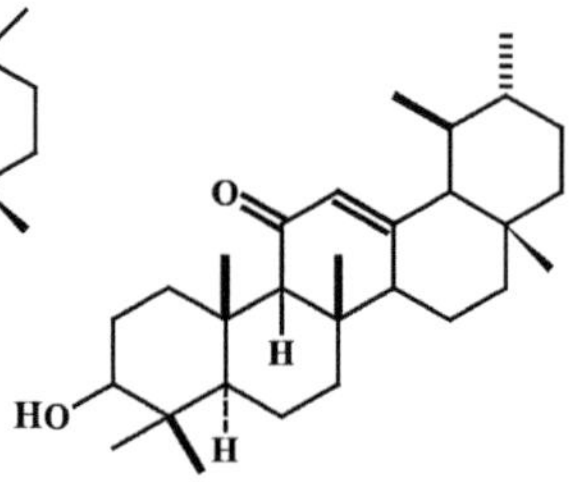

Neoilexonol

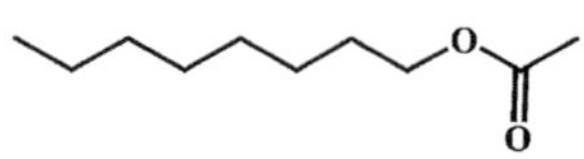

Octyl Acetate

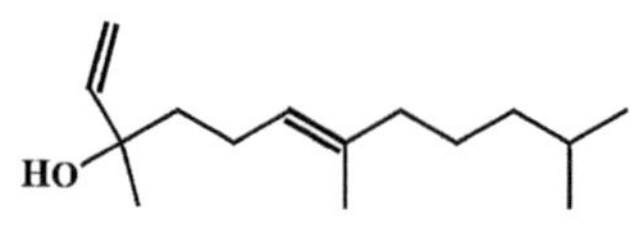

trans-Nerolidol

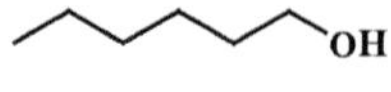

1-Hexanol

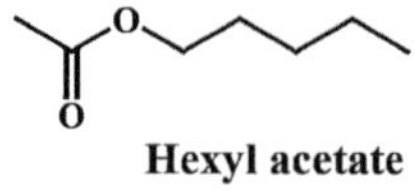

Hexyl acetate

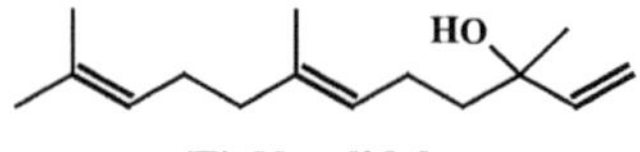

(E)-Nerolidol

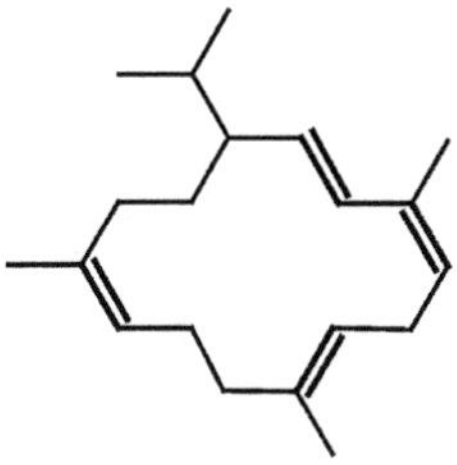

Cembrene

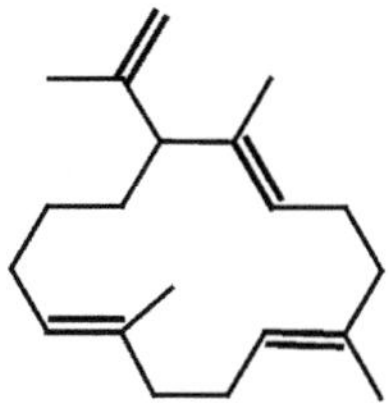

Cembrene A

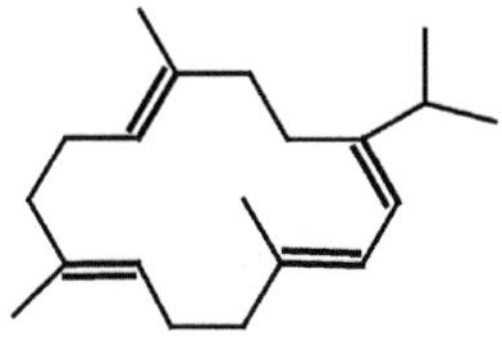

Cembrene C

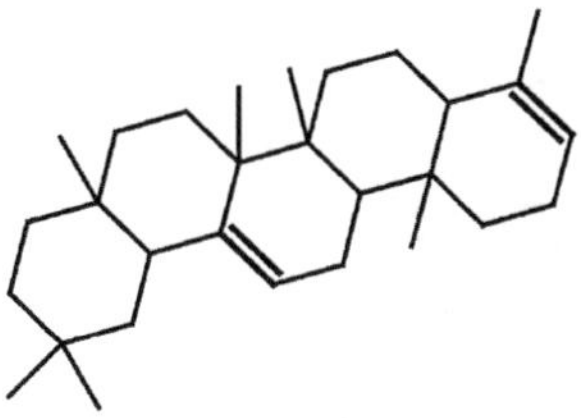

24-Norursa-3,12-diene

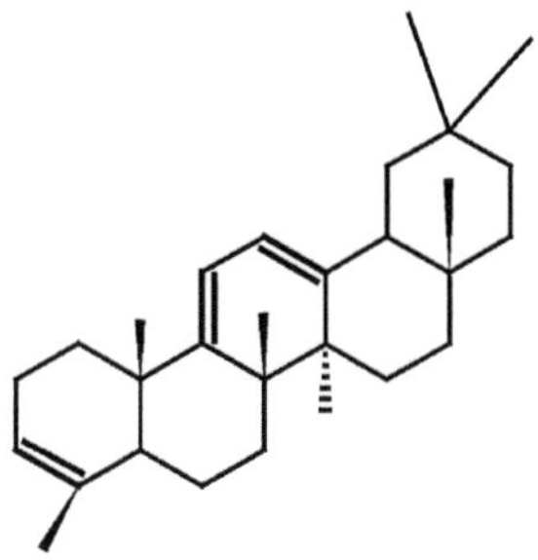

24-noroleana-3,9(11),12-triene

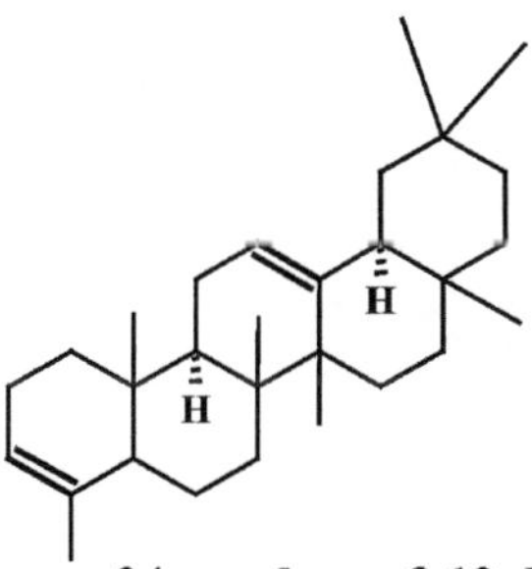

24-noroleana-3,12-diene

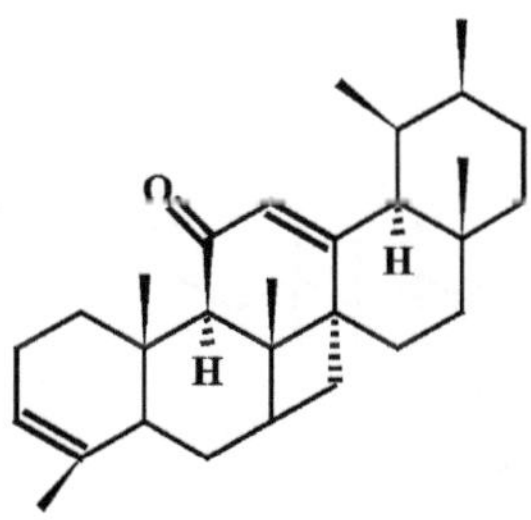

24-noroleana-3,12-diene -11-one

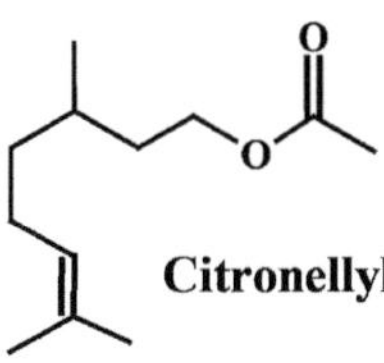

Citronellylacetate

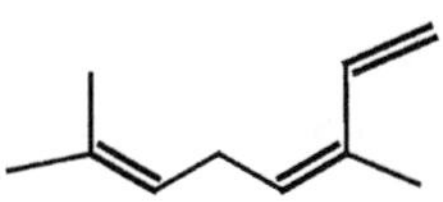

(Z)-b-Ocimene

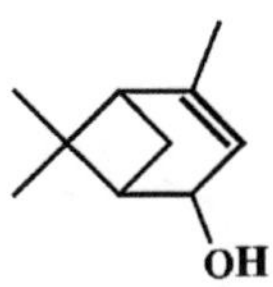

trans-Verbenol

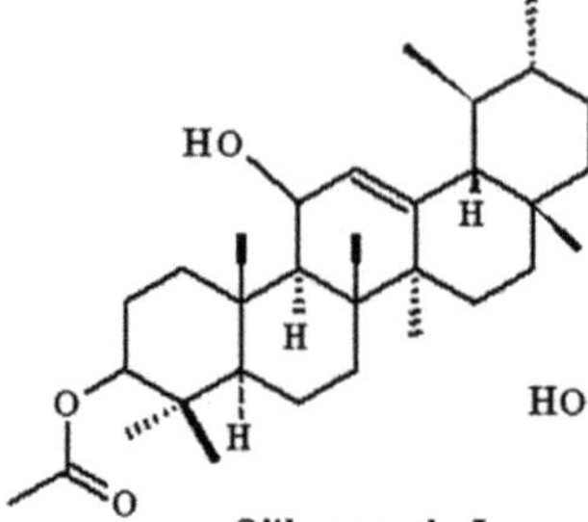

Olibanumol - L

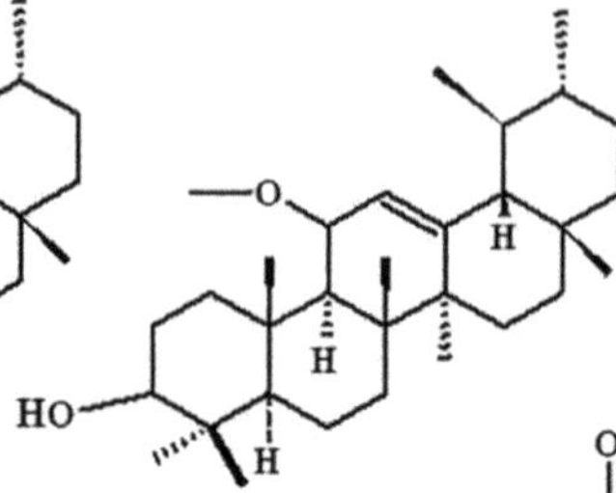

Olibanumol - M

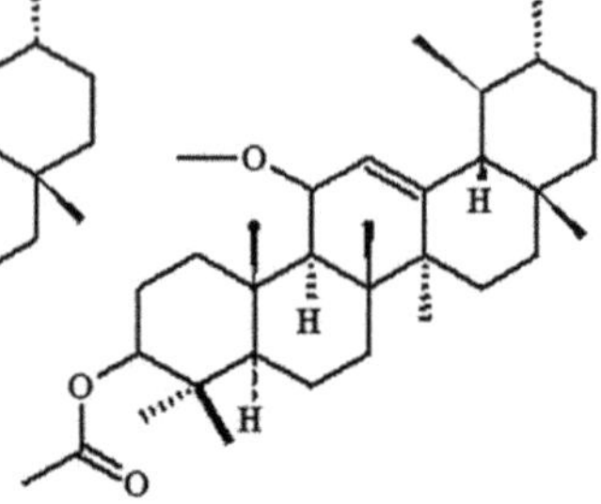

Olibanumol - N

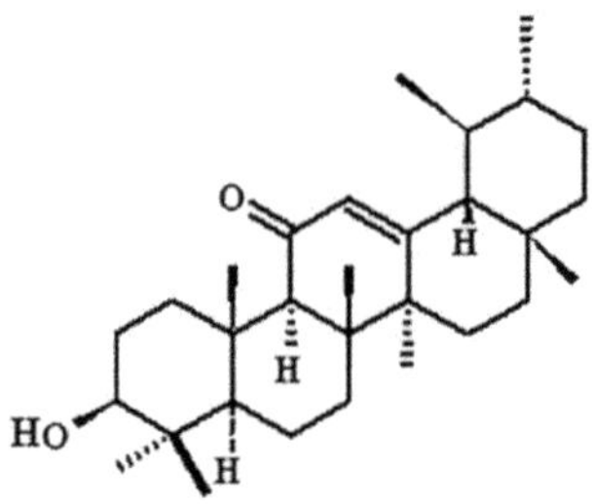

Neoilexonol

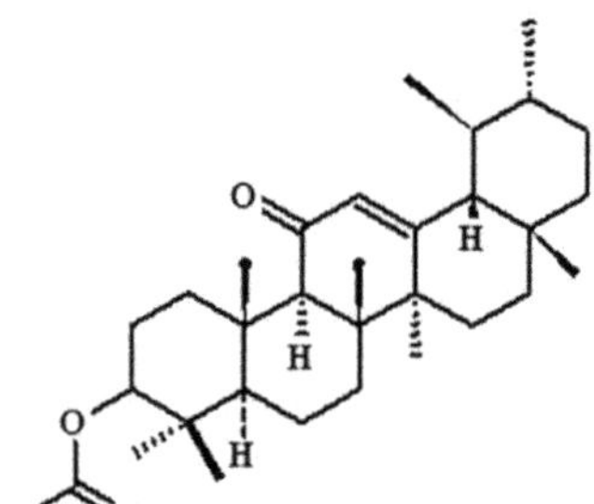

Neoilexinol Acetate

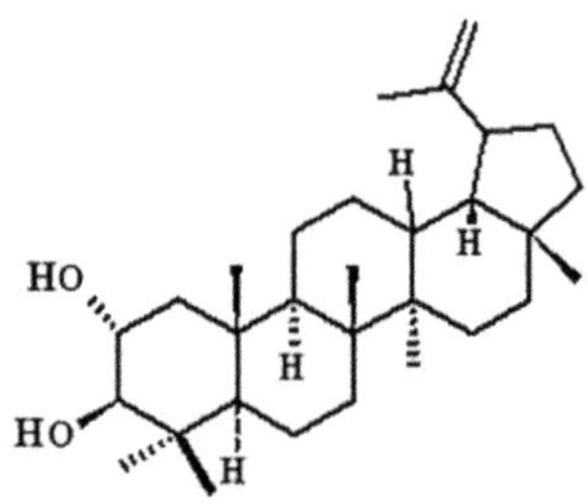

Lup-20(29)-ene-2α,3β-diol

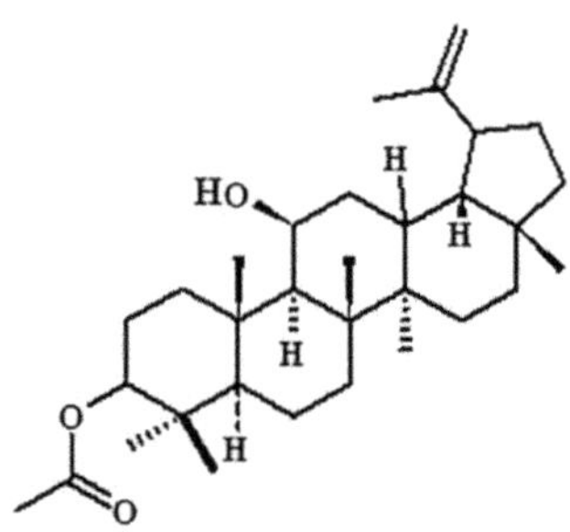

3β-acetoxylup-20(29)-en-11β-ol

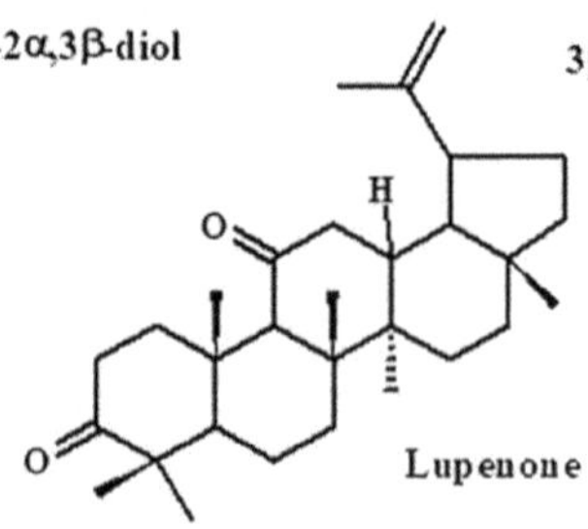

Lupenone

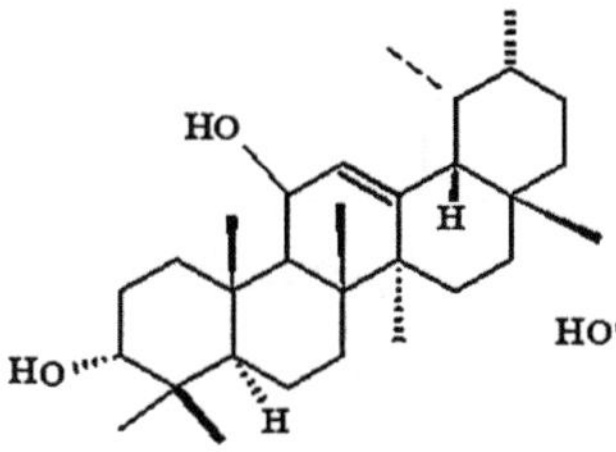

Urs-12-ene-3α 11α-diol

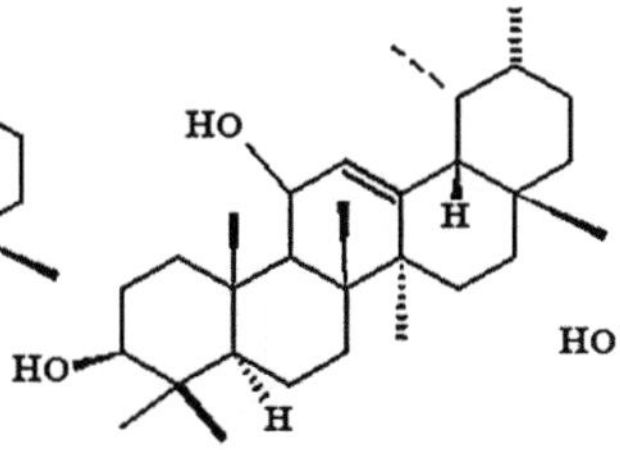

Urs-12-ene-3β, 11α-diol

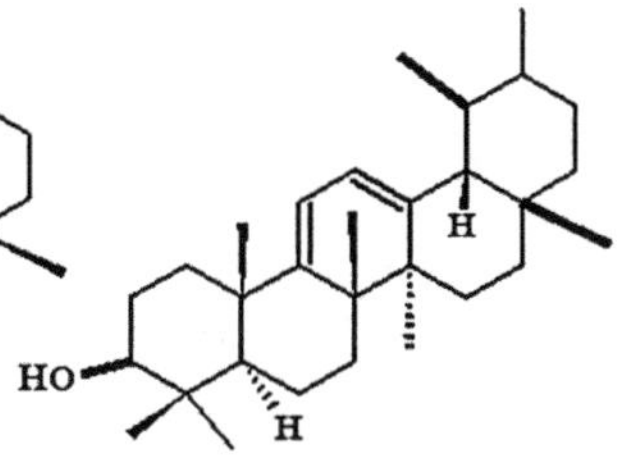

Urs-9(11),12-dien-3β-ol

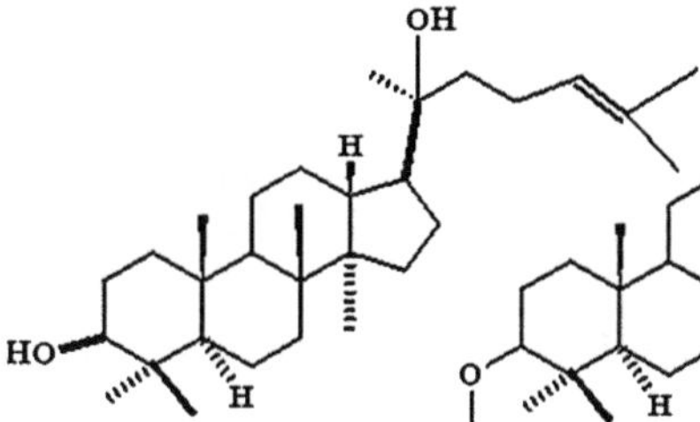

Dammarenediol II

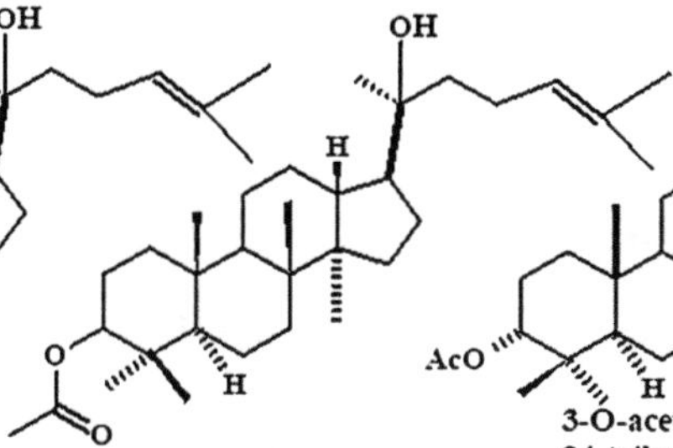

Dammarenediol II Acetate

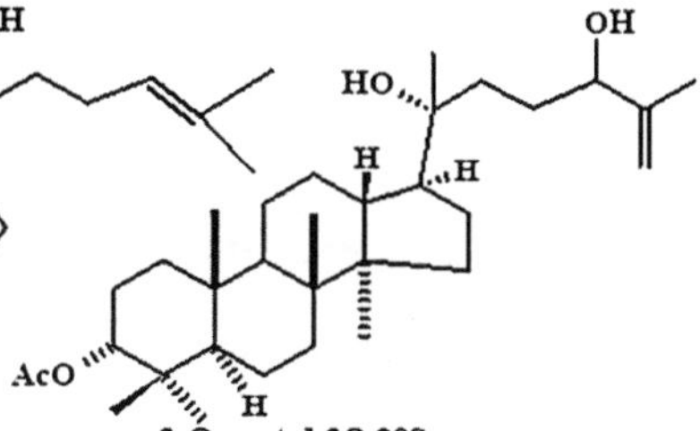

3-O-acetyl-3β-20S, 24-trihydroxydammar-25-ene

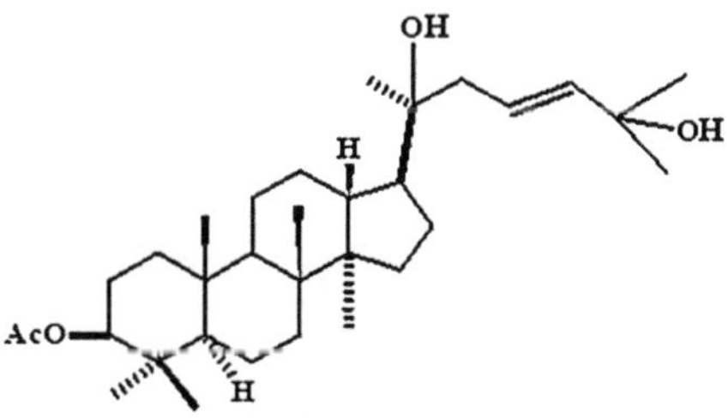

Isofouquierol acetate

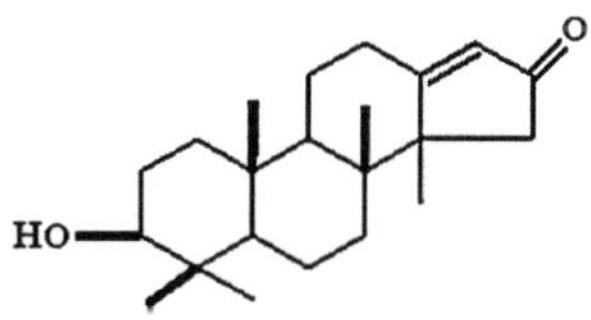

3β-hydroxymansumbin-13(17)-en-16-one

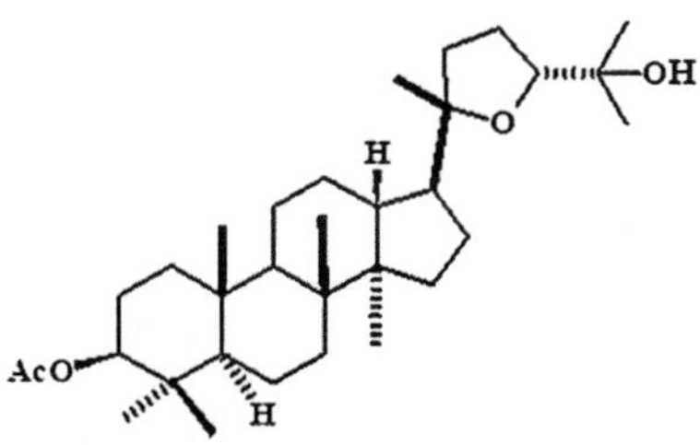

Ocotillol acetate

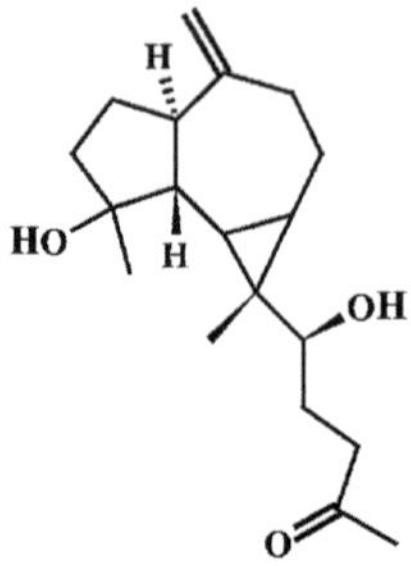

Boscartol - M

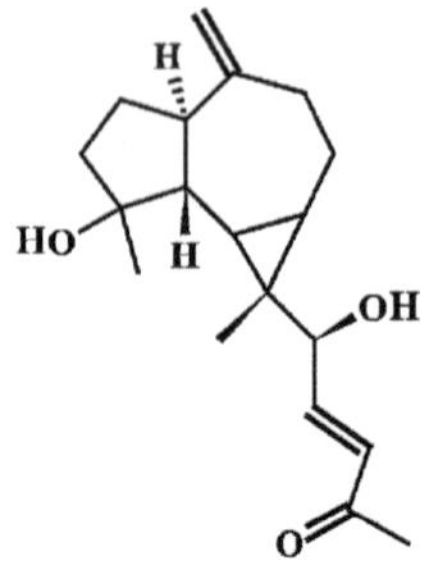

Boscartol - L

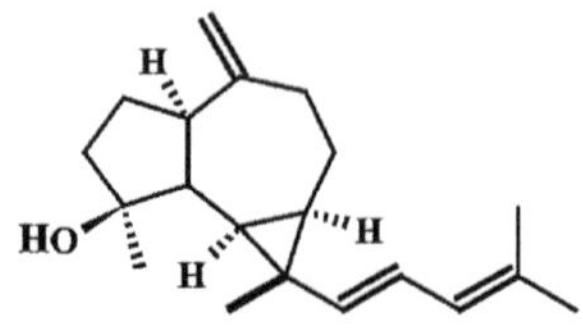

Olibanumol-D

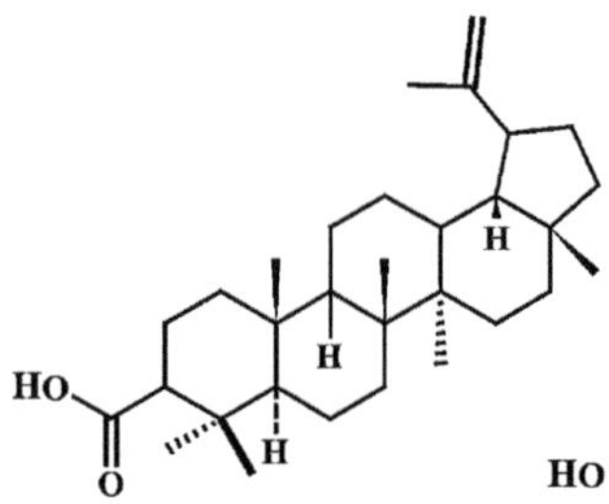

Olibanumol - F

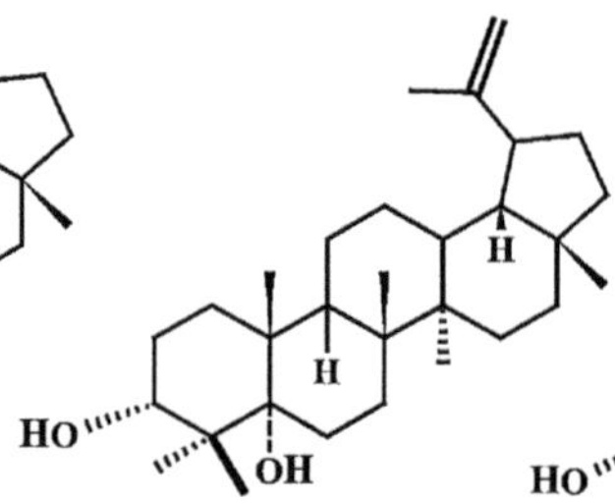

Olibanumol - G

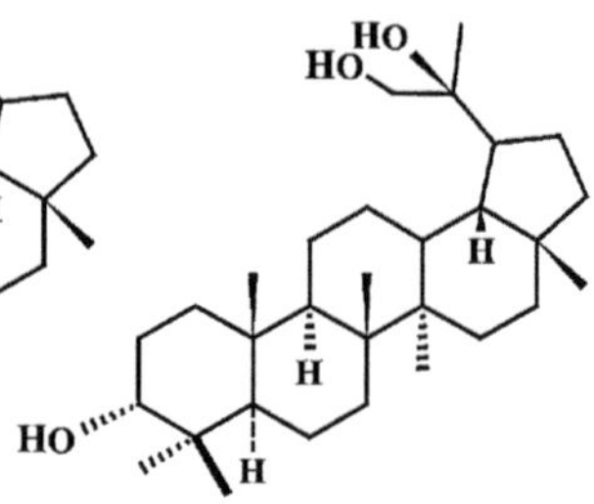

Olibanumol - H

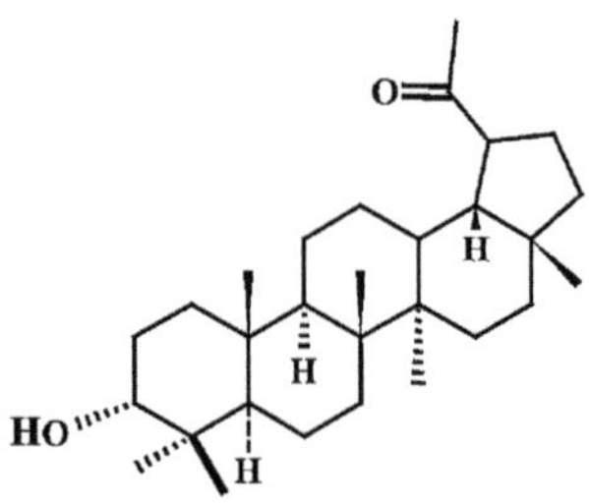

Olibanumol - I

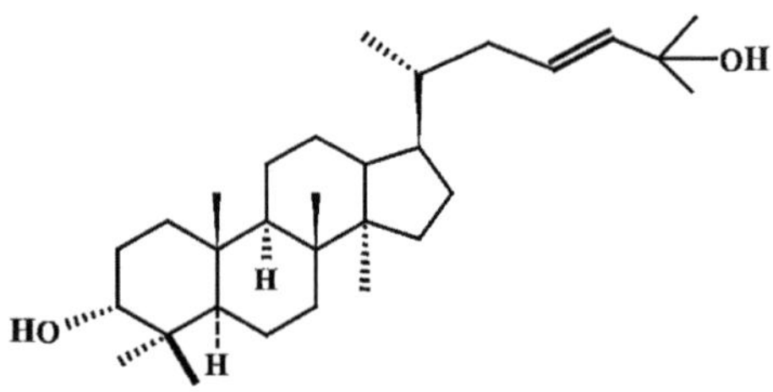

Olibanumol - J

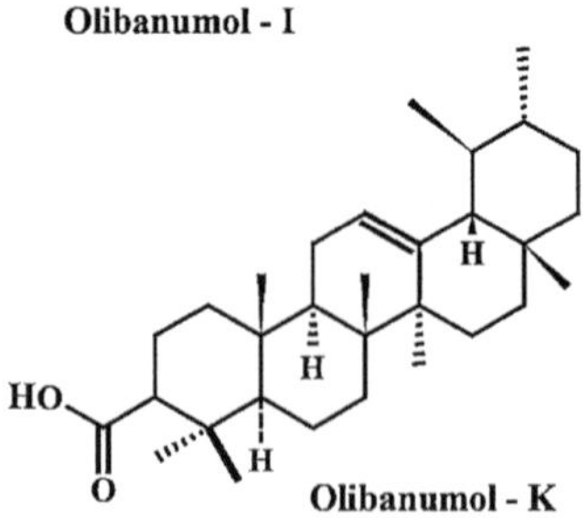

Olibanumol - K

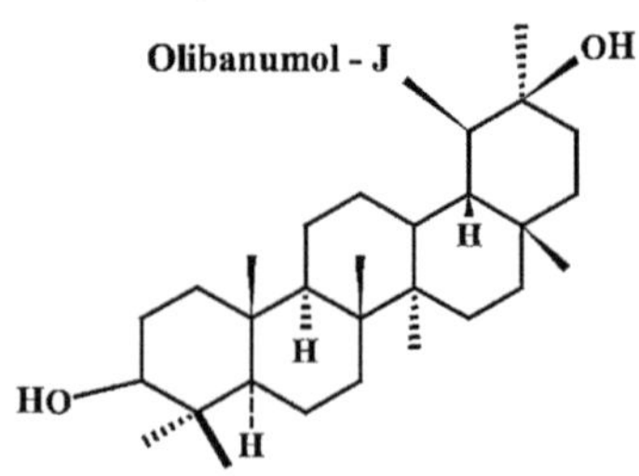

Olibanumol - L

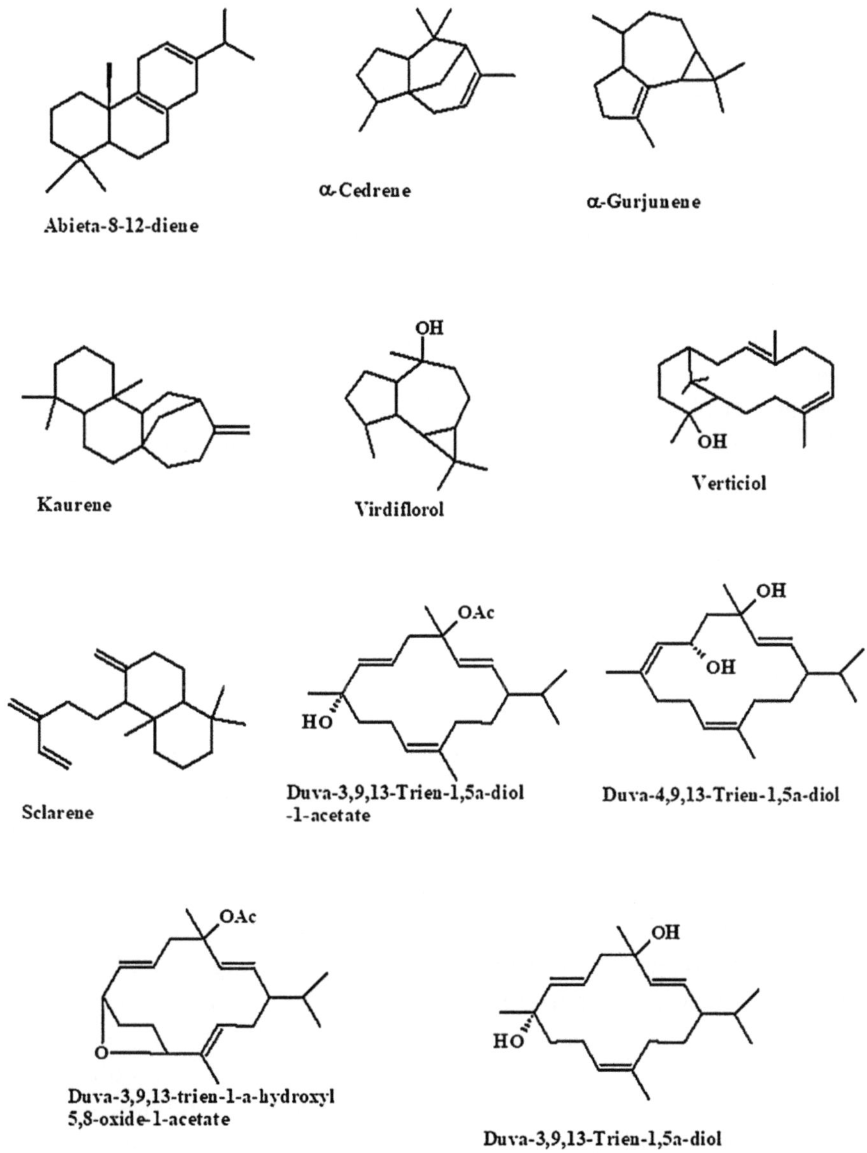
Abieta-8-12-diene
α-Cedrene
α-Gurjunene
Kaurene
OH
Virdiflorol
OH
Verticiol
Sclarene
OAc
HO
Duva-3,9,13-Trien-1,5a-diol
-1-acetate
OH
OH
Duva-4,9,13-Trien-1,5a-diol
OAc
O
Duva-3,9,13-trien-1-a-hydroxyl
5,8-oxide-1-acetate
OH
HO
Duva-3,9,13-Trien-1,5a-diol

FIGURE 5.1(T).

Yu et al. (2020) isolated eight new cembrane-type diterpenoids (boscartins AP-AW) such as 1:12-epoxy-11-acetoxy-2,7-cembrandine-4,6-diol.; 1:12-epoxy-11-acetoxy-2,7-cembrandine-4-ol; 1:12-epoxy-11-acetoxy-3,7-cembrandine-6-ol; (1S,9R,11R,12R,3Z,7Z)-1:12-epoxy-11-acetoxy-3,7-cembrandine-9-ol (boscartins AS); (1S,7S,11R,12R,3E,8Z)-1:12-epoxy-11-acetoxy-3,8-cembrandine-6-ol (boscartins AU); 1:12-epoxy-11-acetoxy-3,6-cembrandine-8-ol; (1S,11R,12R,3Z,5E)-1:12-epoxy-6-oxo-11-acetoxy-3,7-cembrandine-9-ol (boscartins AW); (1S,7S,11R,12R,3E,8Z)-1:12-epoxy-11-acetoxy-3,8-cembrandine-6-ol (boscartins AU) from the gum resin of *B. sacra* (Syn.: *B. carterii*).

Zhang et al. (2021) have isolated seven sacraoxides A–G from gum resin of *B. sacra* which are cembranoids; a class of diterpenes. Moreover, sacraoxide E and sacraoxide F demonstrated inhibitory activities on nitric oxide (NO) production induced by lipopolysaccharide (LPS) in RAW264.7 mouse monocyte-macrophages with the IC_{50} values of 36.4 and 24.9 μM, respectively.

5.17 *BOSWELLIA SERRATA* ROXB. (INDIAN FRANKINCENSE)

Oleo-gum-resin (Gum Olibanum, Indian frankincense) is an exudate, which comes out from the cortex after an injury or natural crack in the bark of *B. serrata*. It is bitter, fragrant, transparent, and golden yellow in color. The oleo gum-resin contain 30–60% resin, 5–10% essential oils, which are soluble in the organic solvents, and the rest is made up of polysaccharides (~ 65% arabinose, galactose, xylose) which are soluble in water (Sharma and Verma, 1980; Bhuchar et al., 1982; Gangwal and Vardhan, 1995). According to Imperial Institute London, *B. serrata* oleo-gum-resin has the following composition, i.e., moisture (10–11%), volatile oil (8–9%), resin (55–57%), gum (20–23%) and 4–5% of insoluble matter (Alam et al., 2012). Its essential oil is a mixture of mono-, di-, and sesquiterpenes, while gum fraction is composed of pentose and hexose sugar with some digestive enzymes. Resin is the most important fraction of Indian frankincense, comprising mainly of pentacyclic triterpenic acids, namely BAs (Sharma et al., 2009). Steam distillation of *B. serrata* oleo-gum-resin yielded colorless volatile oil with a characteristic odor. The major volatile components present in the oil are α- and β-pinene

(Van Vuuren et al., 2010). Basar et al. (2005) investigated the chemical composition of hydro-distilled essential oil obtained from oleo-gum-resin of *B. serrata* by GC and GC-MS analysis. The oil consists of the following components, i.e., α-thujene (12%), α-pinene (8%), sabinene (2.2%), β-pinene (0.7%), myrcene (38%), α-phellandrene (1%), p-cymene (1%), limonene (1.9%), linalool (0.9%), perillene (0.5%), methylchavicol (11.6%), methyleugenol (2.1%), germacrene D (2.0%), kessane (0.9%), cembrene A (0.5%) and cembrenol (1.9%). Girgune and Garg (1979) have reported the presence of α-thujene (50%), α-pinene (6.2%), d-limonene (4.5%), *p*-cymene (14%), cadinene (4%), geraniol (0.8%) and elemol (1.3%) as the main constituents of the essential oil. Wahab et al. (1987) determined the chemical composition of essential oil of the oleo gum resin of *B. serrata*, prepared by direct steam distillation and steam distillation of an n-hexane extract, by GC-MS analysis. Around 33 components were identified in the steam-distilled oil by their Kováts indices on stabilized OV-1 columns coupled with mass spectral data. β-eudesmene (3.3%), γ-murolene (3.0%), γ-cadinene (2.9%), α-copaene (2.8%), α-murolene (2.7%), α-cubebene (1.9,%), δ-cadinene (1.9%), 3,5-dimethoxytoluene (1.6%), germacrene d (1.4%), allo-aromadendrene (1.3%), β-caryophyllene (1.3%), o-methyl anisole (1.2%), linalool (1.0%), β-gurjunene (1.0%) β-bourbonene (0.9%), camphene (0.9%), eucalyptol (0.9%), 1-hydroxy-1,7-dimethyl-4-isopropyl-2,7-cyclodecadiene (0.9%), valencene (0.8%), s-cis-sabinol (0.6%), α-phellandrene (0.5%) trans-carveol (0.5%), β-pinene (0.4%), terpinyl acetate (0.4%), (–)-trans-pinocarveol (0.4%), (+)-α-terpineol (0.4%), aromadendrene (0.3%), bornyl acetate (0.3%), carvone (0.3%), δ-guaiene (0.3%), sabinene (0.3%) δ-elemene (0.2%), α-guaiene (0.2%), cis-carveol (0.1%), cumaldehyde (0.1%), cuminol (0.1%), cis-verbenol (0.1%), γ-terpinene (0.1%), (–)-4-terpineol (0.1%), α-thujene (0.1%), 1,2,3,4,6,8a-hexahydro-1-isopropyl-4,7-dimethyl- (0.6%) and verbenone (0.1%).

Kasali et al. (2002) reported the chemical composition of essential oil separated from *B. serrata* stem bark, obtained by hydrodistillation. The essential oil is rich in monoterpenoids, of which α-pinene (73.3%) was the major constituent. Other monoterpenoids identified included β-pinene (2.05%), cis-verbenol (1.97%), trans-pinocarveol (1.80%), borneol (1.78%), myrcene (1.71%), verbenone (1.71%), limonene (1.42%), thuja-2,4 (10)-diene (1.18%) and p-cymene (1.0%), while α-copaene (0.13%) was the only sesquiterpene identified in the oil. Kumar and Saxena (1979) identified 13 components such as d-α-Thujene (32.0%), bornyl acetate (20.0%), α-terpineol (13.6%), un-identified (5.0%), un-identified (5.0%), d-limonene (3.9%), α-terpinolene (1.9%), methyl chavicol (4.0%), un-identified (5.3%), α-pinene (2.5%), p-cymene (2.2%) and α-phellandrene (1.5%) in *B. serrata* leaf essential oil.

The major chemical components of nonvolatile part of oleo gum resin of *B. serrata* are higher terpenoids, i.e., pentacyclic triterpenes and tetracyclic triterpenes. Pardhy and Bhattacharyya (1978), isolated BAs from the nonvolatile fraction of the resin from *B. serrata* along with other natural products including methyl chavicol, α- and 3-amyrins and a new diterpene alcohol 'serratol.' Of the isolated terpene acids, four are tetracyclic triterpene acids viz. 3-α-acetoxytirucall-8, 24-dien-2l-oic acid, 3-ketotirucall-8, 24-dien-21-oic acid, 3-α-hydroxytirucall-8, 24-dien-21-oic acid, 3-β-hydroxytirucall-8, 24-dien-21-oic acid and four are pentacyclic triterpene acids, βBA, acetyl-β-boswellic acid (AβBA), acetyl-11-keto-β-boswellic acid (AKBA) and 11-keto-β-boswellic acid (KBA). Mahajan et al. (1995) elucidated structure of two new triterpenoids from acidic and neutral fractions of the gum extract of *B. serrata*, i.e., 2α, 3α-dihydroxy-urs-12-ene-24-oic acid and urs-12-ene-3α,24-diol. Pardhy and Bhattacharyya (1978) isolated another isomeric diol, viz. urs-12-ene-3β, 24-diol, from neutral fraction of *B. serrata* gum. Two lupine type triterpene components, i.e., 3α-hydroxy-lup-20(29)-en-24-oic acid (Culioli et al., 2003) and 3α-acetoxy-lup-20(29)-ene-24-oic acid were isolated from *B. serrata* gum (Belsner et al., 2003).

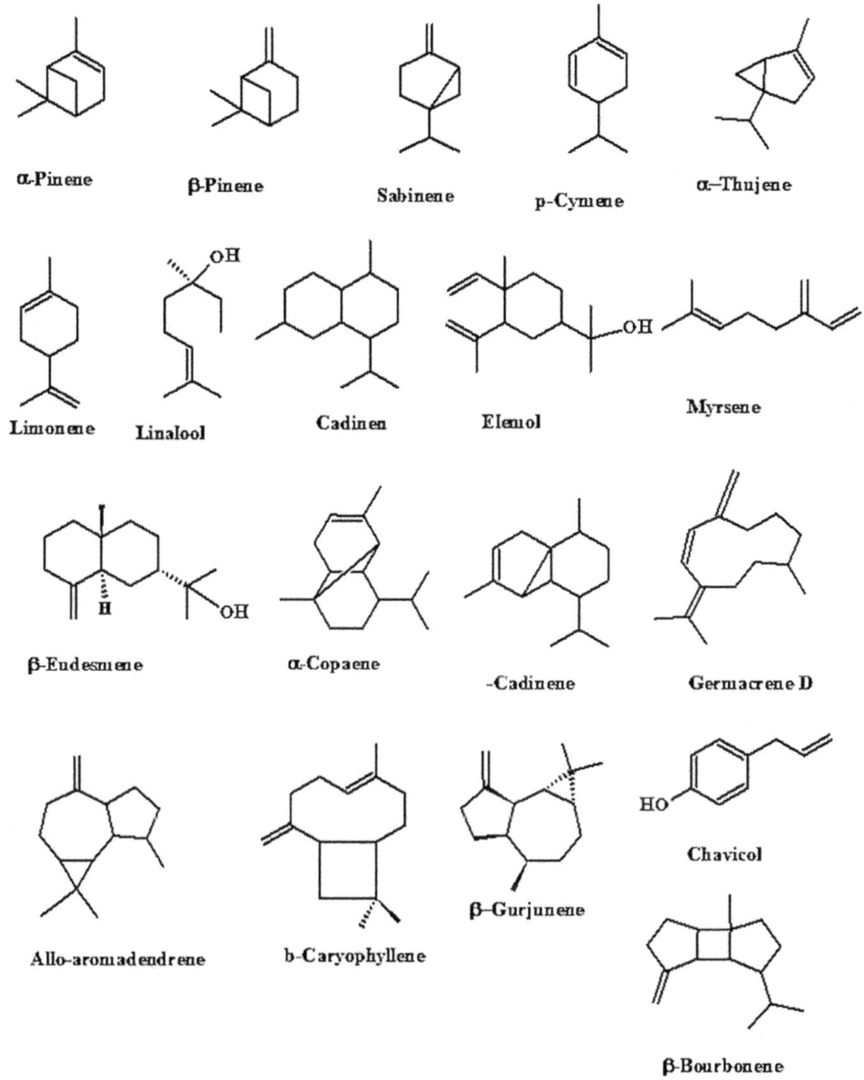

FIGURE 5.1(U).

Some additional chemical constituents were identified in *Boswellia* gum resin include lupeolic acid, acetyl lupeolic acid, incensole acetate, incensole oxide, isoincensole oxide (Siddiqui, 2011). Several added BAs were isolated from *Boswellia* gum resin are 9,11-dehydro-α-boswellic acid, acetyl 9,11-dehydro-α-boswellic acid, 9,11-dehydro-β-boswellic acid and acetyl 9,11-dehydro-β-boswellic acid.

Hanus et al. (2007) identified chemical components such as, 24-noroleana-3,9(11),12-triene (1.3%), 24-norursa-3,9(11),12-triene (5%), 24-noroleana-3,12-diene (14.2%), 24-norursa-3,12-diene (50.8%) and 24-norursa-3,12-diene-11-one (7.6%) in n-hexane fraction of *B. serrata* gum resin by GC-MS analysis. Further, the components such as trimethylsilyl 3-*O*-trimethylsilyl 8,24-diene-tirucallate (7.1%), trimethylsilyl 3-*O*-trimethylsilyl-9,11-dehydro-β-boswellate (2.7%), trimethylsilyl 3-*O*-trimethylsilyl-α-boswellate (6.9%), trimethylsilyl 3-*O*-trimethylsilyl-β-boswellate (25%), trimethylsilyl 3-oxo-8,24-dienetirucallate (10.5%), trimethylsilyl 3-*O*-acetyl-11-methoxy-β-boswellate (2.6%), trimethylsilyl 3-*O*-acetyl-α-boswellate (1.2%), trimethylsilyl 3-*O*-acetyl-β-boswellate (3.0) identified in n-hexane fraction after silylation with N,*O*-bis (trimethylsilyl) trifluoroacetamide (BSTFA).

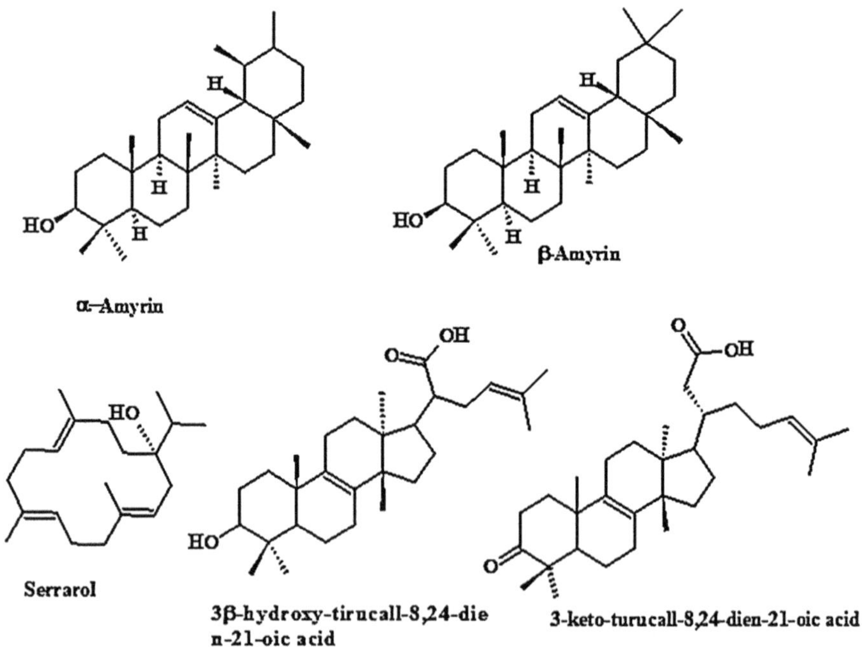

FIGURE 5.1(V).

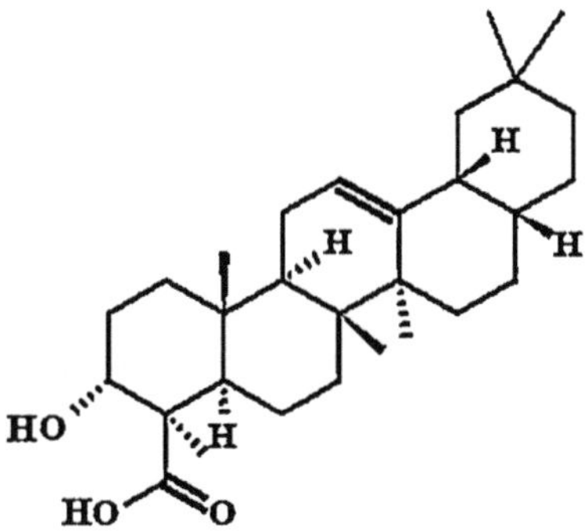

α-Boswellic acid

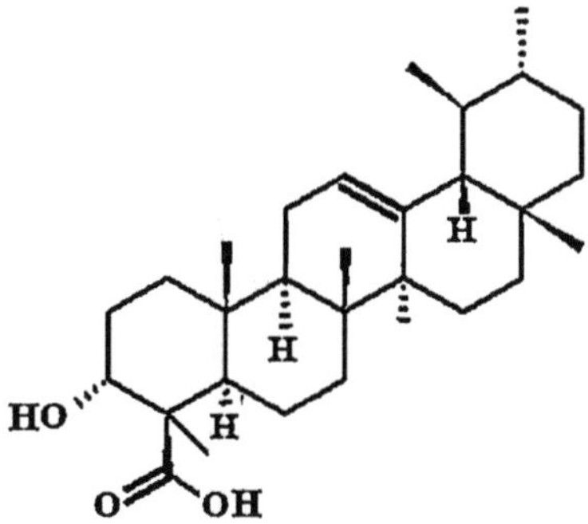

β-Boswellic acid

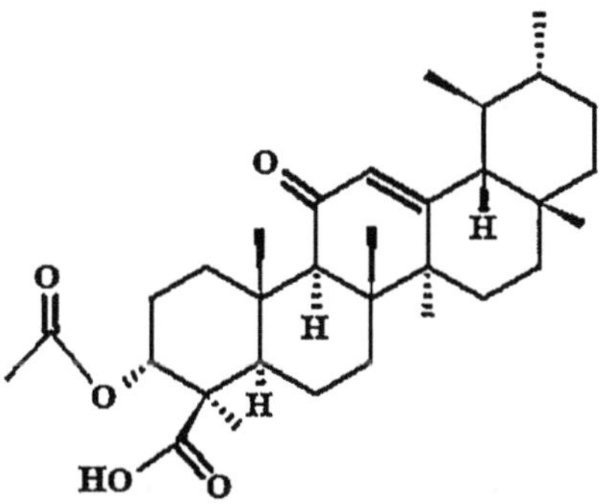

acetyl-11-keto-β-Boswellic acid

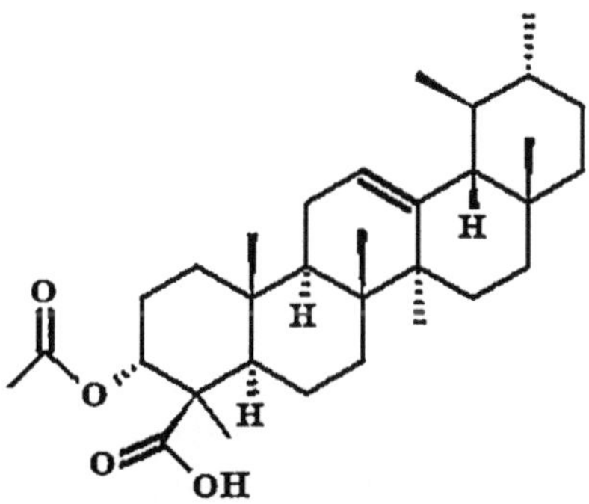

acetyl-β-Boswellic acid

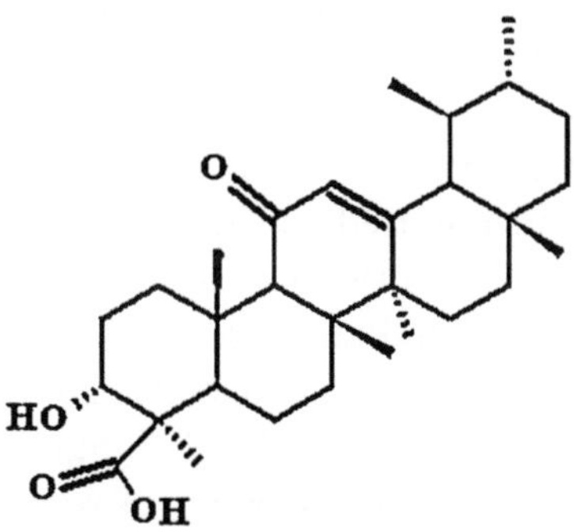

11-keto-β-boswellic acid

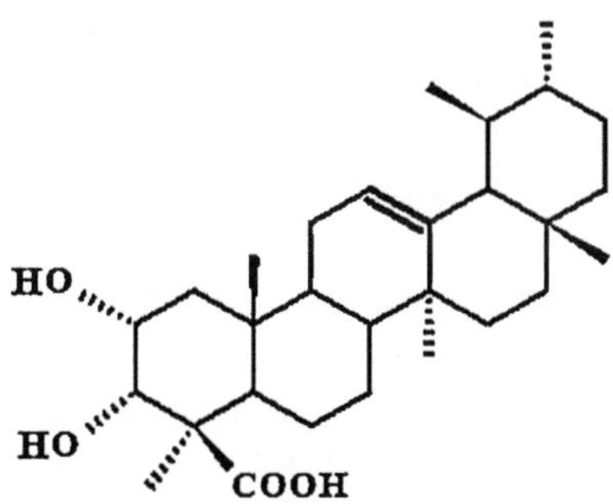

2α,3α-dihydroxy-urs-12-ene
-24-oic acid

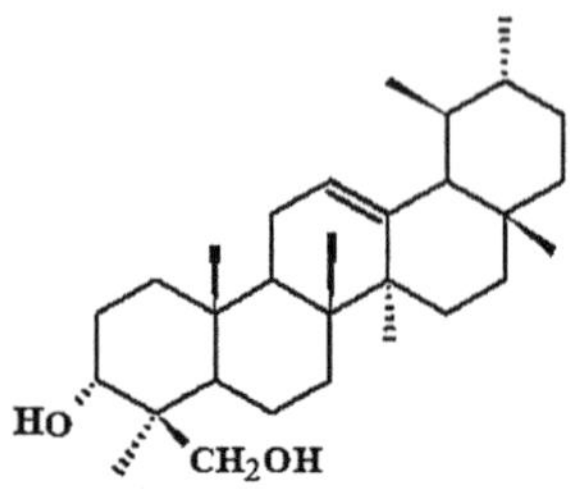

urs-12-ene-3α,24-diol

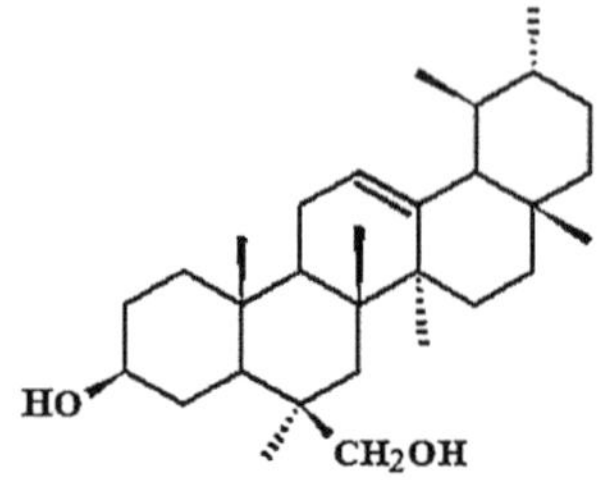

urs-12-ene-3β, 24-diol

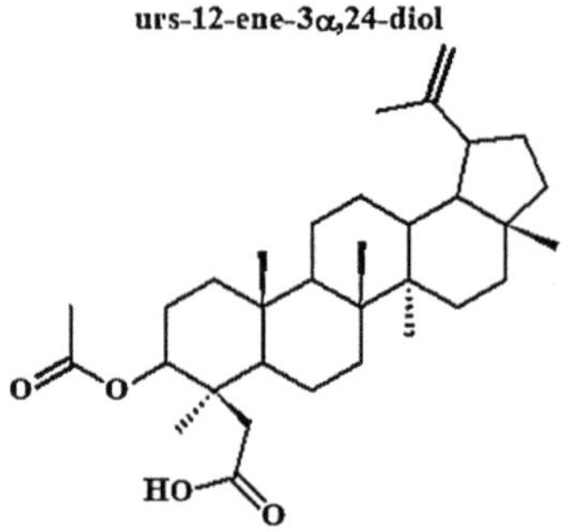

3α–hydroxy-lup-20(29)-en-24-oic acid

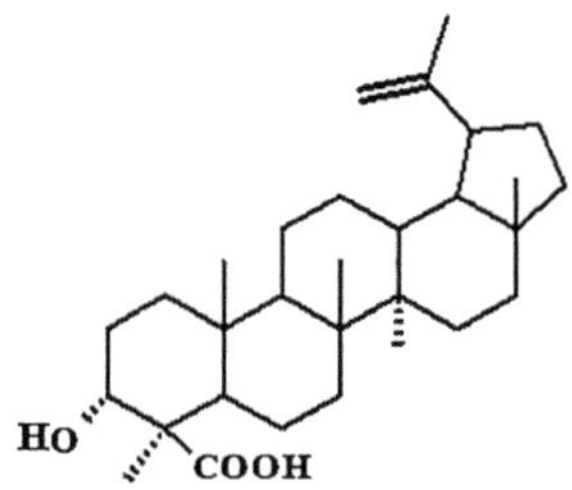

3α-acetoxy-lup-20(29)-ene-24-oic acid

FIGURE 5.1(W).

B. serrata (frankincense) triterpenoids of extract of commercial samples, including αBAs, βBAs, 11-keto-β-boswellic acids (KBAs), acetyl-11-keto-β-boswellic acids (AKBAs), and their esters were analyzed using matrix-assisted laser desorption ionization time-of-flight mass spectrometry (MALDI-TOF MS)-based method for rapid and high-throughput screening of compounds can be used as indicators for determining the authenticity (Lai et al., 2020).

However, the oleoresin of gum or essential oil components were important resources of various terpenes and reported by several scientists from different parts of the globe and have a difference in the major components. For example, α-thujane (12%), methyl chavicol (11.6%) in the total essential oil from the stembark, whereas, borneol acetate (20%), α-terpineol (13.6%) as major components (Kasali et al., 2002; Singh et al., 2007). Wahab et al. (1987) reported diversified components, of which none exceeded 10%. Girgune and Garg (1979) noticed α-thujene 50% and para-cymene 14% in the essential oil. Interestingly Kasali et al. (2002) reported the essential oil from stem bark with α-pinene (73.3%). Kumar and Saxena (1979) reported d-α-thujene at 32% borneol acetate 20% and α-terpineol 13.6% from the leaves. Hence, it is known from the aforementioned phytochemical constituents *B. serrata* may be an important natural resource for the volatile incense components, which were reported in differential concentrations and require more studies on the effect of ecoclimatic impact on their biosynthesis.

5.18 *B. SOCOTRANA* BALF. F.

Essential oil obtained from oleo gum resin of *B. socotrana* was dominated with (E)-2, 3-epoxycarene (51.8%), 1,5-isopropyl-2-methylbicyclo [3.1.0] hex-3-en-2-ol (31.3%). The minor components identified in the oil were α-cymene (7.1%), 4-terpinenylacetate (3.9%), 4-methyl-benzoic acid (2.7%), p-menth-1(7)-en-2-one (2.6%) and α-terpinyl acetate (0.1%) (Ali et al., 2008). The volatile components identified in methanolic extract of *B. socotrana* gum resin as α-thujene, α-pinene, α-fenchene, sabinene, β-pinene, β-myrcene, p-cymene, limonene, cyclodecyne, terpinolene, 1,2,3,4-tetrahydronaphthalene, fenchol, borneol, 4-terpineol, α-terpineol, bornyl acetate, α-terpinenyl acetate, β-caryophyllene, neocembrene (Madera et al., 2017).

Phytochemical composition of essential oil isolated from the stem bark of *B. socotrana* was analyzed by (Mothana et al., 2011). The oil consists about 67 components and it was found to be predominated with monoterpene hydrocarbons (26.4%) and oxygenated monoterpenes (30.7%) with p-cymene (13.0%), 2-hydroxy-5-methoxy-acetophenone (16.3%) and camphor (11.6%).

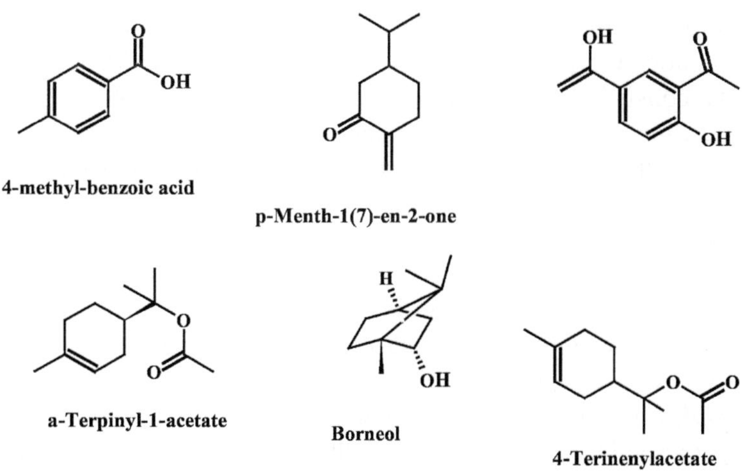

FIGURE 5.1(X).

5.19 CHEMOTAXONOMIC APPREHENSIONS

The study on the marker chemical agents in the plants may serve as a solution in providing the taxonomical gradiences and revealed the controversies existing on *Boswellia* species especially on *B. carterii* and *B. sacra* which are treated as same species, in addition "The Plant List" also denoting as synonyms (Paul et al., 2011; Woolley et al., 2012). Progressively, the identification of the correct botanical name of certain frankincense species, Paul et al. (2011) have proposed a TLC protocol for three resins (olibanum) on the basis of biomarker agents. These findings demonstrated the occurrence of incensole and incensyl acetate confirms the species of *B. papyrifera*, caryophyllene oxide for *B. carterii* and/or *Boswellia sacra*, whereas *B. serrata* has neither incensole acetate nor caryophyllene oxide, but has a notable concentration of serratol in addition to traces of incensole. However, Satyal and Pappas (2016) reported a chemotaxonomic marker as a unique chemical composition (alkyl methyl ether as the natural volatile component in an essential oil), enantiomeric distribution, and biological activities of the oleogum resin essential oil of *Boswellia* spp. from Somalia.

5.20 CONCLUSION

The commercial products of Frankincense/Olibanum have multifaceted benefits to human. They are used as immune boosters, anti-inflammatory, antirheumatic, and pain reliever. Recent studies have evidenced that the products consisting of Frankincense exhibit several pharmacological properties, for instance, anticancer and antiarthritic activities. The phytochemical studies were helpful in understanding the mechanism of pharmacological effects, and deriving the active principle in developing molecular medicine. Species of *Boswellia* were reported to be one of the rich sources of terpenoids that majorly occur in the essential oils of gum resin.

Although, there are 24 species in *Boswellia,* the phytochemical studies are mainly restricted to only one species, i.e., *B. sacra*. However, there is a lot of scope for the phytochemical studies in *B. asplenifolia, B. globosa, B. microphylla, B. ogadensis, B. samahensis*, and *B. scopulorum*. Further, these species to be explored scientifically to disclose the hidden beneficial effects of unexplored phytocompounds. Moreover, many of these species are globally endemic as per the IUCN regulations, and thus needs conservation strategies and sustainable utilization of natural resources through safe extraction methods. Alternatively, biotechnological protocols should be explored in the near future for extracting high yields of their active principles. Besides, the phytochemical investigation strongly supports the pharmacological effects, indeed might be taken as a key for the identification of morphologically similar *Boswellia* species. Likewise, high throughput molecular studies can be engaged to confirm the chemotypic species of *Boswellia* in addition to conservation strategies.

KEYWORDS

- ***Boswellia***
- **Burseraceae**
- **essential oil**
- **frankincense**
- **oleo-gum resin**
- **olibanum**
- **olibanumol**
- **phytochemicals**
- **verticiol**

REFERENCES

Abdurrahman, I. A., Hagr, T. E., & Xia, Y. C., (2020). Isolation and structural elucidation of new steroid from stem bark of *Boswellia papyrifera* (Del.) Hochst. *Adv. J. Chem. Section B, 2*(4), 172–178.

Abdurrahman, I., Lai, H. Y., Xia, Y. C., & Hagr, T. E., (2019). Identification and characterization of diterpenes from ethyl acetate fraction of stem barks of *Boswellia papyrifera* (Del) Hochst, Sudanese medicinal plant. *Asian J. Green Chem., 3*, 322–335.

Ahmad, U. V., & Rahman, A., (1994). In: *Handbook of Natural Products Data, Pentacyclic Triterpenoids* (Vol. 2, p. 718). Elsevier, Amsterdam, The Netherlands.

Alam, M., Khan, H., Samiullah, L., & Siddique, K. M., (2012). A review on phytochemical and pharmacological studies of Kundur (*Boswellia serrata* Roxb. ex Colebr.): A Unani drug. *J. Appl. Pharm. Sci., 2*, 148–156.

Alemika, T. E., Onawunmi, G. O., & Olugbade, T. A., (2004). Isolation and characterization of incensole from *Boswellia dalzielii* stem bark. *J. Pharm. Biores., 1*, 7–11.

Alemika, T. E., Onawunmi, G. O., & Olugbade, T. A., (2006). Antibacterial phenolics from *Boswellia dalzielii. J. Nat. Prod. Med., 10*, 108–110.

Al-Harrasi, A., & Al-Saidi, S., (2008). Phytochemical analysis of the essential oil from botanically certified oleogum resin of *Boswellia sacra* (Omani Luban). *Molecules, 13*, 2181–2189.

Al-Harrasi, A., Ali, L., Ceniviva, E., Al-Rawahi, A., Hussain, J., Hussain, H., et al., (2013b). Antiglycation and antioxidant activities and HPTLC analysis of *Boswellia sacra* oleogum resin: The sacred frankincense. *Trop. J. Pharm. Res., 12*, 597–602.

Al-Harrasi, A., Rehman, N. U., Khan, A. L., Al-Broumi, M., Al-Amri, I., Hussain, J., et al., (2018). Chemical, molecular and structural studies of *Boswellia* species: β-boswellic aldehyde and 3-*epi*-11β-dihydroxy BA as precursors in biosynthesis of boswellic acids. *PLoS One, 13*(6), e0198666.

Ali, N. A. A., Wurster, M., Arnold, N., Teichert, A., Schmidt, J., Lindequist, U., & Wessjohann, L., (2008). Chemical composition and biological activities of essential oils from the oleogum resins of three endemic Soqotraen *Boswellia* species. *Rec. Nat. Prod., 2*(1), 6–12.

Al-Saidi, S., Rameshkumar, K. B., Hisham, A., Sivakumar, N., & Al-Kindy, S., (2012). Composition and antibacterial activity of the essential oils of four commercial grades of Omani Luban, the oleo-gum resin of *Boswellia sacra. Chemistry and Biodiversity, 9*(3), 615–624.

Atta-Ur-Rahman, Naz, H., Fadimatou, Makhmoor, T., Yasin, A., Fatima, N., Ngounou, F. N., et al., (2005). Bioactive constituents from *Boswellia papyrifera. J. Nat. Prod., 68*(2), 189–193.

Ayubova, M., Guelleh, Z. O., Guelleh, M. O., Brevard, H., & Baldovini, N., (2019). Analytical investigations on *Boswellia occulta* essential oils. *Phytochemistry, 164*, 78–85.

Basar, S., (2005). *Phytochemical Investigations on Boswellia Species: Comparative Studies on the Essential Oils, Pyrolysates and Boswellic Acids of Boswellia carterii Birdw., Boswellia serrata Roxb., Boswellia frereana Birdw., Boswellia neglecta S. Moore and Boswellia rivae Engl.* PhD Thesis, Institute of Organic Chemistry, University of Hamburg.

Basar, S., Koch, A., & König, W. A., (2001). A verticillane-type diterpene from *Boswellia carterii* essential oil. *Flavor and Fragrance Journal, 16*(5), 315–318.

Baser, K. H. C., Demirci, B., Dekebo, A., & Dagne, E., (2003). Essential oils of some *Boswellia* spp., myrrh and opopanax. *Flavor Fragrance Journal, 18*(2), 153–156.

Bekana, D., Kebede, T., Assefa, M., & Kassa, H., (2014). Comparative phytochemical analyses of resins of *Boswellia* species (*B. papyrifera* (Del.) Hochst., *B. neglecta* S. moore, and *B. rivae* Engl.) from Northwestern, Southern, and Southeastern Ethiopia. *Analytical Chemistry*, 1–9.

Belsner, K., Buchele, B., Werz, U., & Simmet, T., (2003). Structural analysis of 3-α-acetyl-20(29)-lupene-24-oic acid, a novel pentacyclic triterpene isolated from the gum resin of *Boswellia serrata*, by NMR spectroscopy. *Magnetic Resonance in Chemistry, 41*, 629–632.

Benelli, G., Rajeswary, M., Vijayan, P., Senthilmurugan, S., Alharbi, N. S., Kadaikunnan, S., Khaled, J. M., & Govindarajan, M., (2018). *Boswellia ovalifoliolata* (Burseraceae) essential oil as an eco-friendly larvicide? Toxicity against six mosquito vectors of public health importance, non-target mosquito fishes, backswimmers, and water bugs. *Environ. Sci. Pollut. Res. Int., 25*(11), 10264–10271.

Bhuchar, V. M., Agarwal, A. K., & Sharma, S. K., (1982). Constituents of gum obtained from *Boswellia serrata* exudates. *Indian J. Technol., 20*, 38.

Bothon, F. T. D., Abou, L., Atindehou, M., Yarou, K. A., Ewedje, E. B. K., Agbangnan, P. S. D., Latifou, L., et al., (2019). Evaluation of antibacterial, free radical scavenging activities and phytochemical composition of *Boswellia dalzielii* Hutch extracts. *J. Pharmacogn. Phytochem., 8*(4), 414–418.

Camarda, L., Dayton, T., Stefano, V. D., Pitonzo, R., & Schillaci, D., (2007). Chemical composition and antimicrobial activity of some oleogum resin essential oils from *Boswellia* spp. (*Burseraceae*). *Annali di Chimica, 97*(9), 837–844.

Chiavari, G., Galletti, G. C., Piccaglia, R., & Mohamud, M. A., (1991). Differentiation between resins from *Boswellia carterii* and *Boswellia frereana* (frankincense) of Somali region. *J Essential Oil Res., 3*, 185–186.

Culioli, G., Mathe, C., Archier, P., & Vieillescazes, C., (2003). A lupane triterpene from frankincense (*Boswellia* sp., Burseraceae). *Phytochemistry, 62*, 537–541.

DeCarlo, A., Johnson, S., Ouedraogo, A., Dosoky, N. S., & Setzer, W. N., (2019). Chemical composition of the oleogum resin essential oils of *Boswellia dalzielii* from Burkina Faso. *Plants, 8*, 223.

Dekebo, A., Dagne, E., Gautun, A., & Aasen, J., (2002). Triterpenes from the resin of *Boswellia neglecta. Bull. Chem. Soc. Ethiopia, 16*, 87–90.

Dekebo, A., Zewdu, M., & Dagne, E., (1999). Volatile oils of frankincense from *Boswellia papyrifera. Bull. Chem. Soc. Ethiopia, 13*(1), 93–96.

Dounchaly, A. Z., Jaimand, K., & Mozaffari, S., (2016). Comparison of essential oils compositions of *Boswellia carteri* Birdwood as a food and non-food in different distillation from Iranian market. *J. Med. Plants and By-Products, 1*, 39–44.

Gangwal, M. L., & Vardhan, D. K., (1995). Carbohydrate contents of *Boswellia serrata. Asian J. Chem., 7*, 677.

Geetha, V., & Chakravarthula, S. N., (2018). Chemical composition and anti-inflammatory activity of *Boswellia ovalifoliolata* essential oils from leaf and bark. *J. For. Res., 29*, 373–381.

Girgune, J. B., & Garg, B. D., (1979). Chemical investigation of the essential oil from *Boswellia serrata* Roxb. *J. Sci. Res., 1*, 119–122.

Gonzalez, G. I., & Zhu, J., (1997). First total synthesis of acerogenin C and aceroside IV. *J. Org. Chem., 62*, 7544, 7545.

Halim, S. A., Khan, A., Csuk, R., Al-Rawahi, A., & Al-Harrasi, A., (2020). Diterpenoids and triterpenoids from frankincense are excellent anti-psoriatic agents: An *in-silico* approach. *Front. Chem., 8*, 486.

Hamm, S., Bleton, J., Connan, J., & Tchapla, A., (2005). A chemical investigation by headspace SPME and GC-MS of volatile and semi-volatile terpenes in various olibanum samples. *Phytochemistry, 66*(12), 1499–1514.

Hanuš, L. O., Moussaieff, A., Řezanka, T., Abu-Lafi, S., & Dembitsky, M. V., (2007). Phytochemical analysis and comparison for differentiation of *Boswellia carterii* and *B. serrata. Nat. Prod. Communications, 2*(2), 139–142.

Hussain, H., Al-Harrasi, A., Al-Rawahi, A., & Hussain, J., (2013a). Chemistry and biology of essential oils of genus *Boswellia. Evid Based Complement Alternat Med.,* 140509.

Johnson, S., DeCarlo, A., Satyal, P., Dosoky, S. N., Sorensen, A., & Setzer, N. W., (2019). The chemical composition of *Boswellia occulta* oleogum resin essential oils. *Nat. Prod. Communications, 15*(7), 1–7.

Kasali, A., Adio, A. M., Oyedeji, A. O., Eshilokun, A. O., & Adefenwa, M., (2002). Volatile constituents of *Boswellia serrata* Roxb. (Burseraceae) bark. *Flav. Frag. J., 17*, 462–464.

Khan, G., Farooqi, M. I. H., (1991). Chemical characterization of resins sold in Indian Markets. Part I. Luban and Kundur. *Int. J. Pharmacogn., 29*, 302–305.

Kohoude, M. J., Gbaguidi, F., Agbani, P., Ayedoun, M. A., Cazaux, S., & Bouajila, J., (2017). Chemical composition and biological activities of extracts and essential oil of *Boswellia dalzielii* leaves. *Pharm. Biol., 55*, 33–42.

Kubmarawa, D., Ogunwande, I., Okorie, D., Olawore, N., & Kasali, A., (2006). Constituents of the essential oils of *Boswellia dalzielii* Hutch. from Nigeria. *J. Essent. Oil Res., 18*, 119–120.

Kumar, A., & Saxena, V. K., (1979). TLC and GLC studies on the essential oil from *Boswellia serrata* leaves. *Indian Drugs, 16*, 80–83.

Lai, S. C., You, R. I., Chen, T. T., Chang, Y., Liu, C. Z., Chen, H. P., & Wu, C., (2021). Rapid identification of commercial frankincense products by MALDI-TOF mass spectrometry. *Comb. Chem. High Throughput Screen.,* 28. doi: 10.2174/13862073246 66210301092111.

Madera, P., Paschova, Z., Ansorgova, A., Vrskovy, B., Lvoncik, S., & Habrova, H., (2017). Volatile compounds in oleo-gum resin of socotran species of Burseraceae. *Acta Universitatis Agriculturae et Silviculturae Mendelianae Brunensis, 65*(1), 73–90.

Mahajan, B., Sethi, V. K., Taneja, S. C., & Dhar, K. L., (1995). Two triterpenoids from *Boswellia serrata* gum resin. *Phytochemistry, 39*, 453–455.

Manguro, L. O. A., & Wagai, S. O., (2016). Ursane and tirucallane-type triterpenes of *Boswellia rivae* oleo-gum resin. *J. Asian Nat. Prod. Res., 18*, 854–864.

Manguro, L. O. A., Wagai, S. O., & Onyango, J. O., (2016). Terpenoids of *Boswellia neglecta* oleo-gum resin. *Bull. Chem. Soc. Ethiop., 30*(2), 317–323.

Marongiu, B., Piras, A., Porcedda, S., & Tuveri, E., (2006). Extraction of *Santalum album* and *Boswellia carterii* Birdw. volatile oil by supercritical carbon dioxide: Influence of some process parameters. *Flavour Fragr. J., 21*, 718–724.

Mikhaeil, B. R., Maatooq, G. T., Badria, F. A., & Amer, M. M., (2003). Chemistry and immunomodulatory activity of frankincense oil. *Zeitschrift fur Naturforschung. C, J. Biosci., 58*(3, 4), 230–238.

Mohamed, A. A., Ali, S. I., Kabiel, H. F., Hegazy, A. K., Kord, M. A., & EL-Baz, F. K., (2015). Assessment of antioxidant and antimicrobial activities of essential oil and extracts of *Boswellia carterii* resin. *Intern. J. Pharmacogn. Phytochem. Res., 7*(3), 502–509.

Morikawa, T., Oominami, H., Matsuda, H., & Yoshikawa, M., (2011). New terpenoids, olibanumols D–G, from traditional Egyptian medicine olibanum, the gum-resin of *Boswellia carterii. J. Nat. Med., 65*, 129–134.

Mothana, A. A. R., Hasson, S. S., Schultze, W., Mowitz, A., & Lindequist, U., (2011). Phytochemical composition and *in vitro* antimicrobial and antioxidant activities of essential oils of three endemic Soqotran *Boswellia* species. *Food Chemistry, 126*(3), 1149–1154.
Mustafa, G. I., (2012). *Phytochemical Composition and Bioassay of Oleogum Resins of Boswellia Papyrifera (Tarak tarak), Grown in Some Parts of the Sudan*. PhD. Thesis, University of Khartoum, Sudan.
Ni, X., Suhail, M. M., Yang, Q., Cao, A., Fung, K. M., Postier, R. G., et al., (2012). Frankincense essential oil prepared from hydrodistillation of *Boswellia sacra* gum resins induces human pancreatic cancer cell death in cultures and in a xenograft murine model. *BMC Complement. Altern. Med., 12*, 253.
Obermann, H., (1977). The chemical and aromatic differences of incense resins. *Dragoco Rep., 24*, 260–265.
Omoregbeee, A., & Idu, M., (2017). GC-MS analysis of ethanolic extract of *Boswellia dalzielii* Hutch. (Burseraceae) root from Nigeria. *Chem. Res. J., 2*(2), 33–38.
Onobrudu, D. A., (2017). Saponins and polyphenolics of methanol leaf extract of *Boswellia dalzielii* Hutch. *Arch. Curr. Res. Intern., 8*(3), 1–6.
Osman, A. M. A., (2012). *A Study on the Chemical Composition of the Oleo Gum from Sudanese Boswellia papyrifera.* Dissertation submitted to Sudan University of Science and Technology.
Pailer, M., Scheidl, O., Gutwillinger, H., Klein, E., & Obermann, H., (1981). Uber die zusammensetzung des pyrolysates von weihrauch "Aden," dem gummiharz von *Boswellia carteri* birdw. *Mitt. Monatsh. Chem., 112*, 341–358.
Pardhy, R. S., & Bhattacharyya, S. C., (1978). Tetracyclic triterpene acids from the resin of *Boswellia serrata* Roxb. *Ind. J. Chem., 16B*, 174, 175.
Paul, M., (2012). *Chemotaxonomic Investigations on Resins of the Frankincense Species Boswellia papyrifera, Boswellia serrata, and Boswellia sacra, Respectively, Boswellia carterii*. Dissertation. Saarland University Saarbrücken, Saarland, Germany.
Prasanna, A. R. L., Venkata, R. K., Bhakshu, Md. L., Narasimha, R. B., & Veeranjaneya, R. L., (2015). Chemical profile, antioxidant and antimicrobial activity of essential oils from *Boswellia ovalifoliolata* Bal. et Henry. *Int. J. Pharmaceut. Cli. Res., 7*(1), 96–101.
Rashan, L., Hakkim, F. L., Idrees, M., Essa, M. M., Velusamy, T., Al-Baloshi, M., Al-Balushi, B. S., et al., (2019). *Boswellia* gum resin and essential oils: Potential health benefits – An evidence-based review. *Int. J. Nutr. Pharmacol. Neurol. Dis., 9*, 53–71.
Rehman, N. U., Halim, S. A., Al-Azri, M., Khan, M., Khan, A., Rafiq, K., Al-Rawah, A., Csuk, R., & Al-Harrasi, A., (2020). Triterpenic acids as non-competitive α-glucosidase inhibitors from *Boswellia elongata* with structure-activity relationship: *In vitro* and *in silico* studies. *Biomolecules, 10*, 751. doi: 10.3390/biom10050751.
Ren, J., Wang, Y. G., Wang, A. G., Wu, L. Q., Zhang, H. J., Wang, W. J., Su, Y. L., & Qin, H. L., (2015). Cembranoids from the gum resin of *Boswellia carterii* as potential antiulcerative colitis agents. *J. Nat. Prod., 78*, 2322–2331.
Satyal, P., & Pappas, R. S., (2016). First reporting on the chemistry and biological activity of a novel *Boswellia* chemotype: The methoxy alkane frankincense. *Global J. Sci. Front. Res. B Chem., 16*(2), 1–10.

Savithramma, N., Linga, R. M., & Venkateswarlu, P., (2014). Isolation and identification of phenolic compound from *Boswellia ovalifoliolata* Bal. & Henry and their radical scavenger activity. *Inter. J. Drug Deli. Tech., 4*(1), 14–21.

Savithramma, N., Venkateswarlu, P., Suhrulatha, D., Basha, S. K. M., & Venkata, R. D. C. H., (2010). Studies of *Boswellia ovalifoliolata* Bal. & Henry, an endemic and endangered medicinal plant. *Bioscan, 5*(3), 359–362.

Sharma, A., Chhikara, S., Ghodekar, S. N., Bhatia, M. D., Kharya, V., Gajbhiye, A., Mann, S., et al., (2009). Phytochemical and pharmacological investigations on *Boswellia serrata*. *Phcog. Rev., 3*(5), 206–215.

Sharma, R. A., & Verma, K. C., (1980). Studies on gum obtained from *Boswellia serrata* Roxb. *Indian Drugs, 17*, 225.

Shen, T., & Lou, H. X., (2008). Bioactive constituents of myrrh and frankincense, two simultaneously prescribed gum resins in Chinese traditional medicine. *Chem Biodivers., 5*, 540–553.

Siddiqui, M. Z., (2011). *Boswellia serrata*, a potential anti-inflammatory agent: An overview. *Indian J. Pharm. Sci., 73*, 255–261.

Singh, B., Kumar, R., Bhandari, S., Pathania, S., & Lal, B., (2007). Volatile constituents of natural *Boswellia serrata* oleo-gum-resin and commercial samples. *Flavor and Fragrance J., 22*, 145–147.

Sun, X., Geng, Y., Wang, X., Qin, D., & Yu, J., (2020). Cembrane-type diterpenoids from the gum resin of *Boswellia carterii* and their biological activities. *RSC Advances, 10*, 746–755.

Tegasne, T., Kapche, G. D. W. F., Kamga, I., et al., (2020). Bioguided chemical study of *Boswellia dalzielii* Hutch. (Burseraceae) for antibacterial agents and a new glucopyranoxyl methoxybenzyle. *Nat. Prod. Res*. https://doi.org/10.1080/14786419.2020.1794863.

Thulin, M., & Warfa, M., (1987). The frankincense trees (*Boswellia* spp., Burseraceae) of northern Somalia and southern Arabia. *Kew Bull., 42*, 487–500.

Thulin, M., (2020). The genus *Boswellia* (Burseraceae). The frankincense trees. *J. Plant Taxonomy and Geography, 75*(1), 133–134.

Thulin, M., DeCarlo, A., & Johnson, S. P., (2019). *Boswellia occulta* (Burseraceae), a new species of frankincense tree from Somalia (Somaliland). *Phytotaxa, 394*(3), 219–224.

Tihar, M., & Dagne, E., (2014). Quality assessment of essential oils of *Eucalyptus globulus* and *Boswellia rivae*. *Ethiop. J. Sci. & Technol., 7*(1), 27–36.

Tucker, A. O., (1986). Frankincense and myrrh. *Econ. Bot., 40*, 425–433.

Van, V. S. F., Kamatou, G. P., & Viljoen, A. M., (2010). Volatile composition and antimicrobial activity of twenty commercial frankincense essential oil samples. *South Afr. J. Bot., 76*, 686–691.

Vanimireddy, L. N. R., Ravinder, K., Srinivasulu, M. T., Goud, V., Reddy, S. M., Srujankumar, D., Rao, T. P., et al., (2003). Two new macrocyclic diaryl ether heptanoids from *Boswellia ovalifoliolata*. *Chem. Pharm. Bull., 51*, 1081.

Vollesen, K., (1989). In: *Flora of Ethiopia* (Vol. 3, p. 442). National Herbarium, Science Faculty, Addis Ababa University, Addis Ababa.

Wahab, S. M., Aboutabl, E. A., El-Zalabani, S. M., Fouad, H. A., De Pooter, H. L., & El-Fallaha, B., (1987). The essential oil of olibanum. *Planta Med., 53*(4), 382–384.

Wang, J., Zhen, B., Jiawen, H., Shi, M., Wei, C., Wang, X., & Sun, H., (2019). Boscartins L–O: Cembrane-type diterpenoids from the gum resin of *Boswellia sacra* Flueck. *Phytochemistry, 163*, 126–131. 10.1016/j.phytochem.2019.03.005.

Woolley, C. L., Suhail, M. M., Smith, B. L., Boren, K. E., Taylor, L. C., Schreuder, M. F., et al., (2012). Chemical differentiation of *Boswellia sacra* and *Boswellia carterii* essential oils by gas chromatography and chiral gas chromatography-mass spectrometry. *J. Chromatogr. A, 1261*, 158–163.

Yagi, S., Babiker, R., Tzanova, T., & Schohn, H., (2016). Chemical composition, antiproliferative, antioxidant and antibacterial activities of essential oils from aromatic plants growing in Sudan. *Asian Pacific J. Trop. Med., 9*(8), 741–748.

Yang, J., Jin, R., & Wang, A., (2018). Isolation, characterization, and hepatoprotective activities of terpenes from the gum resin of *Boswellia carterii* Birdw. *Phytochemistry Letters, 23*, 73–77.

Yoshikawa, M., Morikawa, T., Oominami, H., & Matsuda, H., (2009). Absolute stereo structures of olibanumols A, B, C, H, I, and J from olibanum, gum-resin of *Boswellia carterii,* and inhibitors of nitric oxide production in lipopolysaccharide-activated mouse peritoneal macrophages. *Chem. Pharm. Bull., 57*, 957–964.

Yu, J. Q., Geng, Y. L., Wang, D. J., Zhao, H. W., Guo, L., & Wang, X., (2018a). Terpenes from the gum resin of *Boswellia carterii* and their NO inhibitory activities. *Phytochemistry Letters, 28*, 59–63.

Yu, J. Q., Geng, Y., Zhao, H., & Wang, X., (2018a). Diterpenoids from the gum resin of *Boswellia carterii* and their biological activities. *Tetrahedron, 74*(40), 5858–5866.

Yu, J. Q., Zhao, L., Sun, X., Geng, Y., & Wang, X., (2020). Bioactive cembrane diterpenoids from the gum resin of *Boswellia carterii. Fitoterapia, 146*, 104699.

Zhang, B., Liu, D., Ji, W., Otsuki, K., Higai, K., Zhao, F., Li, W., et al., (2021). Sacraoxides A–G, bioactive cembranoids from gum resin of *Boswellia sacra. Front. Chem.* https://doi.org/10.3389/fchem.2021.649287.

CHAPTER 6

TRADITIONAL USES AND PHARMACOLOGY OF *BOSWELLIA* SPECIES

K. VENKATA RATNAM[1] and LEPAKSHI MD. BHAKSHU[2]

[1]Department of Botany, Rayalaseema University, Kurnool – 518007, Andhra Pradesh, India

[2]Department of Botany, PVKN Government College, Chittoor – 517002, Andhra Pradesh, India

**Corresponding author. E-mail: drvenkatapkd@gmail.com*

ABSTRACT

The present chapter has focused its attention on the traditional uses and pharmacological effects of *Boswellia* species in various biological systems. It's a well-appreciated plant resource for its aromatic gum resin, which is preferably grown in drought and open woods and restricted in its distribution and constrained to a few countries. These species yield resinoids, which possess high market value and become a source of income in drought-prone areas. Moreover, a few of the species were highly threatened and needed its conservation and sustainability during the gum extraction. Among the 24 authenticated species, 12 species are being used in the preparation of several medicinal products based on the *in vitro*, *in vivo*, or *in silico* (docking), patent studies. However, many other species are not yet fully explored scientifically and needs attention by the emerging scientists. Here in, we have discussed the biological activities of certain species of *Boswellia*, which are published and cited.

Frankincense – Gum Olibanum: Botany, Chemistry, Extraction, Utilization, Propagation, Biotechnology, and Conservation, T. Pullaiah, K. Venkata Ratnam, Mallappa Kumara Swamy, Lepakshi Md. Bhakshu (Eds.)

6.1 INTRODUCTION

The traditional rituals are associated with health beneficial applications, and can provide a rational support for the usage as medicinal ailments and the same has been claimed and proved in case of frankincense/olibanum, which is scientifically recognized as *Boswellia* species. The natural aromatic gum resin when burned gives a pleasant smell which has also been used in medicines (Van Beek, 1960; Groom, 1981). This precious natural product possesses immense medicinal values which are proved in the various biological systems and appreciated by the pharmacologists and traditional practitioners for its usage in the many inflammatory-related rheumatic pains (arthritis), immune-enchantment during the juvenile stages of human being and established its phytochemical profile throughout the globe, which also stands for good economic source for resin collecting agricultural labor besides incense industry. In relation to the discussion on the pharmacological effects of the *Boswellia* genus, a global outlook has revealed that there are few species of *Boswellia* were extensively studied for the pharmacological effects and analyzed for secondary metabolites. The species, such as *B. ameero, B. bullata, B. carterii, B. dalzielli, B. dioscorides, B. ovalifoliolata B. papyrifera, B. rivae, B. sacra, B. serrata, B. socotrana* were scientifically explored for various biological activities and proved as a best anti-inflammatory agent besides anticancer and other health-protective medicaments. Among 24 authenticated species globally (see Chapter 2), only 12 species were investigated and proved many biological effects. *B. carterii, B. serrata, B. sacra* were studied for molecular signaling pathways and established as modern medicines scientifically and published in highly reputed journals/newsletters and even reviews. Certain species of *Boswellia* such as *B. frereana* Birdw., *B. globosa* Thulin, *B. microphylla* Chiov. *B. nana* Hepper, *B. occulta* Thulin, *B. ogadensis* Vollesen, and *B. pirrottae* Chiov. were not yet scientifically explored and required attention by the young scientists. The medicinal properties, traditional uses, and pharmacological effects of species of *Boswellia* are being discussed herein.

6.2 TRADITIONAL USES OF *BOSWELLIA* SPECIES

6.2.1 *BOSWELLIA AMEERO*

Gum oleoresin obtained from stem bark is used as an incense (Uphof, 1959). The plant has been used to treat asthma, bronchitis, rheumatism, and common cold by local peoples of Socotra Island (Mothana and Lindequist, 2005).

6.2.2 *BOSWELLIA DALZIELII*

6.2.2.1 ETHNOBOTANY

In Nigeria, gum resin is used to cure venereal diseases. The bark boiled in water, and the decoction was used to wash for fever, and rheumatism. The root decoction boiled along *Hibiscus sabdariffa* is used for the treatment of syphilis (Nwinyi et al., 2004). The leaf is used in the treatment of bilharziasis and given as an oxytocic to pregnant women (Kafuti et al., 2017). The stem bark is used as an antidote for snakebite. Decoction of the aerial parts was given orally for tuberculosis. Root bark decoction is applied externally to treat scabies (Kubmarawa et al., 2013). In Ivory Coast the bark decoction is used as an antiseptic solution to wash sores and as an ingredient to treat leprosy (Danlami et al., 2015). In Cameroon, local people use the leaves as insecticide to protect maize, millet, and sorghum against weevils' attack (Younoussa et al., 2016).

The root bark decoction is traditionally used to treat diabetes by the Hausa-Fulanis in Sokoto, Nigeria. The decoction prepared from bark used to bath during fever and rheumatism conditions. The fluid prepared from bark given orally for gastrointestinal trouble. The Fulanis used cold infusion prepared from bark as an antidote for snake bite. Fresh root bark was eaten in Adamawa State, Nigeria, to cause vomiting in case of giddiness, palpitations as well as antidotes to arrow-poison (Shinkafi et al., 2015; Burkill, 1985). Gwandara people of Northern Nigeria used stem bark decoction as a tranquilizer, in the treatment of convulsions and mental derangement (Ibrahim et al., 2007).

Lala people of Nigeria used different parts of *B. dalzeilli* to cure different ailments such as, decoction of the root bark is applied on the spot of scabies. The stem bark paste applied externally on the spot of the arrow wound as an antidote. The decoction of stem bark given orally in laxative conditions. The stem bark paste applied externally on bitten area of python bite. Aerial parts boiled in water; decoction given orally for tuberculosis. The leaves are used as scent (Kubmarawa et al., 2013). Root and stem bark reported to treat gingivitis (Henley-Smith et al., 2013). A bark-decoction is used as an antiseptic wash for sores, and is an ingredient of a complicated prescription for leprosy (Burkill, 1985).

6.2.2.2 AYURVEDIC USES

In the Ayurveda system of medicine, the gum resin of *B. dalzielii* was recommended for arthritic and inflammatory conditions, gastric disorders, pulmonary diseases, and skin ailments. It is also reported to have strong action on the nervous system and reduces phlegm, asthmatic conditions, and as antiemetic drug (Etuk et al., 2006b).

6.2.2.3 VETERINARY USES

The local herbal practitioners of Kano and Sokoto states of north-western region of Nigeria the aqueous stem bark extract of *B. dalzielii* is used in the treatment of gastrointestinal problems (diarrhea and abdominal cramp) and as an antipyretic, analgesic, and anti-inflammatory agent either alone or in combination with other herbs (Nwude and Ibrahim, 1980).

6.2.3 BOSWELLIA ELONGATA

The plant has been used to treat asthma, bronchitis, rheumatism, and common cold by local people of Socotra island (Mothana and Lindequist, 2005).

6.2.4 BOSWELLIA OVALIFOLIOLATA

6.2.4.1 ETHNOBOTANY

The fresh leaf juice given orally to cure throat ulcers (Savithramma and Sulochana, 1998). Stem bark decoction about 10–25 mL per day given orally to reduce rheumatic pains (Nagaraju and Rao, 1990). Dried gum dissolved in water and the formed gel along with curd and given orally to treat amoebic dysentery (Sudhakar, 1998). Gum powders of *B. ovalifoliolata* and *B. serrata* along with fruit powder of *Pedalium murex* mixed in equal parts and ground, paste applied externally on the affected part of the testicle to cure hydrocele. *B. ovalifoliolata* gum mixed with that of stem powder of *Tinospora cardifolia* and honey given orally two times a day to cure hydrocoel (Vedavathy et al., 1995).

6.2.5 *BOSWELLIA PAPYRIFERA*

Leaves and roots are used against lymphadenopathy and resin used as febrifuge (Fitchl and Admasu, 1994). In China gum resin was considered as a stimulant and also used to treat leprosy (Tucker, 1986). The stem bark is chewed to treat stomach problems. The bark fumes used as mosquito repellent in the tropics. The lowlanders chewed stem bark to quench thirst during hot days (Tilahun, 1997). Gum resin was widely used as incense at home and for the treatment of cough and respiratory infections (El Ghazali, 1997).

6.2.6 *BOSWELLIA SACRA* (SYN.: *B. CARTERII*)

6.2.6.1 *TRADITIONAL CHINESE MEDICINE (TCM)*

Oleo gum resin of *Boswellia sacra* was commonly used as a remedy for improving blood circulation, traumatic injuries and as relieving pain in leprosy, gonorrhea, and cancer patients (Zhao et al., 2003), chest pain and stuffiness and rheumatism (Committee of the Chinese Pharmacopoeia, 2015).

6.2.6.2 *AFRICAN TRADITIONAL MEDICINE*

Ancient Somalians used the gum resin to prepare cosmetics, skin care lotions, and perfumes. Ancient Egyptians used gum resin as incense in religious ceremonies and rituals (Basar, 2005).

6.2.6.3 *INDIAN TRADITIONAL MEDICINE*

In Ayurvedic medicinal systems, the oleo gum resin is used to treat chronic lung diseases, diarrhea, dysentery, pulmonary diseases, menorrhea, dysmenorrhea, syphilis, piles, and liver disorders (Kapoor, 1990). The gum resin oil is prescribed with a demulcent drink to treat gonorrhea. Gum resin paste prepared along with coconut oil or lemon juice is applied to ulcers, swellings, boils, and ringworm (Chopra et al., 1956).

6.2.7. *BOSWELLIA SERRATA*

6.2.7.1 UNANI SYSTEM OF MEDICINE

1. **Whitlow:** Gum powder mixed with honey and applied externally (Ghani, 2002).
2. **Skin Scares/Wounds:** Bark powder mixed with honey and applied externally (Ghani, 2002; Sina, 1998).
3. **Anthelmintic:** Gum along with duck fat or vinegar or olive oil given orally to control ringworm infection (Ibn Baiter, 2003).
4. **Ear Ache:** Resin dissolved in ethanol and used as ear drops (Sina, 1998; Alam et al., 2012).

6.2.7.2 AYURVEDIC SYSTEM OF MEDICINE

In Ayurvedic text, *B. serrata* gum resin referred to cure different human ailments like rheumatic-arthritis, fevers, ringworm, diarrhea, dysentery, boils, skin, and blood diseases, cardiovascular diseases, mouth sores, bad throat, lung diseases, cough, gynic diseases, irregular menses, hair-fall, liver diseases, hemorrhoids, syphilitic diseases, and stimulation of liver (Kirtikar and Basu, 1935; Chatterjee and Pal, 1984; Khare, 2004).

6.3 PHARMACOLOGY OF *BOSWELLIA* SPECIES

Poeckel and Werz (2006) reviewed the biological actions and molecular targets of boswellic acids (BAs). Henkel et al. (2012) reported that BAs from gum resin of *Boswellia* species inhibit lipopolysaccharide (LPS) functionality through direct molecular interference. Tausch et al. (2009) identified human cathepsin G as a functional target of BAs from the anti-inflammatory remedy frankincense.

6.3.1 *BOSWELLIA AMEERO*

6.3.1.1 ANTIVIRAL ACTIVITY

Antiviral efficacy of methanol and hot water extracts of 25 medicinal plant species, which includes *B. ameero*, used in Yemeni traditional medicine was investigated against two viral strains, i.e., influenza virus type A/ MDCK cells and herpes simplex virus (HSV) type 1/Vero cells. *B. ameero* showed antiviral effects against anti-influenza virus type A, along with other medicinal plants such as *Boswellia elongata, Buxus hildebrandtii, Cissus hamaderohensis, Cleome socotrana, Dracaena cinnabari, Exacum affine, Jatropha unicostata* and *Kalanchoe farinacea* (Mothana et al., 2006).

6.3.1.2 ANTIBACTERIAL STUDIES

Antibacterial efficacy of methanol and hot water extracts of 25 medicinal plant species, which includes *B. ameero*, used in Yemeni traditional medicine was determined by using the agar diffusion method. The methanol extract of *B. ameero* exhibited good antibacterial activity against the test pathogens along with other medicinal plants tested, i.e., *Boswellia elongata, Buxus hildebrandtii, Commiphora parvifolia, Jatropha unicostata, Kalanchoe farinacea, Pulicaria stephanocarpa, Punica protopunica, Withania adunensis*, and *W. riebeckii* (Mothana and Lindequist, 2005).

6.3.1.3 ANTIOXIDANT ACTIVITY

Antioxidant activity of essential oil isolated from *B. ameero* oleo gum resin was evaluated by DPPH (2,2-diphenyl-1-picrylhydrazyl) method (Ali et al., 2008). In results, oil strongly reduced DPPH purple color by expressing lower IC_{50} value 175.2 μg/mL. This may be due to the synergistic effect among the components present in the essential oil. In another study, methanolic extract of *B. ameero* oleo gum resin exhibited concentration-dependent DPPH scavenging activity (Ali et al., 2014).

6.3.1.4 ACETYL CHOLINESTERASE (ACHE) INHIBITION

Acetylcholinesterase (AChE) inhibition activity of essential oil isolated from *B. ameero* oleo gum resin was assessed by using *in vitro* method. Essential oil at 200 μg/mL concentration showed 41.5% enzyme inhibition (Ali et al., 2008).

6.3.2 *BOSWELLIA DALZIELII*

6.3.2.1 ANTIBACTERIAL STUDIES

Antibacterial effect of ethanol extract and essential oil isolated from leaves of *B. dalzielii* against *E. coli, Enterococcus faecalis, S. aureus* and Methicillin-resistant *S. aureus* (MRSA) *in vitro* microdilution method. Among the tested four bacterial strains, MRSA, and *Enterococcus faecalis* were strongly inhibited by both ethanol extract and essential oil (Bothon et al., 2019). Antibacterial activity of incensole, gallic acid, protocatechuic acid, 4'-methoxy-(E)-resveratrol-3-o-rutinoside, phytosterol, β-sitosterol, and chlorocresol from ethyl acetate (EA) fraction of *B. dalzielii* stem bark was studied by Alemika et al. (2006) against four bacterial species, i.e., *B. subtilis, E. coli, P. aeruginosa* and *S. aureus*, and yeast *Candida pseudotropicalis*. Among the tested components, gallic acid and protocatechuic acid strongly inhibited *S. aureus*, moderate activity against *B. subtilis* and very feeble activity on remaining bacterial strains. No activity was found against *Candida*.

In another study, Gram-positive bacterial species, i.e., *B. megaterium, B. cereus, S. epidermidis, S. aureus* NCTC 6570 and *S. aureus* were strongly inhibited by ethanol extract of *B. dalzielii* stem bark determined in minimum inhibitory concentration (MIC) and minimum bactericidal concentration (MBC) methods. Moderate activity was observed against gram-negative bacterial strains, i.e., *P. aeruginosa, Proteus mirabilis, Escherichia coli* NCTC 10418, *E. coli*, and *K. pneumoniae* (Olukemi et al., 2005). *P. mirabilis, E. coli*, and *S. aureus* exhibited higher inhibition zones (12 mm for each organism) to ethanol extract of *B. dalzielii* stem bark. *P. aeruginosa, and K. pneumoniae* exhibited less inhibition zones to the test extract (Ohadoma et al., 2016).

6.3.2.2 ANTIOXIDANT STUDIES

Antioxidant activity of ethanol extract and essential oil isolated from leaves of *B. dalzielii* was evaluated by DPPH quenching assay. Essential oil strongly reduced DPPH purple color than ethanol extracts and standard compound ascorbic acid (Bothon et al., 2019). The antioxidant capacity of ethanol extract of *B. dalzielii* gum resin was determined by using different *in vitro* models, including 2,2-diphenyl-1-picrylhydrazyl (DPPH), and ferric ion reducing antioxidant power (FRAP) assays. The extract exhibited very strong activity to DPPH scavenging activity with an IC_{50} value of 178.12 mg/mL and a FRAP was found to be 147.85 mg/mL (Védékoi et al., 2019). The methanolic extract, EA, and aqueous fractions of *B. dalzielii* stem bark exhibited significant DPPH quenching potential than standard vitamin C, gallic acid, and rutin. In the FRAP assay, aqueous fraction exhibited the highest FRAP antioxidant capacity with EC_{50} 1.00 μg/ml followed by methanolic 4.78 μg/ml, EA 5.62 μg/ml and hexane fraction 120 μg/ml, respectively (Kafuti et al., 2019).

The DPPH scavenging potential of essential oil, methanol, EA, dichloromethane, and cyclohexane extracts *B. dalzielii* leaves was determined (Kohoude et al., 2017). Of the test extracts, methanol extract exhibited lower IC_{50} value (6.10 ± 0.01 mg/L), followed by the EA extract (15.20 ± 0.01 mg/L), dichloromethane extract (78.10 ± 0.10 mg/L) and cyclohexane extract (IC_{50} > 100 mg/L). Antioxidant capacity of methanolic extract of *B. dalzielii* was determined in streptozotocin-induced diabetic mice by catalase, SOD, MDA, and NO methods. The extract significantly reduced the levels of MDA and NO and enhanced the activity of antioxidant enzymes, i.e., SOD and catalase (Mbiantcha et al., 2020).

6.3.2.3 ANALGESIC EFFECT

Analgesic effect of aqueous extract of *B. dalzielii* stem bark was evaluated in animal models using acetic acid-induced writhing test and formalin-induced nociception methods (Adebisi et al., 2018). The extract significantly inhibited acetic acid-induced writhing in a concentration-dependent manner. At 50 and 100 mg/kg concentration the extract exhibited strong inhibition (69.56 and 71.73%) higher than that of the standard drug piroxicam (59.78%). Regarding to formalin-induced nociception, the extract produced a significant inhibition in a dose-dependent manner at the first and last phase of drug administration.

6.3.2.4 ANTI-INFLAMMATORY ACTIVITY

Anti-inflammatory activity of aqueous extract of *B. dalzielii* stem bark was evaluated in animal models using formalin-induced inflammation (Adebisi et al., 2018). At 100 mg/kg concentration the extract exhibited strong inhibition (48.78%) higher than that of the standard drug piroxicam (43.90%). Anti-inflammatory activity of essential oil and methanol, EA, dichloromethane, and cyclohexane extracts of *B. dalzielii* leaves was evaluated by using soybean 5-lipoxygenase (Kohoude et al., 2017) *in vitro* study. Except the cyclohexane extract, all extracts exhibited good inhibitory activity against 5-lipoxygenase. Anti-inflammatory potential of methanol extract of *B. dalzielii* was investigated against carrageenan, arachidonic acid, histamine, serotonin, prostaglandin, and bradykinin-induced inflammation in rat models (Mbiantcha et al., 2018). Methanol extract significantly reduced edema conditions caused by carrageenan, arachidonic acid, histamine, serotonin, prostaglandin, and bradykinin in rat models.

6.3.2.5 ANTI-ARTHRITIS ACTIVITY

Anti-arthritis activity of methanol extract of *B. dalzielii* was carried out using complete Freund's adjuvant (CFA) rat model (Mbiantcha et al., 2018). The extract exhibited a noticeable systematic reduction in joint diameter and a decrease in pain in the treated animals, which were very close to normal rats.

6.3.2.6 ANTIEPILEPTIC STUDIES

Antiepileptic activity of ethanol extracts of *B. dalzielii* stem bark was evaluated using maximal electroshock-induced seizure test in broiler chicks and pentylenetetrazole (PTZ) using Wistar strain mice. The ethanol extract at lower concentrations (100 and 250 mg/kg body weight) did not protect chicks against tonic hind limb extension (THLE) in maximal electroshock test. But it showed 20% protection at 500 mg/kg body weight. The extract exhibited concentration-dependent protection to PTZ induced seizure in Wistar mice at 100 and 250 mg/kg body weight (Medugu et al., 2020).

6.3.2.7 ANTICONVULSANT ACTIVITY

Anticonvulsant activity of methanol extract of *B. dalzielii* stem bark was evaluated using maximal electroshock-induced convulsion test in chicks and PTZ, strychnine, picrotoxin, and 4-aminopyridine (4-AP) induced convulsion in mice (Nazifi et al., 2017). The extract at 20 mg/kg protected 40% of chicks against HLTE induced by the electroshock and it significantly increased the mean onset of seizure compared to normal saline group. The extract exhibited dose-dependent anticonvulsant activity against PTZ induced convulsion in mice. The extract at 40 and 80 mg/kg protected 33% against seizures induced by PTZ and strychnine. While the extract at 40 mg/kg dose protected 50% of the mice against picrotoxin induced convulsion. No effect was observed against 4-AP-induced seizures.

6.3.2.8 ANTIULCER ACTIVITY

Antiulcer activity of ethanol extract of *B. dalzielii* stem bark was investigated in ethanol-induced ulcers in rat models (Yusuf et al., 2018). The extract significantly reduced ethanol-induced gastric mucosal damage in albino rat models by reducing the gastric ulcer index, gastric tissue level of malondialdehyde (MDA) and TNF-α. The extract also reinforced the level of antioxidant enzymes, i.e., superoxide dismutase (SOD), catalase, total antioxidant capacity, and vascular endothelial growth factor (VEGF).

6.3.2.9 HEPATOPROTECTIVE ACTIVITY

Hepatoprotective potential of aqueous extract of *B. dalzielii* stem bark was investigated on paracetamol-induced hepatotoxicity in adult rabbits. The extract exhibited dose-dependent hepatoprotective activity against paracetamol-induced hepatotoxicity. It was evident by modulating the serum biochemical enzyme levels of alanine aminotransferase (ALT), aspartate aminotransferase (AST), alkaline phosphatase (ALP), total and direct bilirubin (TB and DB) and also improved histopathological liver sections compared to paracetamol-treated animals (Sa'id et al., 2017). Similarly, liver protection activity of aqueous extract of *B. dalzielii* stem bark was studied in rat models by Aliyu et al. (2007). They reported similar results in rats.

6.3.2.10 ANTISPASMODIC STUDIES

Antispasmodic effect of methanol extract of *B. dalzielii* stem bark was determined in acetylcholine and barium chloride-induced contraction of isolated rabbit jejunum and histamine and barium chloride-induced contraction of isolated guinea pig ileum. The extract exhibited dose-dependent inhibitory activity on acetylcholine and barium chloride-induced contraction of isolated rabbit jejunum and histamine and barium chloride-induced contraction of isolated guinea pig ileum (Hassan et al., 2009).

6.3.2.11 ANTICANCER STUDIES

Cytotoxic effects of methanol, EA, dichloromethane, and cyclohexane extracts of *B. dalzielii* leaves was investigated against two ovarian cancer cell lines, OVCAR-3, and IGROV-1 (Kohoude et al., 2017). Except methanol extract, other extracts, i.e., EA, dichloromethane, and cyclohexane extract significantly reduced the proliferation of IGROV-1 cancer cell line. Whereas dichloromethane extract exhibited the highest anticancer activity against OVCAR-3 cancer cell line than other extracts.

6.3.2.12 NEUROPROTECTIVE ACTIVITY

Neuroprotective activity of essential oil and methanol, EA, dichloromethane, and cyclohexane extracts of *B. dalzielii* leaves was evaluated by using AChE inhibitory assay (Kohoude et al., 2017). Essential oil exhibited the strongest inhibition (85.37 ± 0.01%) against AChE. While EA extract expressed lowest IC_{50} value (76.10 ± 0.02 mg/L), followed by the methanol extract (85.01 ± 0.02 mg/L), the dichloromethane extract (45.79 ± 0.01% at 100 mg/L) and the cyclohexane extract (22.79 ± 0.04% at 100 mg/L).

6.3.2.13 ANTI-HYPERURICEMIA STUDIES

Anti-hyperuricemia potential of essential oil and methanol, EA, dichloromethane, and cyclohexane extracts of *B. dalzielii* leaves was determined by antixanthine oxidase assay *in vitro* method. The EA extract expressed high antixanthine oxidase activity by exhibiting lower IC_{50} value (30.20 ± 0.01 mg/L). The essential oil showed very less inhibition to antixanthine oxidase activity (Kohoude et al., 2017).

6.3.2.14 ANTI-DIARRHEA ACTIVITY

Anti-diarrhea effect of aqueous extract of *B. dalzielii* at 100, 200, and 300 mg/kg body weight was evaluated on castor oil-induced diarrhea in albino rats (Etuk et al., 2006a). The extract significantly reduced castor oil-induced diarrhea in albino rats in a dose-dependent manner.

6.3.2.15 ANTI-HYPERNOCICEPTIVE STUDIES

Anti-hypernociceptive effect of the methanolic extract of *B. dalzielii* was investigated in streptozotocin-induced diabetic mice using Von Frey filaments, hotplate, acetone, and formalin tests. The extract at 250 and 500 mg/kg body weight significantly reduced licking/biting behavior during the first and second phase of the formalin test in a dose-dependent manner (Mbiantcha et al., 2020).

6.3.3 *BOSWELLIA DIOSCORIDES*

6.3.3.1 ANTIMICROBIAL ACTIVITY

Antimicrobial activity of essential oil obtained from *B. dioscorides* was evaluated against gram-positive bacteria using the MIC method. The results showed that the oil exhibited moderated antibacterial activity against the test pathogens (Mothana et al., 2011a). In another study antimicrobial activity of methanol and water extracts of *B. dioscorides* was investigated against three gram-positive, two gram-negative, three multi-resistant *S.* strains and one yeast using disc diffusion and MIC methods. It revealed that, the extracts showed >15 mm of inhibition zones and MIC values below 250 μg/ml (Mothana et al., 2009).

6.3.3.2 ANTIOXIDANT ACTIVITY

Antioxidant potential of essential oil obtained from *B. dioscorides* was evaluated using DPPH method. It indicates that the oil exhibited very feeble DPPH quenching activity (Mothana et al., 2011b). Antioxidant potential of methanol and water extracts of *B. dioscorides,* revealed that both extracts strongly reduced DPPH purple color even at lower concentrations (Mothana et al., 2009).

6.3.3.3 ANTICANCER ACTIVITY

Anticancer potential of methanolic and water extracts of *B. dioscorides* was evaluated against three human cancer cell lines (A-427, 5637, and MCF-7) by using *in vitro* methods. The extracts showed notable antiproliferative activity against the tested human cancer cell lines (Mothana et al., 2009).

6.3.4 BOSWELLIA ELONGATA

6.3.4.1 ANTIVIRAL STUDIES

Antiviral activity of methanol and hot water extracts of *B. elongata* along with 24 medicinal against influenza virus type A/MDCK cells and HSV type 1/Vero cells was investigated. Of the tested 24 medicinal plants, 9 plants which includes *B. elongata* showed inhibitions against influenza virus type cells (Mothana et al., 2006).

6.3.4.2 ANTIMICROBIAL ACTIVITY

Antimicrobial activity of essential oil obtained from *B. elongata* was evaluated against gram-positive bacteria using the MIC method. The results showed that the oil exhibited moderated antibacterial activity against the test pathogens (Mothana et al., 2011a). Further, antimicrobial activity of chloroform, methanol, and hot water extracts of 25 medicinal plants (which includes *B. elongata*) was screened against bacterial and yeast species. The results revealed that, *B. elongata* and other 8 plants showed a broad spectrum of antimicrobial activity against the test pathogens (Mothana and Lindequist, 2005).

6.3.4.3 ANALGESIC ACTIVITY

Analgesic activity of methanolic extract obtained from *B. elongata* leaves was investigated in chemical and thermal models induced acute pain in male and female Swiss mice and Wister male rats. The extract showed peripheral analgesic as anti-inflammatory activities, but not central analgesic effect (Ahmed et al., 2012).

6.3.4.4 ANTI-INFLAMMATORY ACTIVITY

Anti-inflammatory effect of methanolic extract obtained from *B. elongata* leaves was investigated in carrageenan and experimental trauma-induced hind paw edema in rat models. The extract exhibited a noteworthy reduction in carrageenan-induced inflammation and inhibited experimental trauma-induced hind paw edema in treated rat models than control and standard drug (Ahmed et al., 2012). Anti-inflammatory and antinociceptive effects of two endemic medicinal plants, i.e., *Boswellia elongata* and *Jatropha unicostata* were assessed in cotton pellet granuloma in rats, carrageenan-induced rat paw edema, acetic acid-induced abdominal writhing, and hot-plate test model in mice. Of the tested two plants *B. elongata* significantly reduced the paw edema and the weight of cotton pellet granuloma, respectively (Mothana et al., 2011b).

6.3.4.5 ACETYLCHOLINESTERASE INHIBITORY ACTIVITY

Acetylcholinesterase inhibitory activity of chloroform and methanol extracts of *B. elongata* along with 13 other medicinal plants was investigated using *in vitro* Ellman method. Among the tested extracts of 13 medicinal plants, chloroform extract of *B. elongata* and four other medicinal plants exhibited more than 50% inhibition (Bakthira et al., 2011).

6.3.4.6 α-GLUCOSIDASE INHIBITION ACTIVITY

α-Glucosidase inhibition activity of 14 triterpene acids isolated from the methanol extract of *B. elongata* resin was evaluated using *in vitro* methods. Of the tested compounds, four compounds, i.e., 3α-Acetoxytirucall-8,24-dien-21-oic acid, 3-O-Acetyl-9,11-dehydro-β-boswellic acid, 9,11-Dehydro-β-boswellic acid and 11α-methoxy-β-boswellic acid demonstrated significant α-glucosidase inhibition activity. The highest inhibition effect on α-glucosidase was observed with the 3-O-scetyl-9,11-dehydro-β-boswellic acid, demonstrated using the *in vitro* assays and the molecular docking studies also confirmed its highest efficacy (Ur Rehman et al., 2020).

6.3.4.7 PEPTIDASES INHIBITORY STUDIES

Peptidase inhibitory effects of methanolic extracts of 20 medicinal plants collected from the island Socotra/Yemen was investigated against angiotensin-converting enzyme (ACE), aminopeptidase N (APN). And neutral endopeptidase (NEP). Of the tested extracts, only three medicinal plants such as *Kalanchoe farinacea, Boswellia elongata*, and *Cissus hamaderohensis* inhibited all three enzymes (Oleski et al., 2006).

6.3.5 BOSWELLIA FREREANA

6.3.5.1 ANTI-INFLAMMATORY ACTIVITY

Anti-inflammatory activity of *B. frereana* extracts was assessed against IL-1α and on costatin-induced cartilage degradation using *in vitro* model for a period of 28 days. *B. frereana* extracts significantly inhibited the breakdown of cartilage and suppressed the expression levels of mRNA of reduced MMP13 and MMP9. Further, the extracts altered the production of proinflammatory markers cycloxygenase-2 and prostaglandin E2 (Blain et al., 2010).

6.3.5.2 ANTICANCER ACTIVITY

Anticancer efficacy of *B. frereana* extract was investigated against BT549 and MDA-MB-231 breast cancer cell lines. Results demonstrate that *B. frereana* extract significantly reduced the migration of Hepatocyte growth factor and strongly inhibited c-Met receptor tyrosine kinase phosphorylation within the tested cancer cell lines (Parr and Ali, 2018). Michael et al. (2019) evaluated the phytochemical composition and anticancer potential of *B. frereana* gum resin. The extract strongly inhibited MDA-MB-231 breast cancer cell lines under *in vitro* conditions (IC_{50} value 16 μg/ml).

6.3.5.3 ANTIMICROBIAL ACTIVITY

Antimicrobial effect of 20 essential oils obtained from six *Boswellia* species (including *B. frereana*) collected from different areas were evaluated for its antimicrobial activity against four bacterial species and one yeast. It revealed that of the tested organisms, *B. cereus* was strongly inhibited by *B. frereana* essential oil (Van Vuuren et al., 2010). Bactericidal activity of hexane, acetone, and methanol crude extracts obtained from oleo gum resins of *Boswellia frereana, Commiphora myrrha* and *Boswellia carterii* were evaluated against *Streptococcus mutans* and *Lactobacillus* strains. Of the tested three plants methanolic extract of *B. frereana* exhibited strong antibacterial activity against the tested pathogens (Barre et al., 2014). Antibacterial and antifungal effect of three *Boswellia* species, including *B. frereana*, collected from Luban, were evaluated against 11 bacterial species and two fungal strains. It revealed that *B. frereana* collected from Somalia exhibited very less activity against the tested pathogens (Hasson et al., 2011).

6.3.6 *BOSWELLIA NEGLECTA*

6.3.6.1 ANTIBACTERIAL ACTIVITY

Antibacterial effects of essential oil isolated from *B. neglecta* was studied against *S. aureus* and *E. coli* using *in vitro* methods. The oil exhibited very strong bactericidal activity against the tested pathogens (Getachew et al., 2013).

6.3.6.2 INSECTICIDAL ACTIVITY

The biocidal efficacy of essential oil isolated from *B. neglecta* was studied against cockroaches, termites, and tics using *in vitro* methods. The essential oil at 10% concentration showed 100% inhibition at different time intervals against termites, cockroaches, and mites (Getachew et al., 2013).

6.3.6.3 ANTICANCER ACTIVITY

Phytochemical composition and anticancer potential of *B. neglecta* gum resin was evaluated (Michael et al., 2019). The extract strongly inhibited

MDA-MB-231 breast cancer cell lines under *in vitro* conditions (IC_{50} value 9.0 μg/ml).

6.3.7 *BOSWELLIA OVALIFOLIOLATA*

6.3.7.1 ANALGESIC ACTIVITY

Analgesic activity of chloroform and methanol extracts of root of *B. ovalifoliolata* was investigated by hot plate nociception method in rat models (Boonsri and Eswaraiah, 2014). Among the test extracts chloroform extract exhibited significant analgesic effect dose-dependently. While analgesic activity of petroleum ether extract of *B. ovalifoliolata* leaves was evaluated by Kumar et al. (2012). The extract significantly inhibited the acetic acid-induced writhing conditions and delayed the tail withdrawal response in dose-dependently.

6.3.7.2 ANTIMICROBIAL ACTIVITY

Antibacterial effect of ovalifoliolatin A, acerogenin C, 3a-hydroxyurs-12-ene and sitost-4-en-3-one isolated from stems of *B. ovalifoliolata* was studied by Reddy et al. (2003) against Gram-negative bacteria, i.e., *Chromobacterium violaceum*, *K. aerogenes*, *P. aeruginosa* and *P. fluorescens* and Gram-positive bacteria *S. aureus*, *B. subtilis* and *B. sphaericus* using disc diffusion method. Ovalifoliolatin A, inhibited strongly *S. aureus* and *C. violaceum* whereas moderate effect was observed against *K. aerogenes*, *B. subtilis*, *P. aeruginosa*, and *B. sphaericus*. In another study conducted by Anitha and Sudarsanam (2012), ethanolic extract of stem bark, chloroform extract of leaf and chloroform, hexane extracts of stem strongly inhibited *Xanthomonas citri*. While, hexane extract of stem, water extract of stem bark and chloroform extract of leaf exerted strong antimicrobial activity against *Salmonella typhimurium.* The methanolic extracts obtained from stem bark and gum resin of *B. ovalifoliolata* exhibited concentration-dependent antibacterial effect against *B. subtilis, S. aureus, P. vulgaris* and *P. aeruginosa.* Extract at higher concentrations exhibited a broad spectrum of antibacterial activity against *P. vulgaris* (Paul and Basha, 2015).

Antibacterial activity of methanolic extracts of *B. ovalifoliolata* stem showed a broad spectrum of antibacterial effect against *P. aeruginosa,* while moderate activity was exhibited against *B. subtilis* (Ramireddy et al., 2017). Further, silver nanoparticles synthesized from *B. ovalifoliolata* stem bark exhibited strong antibacterial effect against *Proteus* species and moderate activity against *E. coli* and *P. aeruginosa* (Savithramma et al., 2011). While silver nanoparticles synthesized from *B. ovalifoliolata* gum resin showed a broad spectrum of antibacterial effect against *Klebsiella* followed by *Bacillus, E. coli* and *P. vulgaris* and lowest activity was observed against *Pseudomonas* species (Savithramma et al., 2014).

Antimicrobial activity of three essential oils isolated from *B. ovalifoliolata* leaf, stem bark, and gum resin were assessed against two Gram-positive, two Gram-negative bacterial and two fungal strains using diffusion method (Prasanna et al., 2015). In results leaf essential oil strongly inhibited *C. albicans* and *C. rugosa* (21.83±1.04 mm and 18.0 ±1.00 mm), *Pseudomonas* and *C. rugosa* were found to be highly sensitive to stem bark essential oil (14.83±0.58 mm and 14.33 ±0.58 mm), whereas gum resin oil strongly suppressed the growth of *E. coli* and *C. rugosa* (15.83±0.29 mm and 15.00 ±1.00 mm). Antimicrobial activity of petroleum ether, EA, and ethanol extracts of *B. ovalifoliolata* leaves against *B. cereus, S. aureus, E. coli, P. aeruginosa, K. pneumoniae,* and *C. albicans* was assessed using disc diffusion method. Except petroleum ether extract, all the test extracts exhibited antimicrobial activity at higher concentrations. Among the tested pathogens *S. aureus* was strongly inhibited by ethanol extract (Venkata Ratnam and Venkata Raju, 2008).

6.3.7.3 ANTI-HYPERLIPIDEMIC ACTIVITY

Anti-hyperlipidemic activity of methanol extract prepared from *B. ovalifoliolata* stem bark in high-fat diet-fed rats was investigated. The extract at 200 mg/kg body weight significantly reduced high-fat diet-induced higher levels of lipids, i.e., LDL-C, VLDL-C, TC, and TG and increased good cholesterol, HDL levels in rat models (Saritha et al., 2014). A similar study was conducted by Kumar et al. (2014). They concluded that methanolic extract of *B. ovalifoliolata* significantly reduced the levels of TC, LDL, VLDL, TG, HDL+VLDL, VLDL+LDL, LDL/TC, AI, SGOT, SGPT, and elevated levels of HDL, HDL/TC in a dose-dependent manner. Further, ethanolic extract of *B. ovalifoliolata* gum resin significantly reduced the levels of AST, ALT, and ALP in atherogenic diet induced rats (Geetha and Ganapaty, 2013a).

6.3.7.4 ANTI-INFLAMMATORY STUDIES

Anti-inflammatory activity of three new terpenoids, i.e., (1R,6S,9S,10R)-2,2,10-trimethyl-7-methylenedecahydro-10H-benzo[4,5]cyclo hepta[9,10] oxirene (1), 6R,7R-epoxy-1-oleanen-3-ol, 6R, 7R-epoxy-1-oleanen-3-one along with known seven components, i.e., 1β,3β,11α-trihydroxy-urs-12-ene, 3α-hydroxy-tirucall-8,24-dien-21-oic acid, 3α-acetoxytirucall-8,24-dien-21-oic acid, serratol, neoilexonol, 3-hydroxy-urs-9,11-dien-24-oic acid and α-amyrin isolated from *B. ovalifoliolata* oleo gum resin was investigated in murine neutrophils (Renu et al., 2014). Among the tested components, 6R, 7R-epoxy-1-oleanen-3-ol, and 3α-hydroxy-tirucall-8,24-dien-21-oic acid significantly attenuated LPS induced activation of inflammatory markers TNF-α, interleukin (IL-6 and IL-8) and nitric oxide (NO) production in murine neutrophils. While anti-inflammatory potential of methanolic extract of *B. ovalifoliolata* leaves was assessed against carrageenan, histamine, and egg white-induced inflammation in rat paw edema (Selvan et al., 2013). The extract at 200 mg/kg body weight exhibited 41.26% inhibition to edema induced by carrageenan, whereas histamine-induced inflammation was reduced dose-dependently.

6.3.7.5 ANTIOXIDANT ASSAY

In vitro antioxidant capacity of ethanol extract of *B. ovalifoliolata* oleo gum resin was investigated using DPPH, NO, and lipid peroxidation (LPO) assays. The extract exhibited concentration-dependent antioxidant activity in tested models (Geetha and Ganapaty, 2013b). Antioxidant potential of essential oils isolated from *B. ovalifoliolata* leaf, stem bark and gum resin were assessed using DPPH quenching assay. Three essential oils and standard compound ascorbic acid exhibited dose-dependent DPPH reducing activity. Leaf essential oil strongly inhibited (99%) over other oil samples and ascorbic acid indicates its significant antioxidant potentiality (Prasanna et al., 2015). Antioxidant activity of hexane, EA, and ethanol extracts of *B. ovalifoliolata* stem bark was assessed using DPPH and ABTS method. Of the tested extracts EA and ethanol expressed concentration dependent DPPH quenching and ABTS reducing activity (Mahesh et al., 2014). Oral administration of aqueous extract of *B. ovalifoliolata* stem bark at a dose of 200 mg/kg body weight/day/for a period of 40 days significantly decreased hepatic and renal thiobarbituric acid reactive substances (TBARS). The extract restored the activity of antioxidant enzymes, i.e., SOD, glutathione (GSH) peroxidase (GPx) and glutathione-s-transferase (GST) in liver and kidney of STZ-induced diabetic rats when compared with untreated diabetic rats (Prabhakar et al., 2013).

6.3.7.6 HEPATOPROTECTIVE STUDIES

Hepatoprotective effect of ethanol extract of *B. ovalifoliolata* was assessed in paracetamol-induced hepatotoxicity in rat models. The extract significantly reduced the paracetamol-induced hepatic LPO and marked depletion in reduced GSH and increased the serum levels of SGPT, SGOT, and LDH at a dose of 400 mg/kg when compared to the paracetamol treated group (Mahesh et al., 2014).

6.3.7.7 ANTI-ADIPOGENESIS ACTIVITY

Ethanolic extract of *B. ovalifoliolata* gum resin and leaves exhibited concentration-dependent cytotoxic effect in 3T3-L1 cells. The ethanolic extract of gum resin significantly reduced the lipid accumulation with 46.1% over control at 100 μg/mL concentration. Whereas the ethanolic extract of leaves did not show significant inhibition of fat accumulation at the tested concentration (150 μg/mL) in 3T3-L1 cells (Geetha et al., 2014).

6.3.7.8 ANTI-ULCER ACTIVITY

Methanolic extract of *B. ovalifoliolata* leaves at 400 mg/kg body weight significantly reduced aspirin-induced gastric ulceration changes, i.e., ulcer index, pH, and acidity when compared with ulcer control group and standard group of albino rats (Dhanalakshmi and Maddilety, 2012). Similarly, the anti-ulcer activity of aqueous extract of *B. ovalifoliolata* gum resin indomethacin, aspirin, and ethanol-induced ulcers in animal models. The water extract significantly reduced ulcers in a dose-dependent manner in the tested animal models (Venkateswarlu et al., 2012).

6.3.7.9 LARVICIDAL STUDY

The mosquito larvicidal potency of essential oil isolated from *B. ovalifoliolata* leaves was investigated against *Anopheles stephensi*, *A. subpictus*,

Aedes aegypti, *Culex quinquefasciatus*, and *C. tritaeniorhynchus*. The essential oil strongly inhibited larvae of *An. stephensi*, *Ae. aegypti*, *C. quinquefasciatus*, *An. subpictus*, *Ae. albopictus*, and *C. tritaeniorhynchus* with LC_{50} values below 100 μg/ml (Benelli et al., 2018).

6.3.7.10 CARDIO PROTECTIVE ACTIVITY

The cardioprotective effect of ethanolic extract of *B. ovalifoliolata* bark and leaf was evaluated against doxorubicin (DOX)-induced cardiotoxicity in mice (Mahesh et al., 2014). Bark extract at 400 mg/kg and leaf extract 250 mg/kg concentrations significantly attenuated DOX-provoked GSH depletion and accumulation of malonaldehyde. Further, the extracts modulated low levels of antioxidant enzymes, i.e., catalase, SOD, and GPx to normal stage in cardiac tissue. Both extracts significantly restored the levels of DOX-induced rise of SGPT, SGOT, serum lactate dehydrogenase, and creatine kinase-MB levels.

6.3.7.11 ANTI-CANCER ACTIVITY

Anticancer potential of ethanolic extract prepared from *B. ovalifoliolata* leaves against triple-negative breast cancer (TNBC) cells, i.e., MDA-MB-231 and MDA-MB-453. Western blot analysis demonstrated that the ethanol extract significantly decreased the enzymes involved in pathogenesis of cancer, such as Phospho-NF-κB (ser536), PCNA, anti-apoptotic protein Bcl-2 expression and increased the expression of pro-apoptotic protein Bax, in MDA-MB-231 and MDA-MB-453 cell lines (Dinesh et al., 2014).

6.3.8 *BOSWELLIA PAPYRIFERA*

6.3.8.1 ANTIMICROBIAL ACTIVITY

A preliminary antibacterial study of methanol, EA, petroleum ether and water extracts of *B. papyrifera* and *Commiphora molmol* was evaluated on *B. subtilis, S. aureus, P. aeruginosa*, and *P. vulgaris* and seven methicillin-resistant bacterial strains and *K. pneumoniae* isolated from patients using

Micro-atmosphere assay. Methanolic extracts of both the tested medicinal plants exhibited very feeble antibacterial effect against the tested pathogens (Abdallah and Khalid, 2012). Antibacterial and antifungal activity of volatile oil and solvent extracts of oleo gum resin of *B. papyrifera* collected from three different localities, was evaluated against *E. coli, S. aureus, Salmonella* spp., *B. subtilis,* and *P. aeruginosa* and fungal species such as *Aspergillus niger, A. flavus* and *C. albicans*. Essential oils of three samples exhibited higher antibacterial and antifungal activity against the tested bacterial and fungal species by exhibiting lower MIC values (5–20 μg/ml) (Mustafa et al., 2020).

6.3.8.2 ANTIOXIDANT ACTIVITY

Antiradical activity of volatile oil and solvent extracts of oleo gum resin of *B. papyrifera* collected from three different localities, was evaluated using DPPH and ion chelating methods. All the tested extracts showed dose-dependent antioxidant activity in the tested antioxidant models. The highest DPPH quenching activity and moderate ion chelating activity was exhibited by essential oils and petroleum ether extract showed strong ion chelating activity (Mustafa, 2012).

6.3.8.3 CYTOTOXIC ACTIVITY

Cytotoxic effects of volatile oil and solvent extracts of oleo gum resin of *B. papyrifera* collected from three different localities, evaluated using brine shrimp assay, revealed that all the tested extracts and essentials oil significantly inhibited the growth of brine shrimp larvae in dose-dependent manner (Mustafa, 2012). Phytochemical composition and anticancer potential of *B. papyrifera* gum resin was evaluated (Michael et al., 2019). The extract strongly inhibited MDA-MB-231 breast cancer cell lines under *in vitro* conditions (IC_{50} value 10.0 μg/ml).

6.3.8.4 EFFECT ON NEURODISEASES

A clinical experiment was carried out effect of *B. papyrifera* gum resin capsule in 80 patients with multiple sclerosis. The results demonstrated that capsules significantly enhanced visuospatial memory in multiple sclerosis patients. But no effect was observed on information processing speed and verbal memory (Sedighi et al., 2014).

6.3.8.5 EFFECT ON SPATIAL MEMORY IN ANIMALS

The effect of *B. papyrifera* gum resin on learning and memory-enhancing capacity in rats and mice was investigated (Amir et al., 2010). The results revealed that *B. papyrifera* gum resin significantly reduced the number of days required to learning and food finding time in both the tested animals. A similar type of experiment with ethanolic extract of *B. papyrifera* gum resin was carried out by Ali et al. (2011). The results demonstrated that ethanol extract exhibited a noteworthy reduction in escape latency and distance traveled. But no effect was observed on swimming speed.

6.3.9 *BOSWELLIA RIVAE*

6.3.9.1 ANTICANCER ACTIVITY

Phytochemical composition and anticancer potential of *B. rivae* gum resin was evaluated (Michael et al., 2019). The extract strongly inhibited MDA-MB-231 breast cancer cell lines under *in vitro* conditions (IC_{50} value 11.0 μg/ml).

6.3.10 *BOSWELLIA SACRA* (SYN.: *B. CARTERII*)

6.3.10.1 ANTIVIRAL ACTIVITY

Antiviral infective effects of chewing gum prepared from *B. sacra* (syn. *B. carterii*) along with *Saccharomyces cerevisiae*, eucalyptus oil, and zinc was evaluated by Ali and Hashim (2020) in influenza vaccinated rat models at 5 mg, 8 mg, 25 ml, and 5 μg/kg dose, given three times per day for a period of 10 days. Chewing gum significantly reduced viral infection by increasing the number of leukocytes in the blood samples of the experiment animals than negative control group animals. Antiviral activity of triterpenoids, i.e., lupeol, β-boswellic acid (βBA), 11-keto-β-boswellic acid, acetyl β-boswellic acid, acetyl 11-keto-β-boswellic acid, acetyl-α-boswellic acid, 3-oxotirucallic acid, 3-hydroxy-tirucallic acid and palmitic acid isolated from *B. sacra* gum resin was evaluated against Herpes simplex type I virus. Among all test compounds, acetyl 11-keto-β-boswellic acid, acetyl β-boswellic acid and 11-keto-β-boswellic acid exhibited strong inhibition (75%) at 20 and 80 μg/mL concentration, respectively (Badria et al., 2003a).

6.3.10.2 ANTIBACTERIAL ACTIVITY

Antibacterial effect of essential oil, methanol, and EA extracts of *B. sacra* (*B. carterii)* gum resin was evaluated against four Gram-positive bacteria, *Streptococcus faecalis*, *Bacillus subtilis*, *B. circulans*, *Listeria monocytogenes*, and two Gram-negative bacteria *Escherichia coli* and *Pseudomonas aeruginosa*. The methanol and EA extracts exhibited the highest antibacterial activity against *S. faecalis*, *B. subtilis*, and *B. circulans* with MIC value 25 μg/mL (Mohamed et al., 2015). Antibacterial activity of methanol, acetone, and hexane extracts obtained from *B. sacra* (Syn.: *B. carterii*) gum resin was evaluated against *Streptococcus mutans* and *Lactobacillus* species (Barre et al., 2014). *S. mutans* exhibited the highest inhibition zone (11±0.4 mm) to acetone extract, while *Lactobacillus* species expressed highest inhibition zone (11±0.4 mm) to hexane extract.

Antibacterial activity of Clarithromycin and *Boswellia sacra* (Syn.: *B. carterii*) gum alone and in combination was studied against five Gram-positive bacteria, i.e., *S. aureus*, *S. faecalis, B. cereus, S. epidermidis, S. saprophyticus*) and three Gram-negative bacteria, i.e., *K. pneumoniae* and *Escherichia coli* and *P. aeruginosa*. *Boswellia sacra (B. carterii)* gum alone strongly inhibited *K. pneumoniae.* All the tested organisms, except *P. aeruginosa* strongly inhibited in combination of *Boswellia sacra (B. carterii)* and Clarithromycin in both agars well diffusion and broth dilution methods (Al-Kuraishy et al., 2012). Antimicrobial efficacy of ethanol, chloroform, and water extracts of *B. sacra* (*B. carterii)*, licorice, and rosemary were carried out using disc diffusion method and minimum inhibition concentration (MIC) method against *S. aureus*, *E. coli, P. aeruginosa, K. pneumoniae* and *B. subtilis.* All the tested organisms exhibited feeble activity to *B. sacra* (El Kichaoui et al., 2017).

Antibacterial efficacy of two commercial essential oil of *B. sacra* gum resin was evaluated against *S. aureus*, *B. cereus*, *E. coli*, *P. vulgaris* and the yeast *C. albicans* using microdilution assay (Van Vuuren et al., 2010). Both the oil samples strongly inhibited Gram-positive organism, *B. cereus*, followed by *P. vulgaris*, *E. coli*, *S. aureus*, and *C. albicans*. While antimicrobial activity of essential oil and methanolic extracts prepared from *B. sacra* branches against 11 bacterial species and four fungal species (which includes 5 clinical isolates) using disc diffusion method (Abdou-Latif et al., 2012). *Enterococcus faecalis* (37 mm), *Shigella dysenteriae* (37 mm), *Salmonella enterica* (35 mm), *B. cereus* (34 mm), *Listeria innocua* (34 mm) were very sensitive to the essential oil. Other bacterial strains exhibited the inhibition zones between 19 mm and 29 mm, whereas methanol extract exhibited good antibacterial effect against *S. enterica* (34 mm) and *Listeria innocua* (30 mm). The other strains had sensitivities between 9 mm and 25 mm.

Further, antibacterial effect of four essential oils obtained from four commercial grade *B. sacra* gum resin samples was determined against *Escherichia coli*, *Pseudomonas aeruginosa*, *S. aureus*, and clinical isolates *B. subtilis*, *Micrococcus luteus*, *S. aureus*, *P. aeruginosa*, *K. pneumoniae*, *Enterobacter aerogenes*, *E. coli*, and *P. vulgaris*. The tested four oils were active both Gram-positive and Gram-negative bacteria. The clinical isolates of *B. subtilis*, *Micrococcus luteus*, *S. aureus*, *K. pneumoniae*, and *Enterobacter aerogenes* were sensitive to the tested oil samples. While *P. aeruginosa*, *E. coli*, and *P. vulgaris* were resistant to the oil samples collected from localities such as Shathari, Najdi, and Hoojri, respectively (Al-Saidi et al., 2012). Antibacterial activity of *B. sacra* gum resin ethanol and chloroform extracts showed that ethanol extract exhibited antibacterial activity against *E. coli* (Soliman et al., 2020).

6.3.10.3 ANTIFUNGAL ACTIVITY

Antifungal potential of essential oil, methanol, and EA extracts of *B. sacra* (*B. carterii*) gum resin was investigated against *Saccharomyces cerevisiae* and *Candida albicans*. *S. cerevisiae* was proved as sensitive organism to the extracts and essential oil. Methanol extract strongly inhibited *S. cerevisiae* (Mohamed et al., 2015). Antifungal activity of the aqueous extract of *B. sacra* (*B. carterii)* against *Alternaria alternata, Helminthosporium rostratum* and *Fusarium solani* using the poisoned food technique. The extract exhibited concentration-dependent antifungal activity against the tested fungal strains. The scanning electron microscope (SEM) photographs of the fungal mycelia treated with the crude aqueous extract of *B. sacra* (*B. carterii),* indicated that, it showed structural alterations in the mycelial and conidial structures compared with the untreated mycelia (Fatimah et al., 2019).

6.3.10.4 IMMUNOMODULATORY ACTIVITY

The immunomodulatory property of *B. sacra* (*B. carterii)* gum resin ethanol extract and purified BAs was studied in murine splenocytes (Chevrier et al., 2005). The resin extract and BAs significantly altered the production of TH1 cytokines (interleukin-2 [IL-2] and gamma-interferon) and TH2 cytokines (IL-4 and IL-10) in a concentration-dependent manner. They concluded that a purified mixture of BAs exhibited carrier-dependent immunomodulatory activity in *in vitro* studies. Similar *in vitro* study on the immunomodulatory effect of *B. sacra (B. carterii),* gum resin extract and isolated BAs was conducted by Badria et al. (2003b) using lymphocyte transformation assay. The results revealed that the extract was more potent than that of individual isolated component.

6.3.10.5 ANTIOXIDANT ACTIVITY

Antioxidant potential of essential oil, methanol, and EA extracts of *B. sacra* (*B. carterii*) gum resin was evaluated by DPPH quenching, ferrous ion chelating and ferric reducing power activity using *in vitro* methods. In results, methanol extract exhibited the highest antioxidant activity against DPPH radical scavenging, ferrous ion chelating and the reducing power assays compared to EA and essential oil (Mohamed et al., 2015).

6.3.10.6 ANTICANCER ACTIVITY

Anti-tumor effect of aqueous and methanolic extracts of *B. sacra* at 50, 25, 12.5, 6.25, and 3.12 mg/ml concentrations was investigated in Mice RD and lymphocyte cell line (Jasim et al., 2019). Water extract exhibited potent anti-tumor activity at 25, 12.5, 6.25, and 3.12 concentrations. Whereas methanolic extract had most inhibition activity at 6.25 mg/ml. Anti-proliferative potential of essential oil and its distillate fractions collected at different durations obtained from *B. sacra* gum resin was investigated in Human pancreatic cancer cells (Ni et al., 2012). Pre-treatment of Human pancreatic cancer cells with fractions III and IV strongly suppressed the cell viability and increased cell death. While essential oil activated the caspase-dependent apoptotic pathway, and down-regulated levels of cyclin D1 and cdk4 expression in Human pancreatic cancer cells. Cytotoxic potential of *B. sacra* gum resin ethanol and chloroform extracts demonstrated moderate activity against the tested cancer cell lines (Soliman et al., 2020). Anticancer activity of essential oil isolated from *B. sacra* against MDA-MB-231 breast cancer cell lines indicated that oil strongly reduced cell viability and increased cell death by preventing cellular network formation (Suhail et al., 2011).

Anti-proliferative activity of eight new cembrane-type diterpenoids, i.e., boscartins AH–AO and two known components boscartin Q and boscartin Z isolated from the gum resin of *B. sacra* was investigated against HCT-116 human colon cancer cell by MTT assay. All the test components exhibited weak anti-proliferative activity (IC_{50} > 100 μM), against HCT-116 human colon cancer cells (Sun et al., 2020). Anticancer effect of essential oil obtained from *B. sacra* oleo gum resin was evaluated in human bladder cancer J82 cells and normal bladder urothelial UROtsa cells. Essential oil inhibited cell viability in bladder transitional carcinoma J82 cells but failed to inhibit UROtsa cells *in vitro* studies. Results on gene expression study revealed that essential oil upregulated genes that are responsible for cell cycle arrest, cell growth suppression and apoptosis in J82 cells (Frank et al., 2009).

Around 15 triterpene acids isolated from methanol extract of the resin of *B. sacra* (*B. carterii*), along with derivative triterpene acid were evaluated for their inhibitory effects on 12-O-tetradecanoylphorbol-13-acetate (TPA) induced Epstein-Barr virus early antigen (EBV-EA) expression in Raji cells and on activation of (–)-(E)-methyl-2[(E)-hydroxyimino]-5-nitro-6-methoxy-3-hexemide (NOR 1), a nitrogen oxide (NO) donor, and cytotoxic activities against IMR-32, NB-39, and SK-N-SH human neuroblastoma cell lines, by *in vitro* methods. Of the tested 15 components, seven compounds, i.e., acetyl β-boswellic acid, lupeolic acid, acetyl lupeolic acid, elemonic acid, 3α-hydroxytirucalla-7,24-dien-21-oic acid,

3α-acetoxytirucalla-7,24-dien-21-oic acid and 3β-hydroxytirucalla-8,24-dien-21-oic acid strongly inhibited (TPA) induced EBV-EA expression in Raji cells. Five compounds, i.e., acetyl 9,11-dehydro-β-boswellic acid, elemonic acid, 3α-hydroxytirucalla-7,24-dien-21-oic acid, 3α-acetoxytirucalla-7,24-dien-21-oic acid and 3β-hydroxytirucalla-8,24-dien-21-oic acid attenuated the activation of NOR 1 in IMR-32, NB-39, and SK-N-SH human neuroblastoma cell lines. Regarding cytotoxic study, 15 compounds, i.e., β-boswellic acid, acetyl β-boswellic acid, 11-keto-β-boswellic acid, acetyl 11-keto-β-boswellic acid, acetyl 11a-methoxy-β-boswellic acid, 9,11-dehydro-β-boswellic acid, acetyl 9,11-dehydro-β-boswellic acid, acetyl α-boswellic acid, lupeolic acid, acetyl lupeolic acid, elemonic acid, 3a-hydroxytirucalla-7,24-dien-21-oic acid, 3a-acetoxytirucalla-7,24-dien-21-oic acid, incensole, and incensole acetate exhibited potent cytotoxic activities in IMR-32, NB-39, and SK-N-SH human neuroblastoma cell lines (Akihisa et al., 2006).

6.3.10.7 ANTI-INFLAMMATORY ACTIVITY

Anti-inflammatory activity of eight new cembrane-type diterpenoids, i.e., boscartins AH–AO and two known components boscartin Q and Z isolated from the gum resin of *B. sacra* (*B. carterii*), was studied against NO production in RAW264.7 mouse peritoneal macrophages by *in vitro* method. Boscartin AO significantly attenuated NO production in RAW264.7 mouse peritoneal macrophages, while other components expressed weak anti-inflammatory activity (Sun et al., 2020). NO production inhibitory activity of three new monoterpenes (olibanumol A, B, C), three triterpenes (olibanumol H, I, J) and isofouquierol was investigated against LPS challenged mouse peritoneal macrophages. Among the test constituents, olibanumols A, H, I, and isofouquierol significantly reduced NO production levels in LPS challenged mouse peritoneal macrophages (Yoshikawa et al., 2009).

NO inhibitory activity of three new terpenes, classified as cembrane-type diterpenoid (rel)-(1S,5R,7E,11E)-1-isopropyl-8,12-dimethyl-4-methylenecyclotetradeca-7,11-diene-1,5-diol, dinortirucallane-type triterpene 20,21-dinortirucalla-8,24-diene-3β-ol-7-one, ursane-type triterpene 21β-hydroxy-3-O-acetyl-11-keto-boswellic acid along with four known triterpenes, i.e., 3-oxo-tirucalla-8, 24-dien-21-oic acid,

3β-hydroxytirucalla-8, 24-dien-21-oic acid, 3α-hydroxytirucalla-8, 24-dien-21-oic acid and 3-O-acetyl-11-keto-boswellic acids isolated from the gum resin of *B. sacra* (*B. carterii*), was evaluated in LPS activated RAW264.7 cells using *in vitro* methods. Among the tested components, (rel)-(1S,5R,7E,11E)-1-isopropyl-8,12-dimethyl-4-methylene-cyclotetradeca-7,11-diene-1,5-diol (IC_{50} value 1.32 μM), 3-oxo-tirucalla-8, 24-dien-21-oic acid (IC_{50} value 3.04 μM), 3β-hydroxytirucalla-8, 24-dien-21-oic acid (IC_{50} value 1.42 μM) and 3-O-acetyl-11-keto-boswellic acid (IC_{50} value 3.25 μM) significantly reduced the LPS induced NO production in RAW264.7 cells (Yu et al., 2018). Anti-inflammatory activity of Boscartins AP to AW isolated from gum resin of *B. carterii,* demonstrated that boscartin AP significantly reduced the LPS induced higher levels of inflammatory markers in RAW 264.7 macrophages. The other components showed moderate activity (Yu et al., 2020).

6.3.10.8 HEPATOPROTECTIVE STUDIES

Hepatoprotective activity of eight new cembrane-type diterpenoids, i.e., boscartins AH–AO and two known components boscartin Q and Z isolated from the gum resin of *B. sacra* (*B. carterii*), was investigated against paracetamol-induced cytotoxicity in HepG2 cells. Boscortin AH, exhibited beneficial effects against paracetamol-induced cytotoxicity in HepG2 cells than positive control bicyclol at 10 μM concentration (Sun et al., 2020). Hepatoprotective activity of Boscartins AP to AW isolated from gum resin of *B. sacra* (*B. carterii*), revealed that boscartin AP significantly altered the APAP-induced changes in HepG2 cell lines (Yu et al., 2020).

6.3.10.9 ANTI-ULCERATIVE COLITIS ACTIVITY

Anti-ulcerative colitis activity of eight new cembrenoids, boscartins A-H and incensole oxide, isolated from the gum resin of *B. sacra* (*B. carterii*), was evaluated by using *in vitro* x-box-binding protein 1 (XBP 1) transcriptional assay using dual luciferase reported detection. Among the test components, boscartin A, D, E, and F significantly activated XBP1 transcription (Ren et al., 2015).

6.3.11 *BOSWELLIA SERRATA*

Abdel-Tawab et al. (2011) in their review on *Boswellia serrata* gave an overall assessment of *in vitro*, preclinical, pharmacokinetic, and clinical data. Safayhi et al. (1992) from their investigation reported that BAs are specific, nonreducing-type inhibitors of the 5-lipoxygenase product formation either interacting directly with the 5-lipoxygenase or blocking its translocation. Inhibition by BAs of human leukocyte elastase was investigated by Safayhi et al. (1997).

6.3.11.1 ANTIMICROBIAL ACTIVITY

Polar and nonpolar solvent extracts, volatile oil and BAs isolated from oleogum resin of *B. serrata* reported to possess potent antimicrobial activity against both bacterial and fungal species. Patel and Patel (2015) reported antibacterial activity of aqueous, acetone, methanol, and petroleum ether extracts of *B. serrata* gum against three gram-positive bacteria (*S. albus, S. aureus,* and *S. citreus*) causing urinary tract infection using disc diffusion method. The results revealed that, aqueous extract strongly inhibited *S. aureus* and *S. citreus* by expressing lower MIC value (12.5 μg/ml), while aqueous extract strongly (MIC value, 12.5 μg/ml), inhibited *S. albus, S. aureus,* and *S. citreus*. A comparative antibacterial activity of ethanol, DMSO, acetone, and chloroform extracts of *B. serrata* gum extracts against two gram-positive and gram-negative bacteria, was reported by Gaurea and Bapat (2016). Among the test extracts, ethanol extract exhibited maximum zone of inhibition against *B. cereus*. Acetyl-11-keto-β-boswellic acid (AKBA), is a potent pentacyclic triterpene found in *B. serrata* gum resin. Essential oil isolated in hydro distillation and steam distillation methods from *B. serrata* oleo gum resin exhibited potential antimicrobial activity of against the tested four bacterial and four fungal species than essential oil extracted in supercritical fluid extraction (SFE) method (Ayub et al., 2018).

Raja et al. (2011a) demonstrated the antibacterial activity of acetyl-11-keto-β-boswellic acid (AKBA) against methicillin-resistant *Staphylococcus aureus* (MRSA) and vancomycin-resistant *Enterococcus faecium* (VRE) and vancomycin-resistant *S. aureus* isolates (VRSA) using time-kill studies, post-antibiotic effect (PAE) and biofilm susceptibility assay. The results revealed that AKBA was found to be the most active compound showing an MIC range of 2–8 μg/ml against the entire gram-positive bacterial pathogens tested and inhibited the formation of biofilms generated by *S. aureus* and *S. epidermidis*. Antibacterial effect of AKBA was evaluated using time-kill studies, mutation prevention frequency, PAE, and biofilm susceptibility assays against oral cavity pathogens *Streptococcus mutans* and *Actinomyces viscosus* (Raja et al., 2011b). *B. serrata* essential oil demonstrated significant antimicrobial activity against *Trichophyton* species, *S. epidermidis* biofilm, and its combination with azoles has shown synergistic activity against azole-resistant strain of *C. albicans* (Sadhasivam et al., 2016). The essential oil exhibited feeble antifungal activity against human pathogens, and highly effective against plant pathogens like *Phytophthora parasitica* (Garg, 1974). Antifungal activity of water, ethanol, and chloroform extracts of *B. serrata* gum resin was investigated against plant pathogenic fungus (red rot disease-causing agent) *Colletotrichum falcatum*. It concluded that the ethanolic extract was much more efficient than chloroform extract (Chaurasia and Gharia, 2017).

Antimicrobial efficiency of different polar solvents obtained from fruits, leaves, and bark parts of *B. serrata* displayed good antimicrobial activity against tested both Gram-positive and negative bacterial strains (Sharma et al., 2016). Further, the antimicrobial activity of petroleum ether, methanol, and n-hexane extracts of *B. serrata* was evaluated by Bhutada et al. (2019), revealed that methanolic extracts exhibited a wide range of antimicrobial activity than the other two extracts.

6.3.11.2 ANTI-INFLAMMATORY ACTIVITY

Inflammation is a complex biological response of the body tissue to harmful pathogens or injury. Pro-inflammatory cytokines such as TNFα, IL-1β and IL-6 activate inflammatory cells and induce production of other inflammatory mediators, which in turn modulate important cellular events, including gene expression, DNA damage and cellular proliferation contributing to various inflammatory disorders. Therefore, cellular modulation of the production of TNFα, IL-1β and IL-6 is important for treating inflammatory diseases (Langmead and Rampton, 2006; Shehata et al., 2011). *B. serrata* gum resin traditionally used in

Indian traditional systems (Ayurvedic/ethnomedicine/Unani) to treat various types of blood disorders, inflammatory health ailments, pain, and cardiac debility, rheumatism, urinary disorders, and corneal ulcer (Paranjpe, 2001). Siddiqui (2011) gave an overview of anti-inflammatory activity of *B. serrata*. Anti-inflammatory activity of *B. serrata* gum dry extract and hydroenzymatic extract and pure BAs characterized by HPLC, on LPS-challenged porcine aortic endothelial cells (pAECs) by *in vitro* methods (Martina et al., 2018). The results revealed that both extracts and BAs expressed dose-dependent anti-inflammatory activity in pAECs. Anti-inflammatory potential of alcoholic extract of *B. serrata* gum resin have been tested against carrageen induced paw edema in rats. The extract administered orally/intera-peritoneal exhibited anti-inflammatory activity in a dose-dependent manner (Atal et al., 1980, 1981).

Crude methanol extract and isolated pure compound 12-ursene 2-diketone of *B. serrata* gum, downregulated the key inflammatory markers, TNF-α, IL-1β and IL-6 in LPS challenged human peripheral blood mononuclear cells (PBMCs). They also inhibited NO production in LPS activated mouse macrophages (RAW 264.7), possibly via suppression of inducible NO synthase mRNA expression (Gayathri et al., 2007). Incensole acetate, a novel diterpenoid isolated from *B. serrata* gum, inhibited nuclear factor-kappa B (NF-κB) activation through tumor necrosis factor (TNF) α-stimulated HeLa cells (Moussaieff et al., 2007). BAs isolated from *B. serrata* gum resin, inhibited the leukotriene synthesis via 5-lipoxygenase, but did not affect the 12-lipoxygenase and the cyclooxygenase activities. Additionally, BAs did not impair the peroxidation of arachidonic acid by iron and ascorbate. It indicates that BAs are to be specific, non-redox inhibitors of leukotriene synthesis either interacting directly with 5-lipoxygenase or blocking its translocation (Ammon et al., 1993). A combination therapy with 3-acetyl-11-keto-β-boswellic acid (AKBA), a potent anti-inflammatory drug and celecoxib, a selective cyclooxygenase-2 (COX-2) inhibitor, reversed the behavioral and molecular changes caused by LPS induced cognitive dysfunction in mice model (Sayed and Sayed, 2016). Two molecules such as methyl boswellate and methyl-11-keto boswellate synthesized, significantly reduced carrageenan-induced paw edema in rat models (Shenvi et al., 2015). Ammon (2002, 2006, 2010, 2016) reviewed the role of boswellic acids in chronic inflammatory diseases.

6.3.11.3 ANTIOXIDANT ACTIVITY

Antioxidant capacity of aqueous and alcoholic extracts of *B. serrata* gum resin was studied by Kokkiripati et al. (2011) using human monocytic (THP-1) cells and rat liver tissue homogenates by *in vitro* methods. Aqueous and alcoholic extracts inhibited ferric chloride-induced LPO in liver (80%) and heart (50%) tissue homogenates of male Wistar rats. Further, both extracts strongly (>60%) inhibited H_2O_2-mediated reactive oxygen species (ROS) generation in THP-1 cells. Water extract significantly enhanced two-fold catalase activity in H_2O_2 treated THP-1 cells, while alcohol extract showed only 40% activity. Alcohol extract sustained cellular reducing power by 80% and water extract by 70% in H_2O_2-treated THP-1 cells. A comparative study on antioxidant capacity of essential oils isolated using hydro distillation, steam distillation and supercritical carbon dioxide methods (Ayub et al., 2018) revealed that, essential oil isolated in supercritical carbon dioxide exhibited better antioxidant and free radical scavenging activity.

6.3.11.4 ANTIPLATELET AGGREGATION ASSAY

B. serrata gum resin alcoholic and water extracts exhibited very high activity in prolonging coagulation time periods of prothrombin time (PT) from ~13 to >60 s and activated partial thromboplastin time (APTT) from ~32 s to >90 s in human plasma samples. While alcohol extract strongly inhibited clotting factors Xa and Xia whereas water extract did not show much effect on FXa and showed 30% inhibition on FXIa (Kokkiripati et al., 2011).

6.3.11.5 ANTI-ATHEROSCLEROTIC STUDIES

Atherosclerosis is a blood vascular disease that occurs due to the formation of plaque inside the blood vessels leading to the thickening of the arteries. Anti-atherosclerotic effect of ethanol extract of *B. serrata* gum resin was investigated on rabbits fed on high lipid diet for three months. The results, monitored at weekly intervals, revealed that rabbits treated with ethanol extracts of *Boswellia* gum significantly reduced serum cholesterol and triglyceride levels by 32–46% and 53–62%, respectively (Atal et al., 1980, 1981). BHUx, is a patented polyherbal formulation consisting of water-soluble extract of five medicinal plants *Commiphora mukul* (Burseraceae), *Terminalia arjuna* (Combretaceae), *B. serrata* (Burseraceae), *Semecarpus anacardium* (Anacardiaceae) and *Strychnos nux-vomica* (Loganiaceae), which are time tested and have been used in the ayurvedic system of medicine to treat oxidative stress and inflammation for many centuries (Pandey and Chunekar, 1967). *In vivo* experiment conducted by Tripathi et al. (2005) BHUx remarkably prevented the formation of thickened intima loaded heavily with fatty vacuoles in the rabbits fed with the atherogenic diet. Several mechanisms can explain this effect. It may be due to the reduced formation of foam cells or induction of their apoptosis, or both. AKBA exhibited potent antiatherogenic effects in LPS-challenged apolipoprotein E-deficient (ApoE+/-) mice, by inhibiting the NF-kB, a key element for the up-regulation and expression of pro-inflammatory cytokines and genes in inflammatory conditions (Cuaz-Perolin et al., 2008). The essential oil of *B. serrata* displayed significant antifungal activity against fluconazole-resistant and susceptible isolates of *C. albicans* isolates *in vitro* (Rasoul et al., 2006).

6.3.11.6 ANTI-ARTHRITIC ACTIVITY

In India, *B. serrata* gum has been used for centuries as potential traditional medicine in the Ayurvedic system of medicine for the treatment of rheumatoid arthritis conditions (Henderson and Panush, 1999). Kumar et al. (2019) reported the effect of *B. serrata* gum extract on CFA-induced arthritis in rat models. It revealed that the extract significantly improved the body weight and reduced ankle diameter and arthritic index in rats. However, there was an insignificant change in paw volume and TNF-α levels in blood plasma. Histopathological results exhibited a reduction in inflammatory parameters. The anti-arthritic activity of *B. serrata* gum resin on the mycobacterial adjuvant-induced poly-arthritis in rat models was studied by Atal et al. (1980, 1981). The results revealed that *B. serrata* gum showed 34% and 49% inhibition of paw swelling with 50 and 100 mg per kg doses, respectively, as compared to controls. Oral administration of commercial BAs at different concentrations significantly reduced the population of leucocytes in a bovine serum albumin-induced arthritis in rabbit model. BAs showed its action by reducing the infiltration of leucocytes into the knee joint and the infiltration of leucocytes into the pleural cavity (Sharma et al., 1989). Etzel (1996) reported that a special extract of gum resin of *B. serrata* is effective in the treatment of rheumatoid arthritis. These findings were obtained in more than 260 patients by using a range of different clinical approaches for evaluation. The criteria for assessment were mainly joint swelling, pain, erythrocyte sedimentation rate (ESR), stiffness, additional use of NSAID, side effects and tolerance.

B. serrata gum resin extract significantly reduced inflammatory markers IL-1β, IL-6, TNF-α, IL-10, IFN-γ and PGE2 up-regulation in collagen-induced arthritic male Wistar rats by modulating the antioxidant biochemical markers, i.e., articular elastase, GSH, LPO, myeloperoxidase (MOP), catalase, SOD, and NO. The protective effect of *B. serrata* gum resin extract on collagen-induced arthritic male Wistar rats was evident from the decrease in arthritis scoring and bone histology (Umar et al., 2014). Sengupta et al. (2011) conducted a comparative anti-inflammatory and anti-arthritic effect of *Boswellia* products, BE-30 (5-Loxin (®)) and Aflapin, a novel compound isolated from non-volatile oil portion of *B. serrata* gum resin, in Freund's Complete Adjuvant (FCA)-induced inflammation in Sprague-Dawley rats. In comparison with BE-30, Aflapin provides more potential benefits in recovering articular cartilage damage or protection from proteolytic degradation due to inflammatory insult in arthritis such as osteoarthritis or rheumatoid arthritis. A comparative study on the effect of *Boswellia* formulation and curcuminoids on osteoarthritis patients. It revealed that both drugs effectively reduced pain and functional improvement in patients than NSAIDs (Bannuru et al., 2018). The effect of herbal mixture containing *Boswellia* spp., *Uncaria tomentosa* and *Lepidium meyenii* along with L-leucine on IL-1β induced osteoarthritis in chondrocytes. It significantly reduced the gene expression levels of inflammatory markers (Akhtar et al., 2011). Further anti-arthritic efficiency of Ayurvedic and other polyherbal formulations of *B. serrata* was evaluated by using different models (Sumantran et al., 2011; Chopra et al., 2013; Natarajan and Majeed, 2012) and standardized *Boswellia* extracts rich in BAs (Majeed et al., 2016a, b, 2019).

Clinical evaluation of *B. serrata* resin in the management of osteoarthritis was carried out by Gupta et al. (2011). After a course of therapy for 2 months, symptomatic improvement was observed at various levels with promising results in the patients. Kimmatkar et al. (2003) investigated the efficacy and tolerability of *B. serrata* extract (BSE) in treatment of osteoarthritis of the knee. They reported that the extract was well tolerated by the subjects. Majeed et al. (2021) in a recent investigation reported that BSE containing 30% 3-acetyl-11-keto-boswellic acid attenuates inflammatory mediators and preserves extracellular matrix in collagen-induced arthritis. Sander et al. (1998) from a double-blind pilot study reported that H15, the resin extract of *B. serrata*, "incense" is a useful supplement to established drug therapy of chronic polyarthritis.

6.3.11.7 ANTI-DIABETIC STUDIES

B. serrata gum resin, its chemical constituents and herbal formulations have been well recognized for its beneficial effects in a large number of diseases, including diabetes mellitus. Multiple low dose-streptozotocin (MLD-STZ) induced diabetes in animals is similar to Type 1 diabetes in humans. MLD-STZ induced diabetic mice increased levels of inflammatory factors like, granulocyte colony-stimulating factor (G-CSF), granulocyte/macrophage colony-stimulating factor (GM-CSF) and pro-inflammatory cytokines (IL-1A, IL-1B, IL-2, IL-6, IFN-γ, TNF-α) in the blood were significantly reduced to normal levels in animals treated with *B. serrata* gum extract (Shehata et al., 2011). Similarly, MLD-STZ induced diabetic mice treated with 11-keto-β-boswellic acid (KBA) and O-acetyl-11-keto-β-boswellic acid (AKBA) strongly inhibited upregulation of pro and anti-inflammatory cytokines (IFN-γ, IL-1A, IL-1B, IL-2, TNF-α, IL-4, IL-10) in the blood. No infiltration of lymphocytes into pancreatic islets and appearance of peri-insular cells were detected (Shehata et al., 2015). Administration of mixture of plant extract comprising of *Nigella sativa*, Myrrh, gum olibanum, gum Asafetida and Aloe, in streptozotocin-induced diabetic rats, significantly inhibited hepatic gluconeogenesis in isolated hepatocytes and activity of key gluconeogenic enzymes, pyruvate carboxylase (PC) and phosphoenol-pyruvate carboxykinase (PEPCK) (Awadi et al., 1991).

Aqueous extract of *B. serrata* gum at 200 and 400 mg/kg dose significantly reduced streptozotocin-induced symptoms, i.e., necrosis of hepatocytes, anarchism of liver plates and lymphocytic inflammation in diabetic female rats. But, the same extract at the dose of 600 mg/kg did not show a protective effect on the liver and also cause cell death, inflammation, and vasodilatation in sinuosoids space (Azemi et al., 2012). Ahangarpour et al. (2013) conducted a clinical study in 60 diabetic patients of male and female between 30 and 48 age group, revealed that, oral administration of *B. serrata* gum extract at the dose of 300 mg for three times a day for a period of six weeks, significantly reduced blood glucose levels and increased plasma insulin levels. Insulin resistance index did not change in the test group. They conducted the similar type of clinical experiment with 900 mg/day of *Boswellia* extract for a period of six weeks, significantly reduced fructosamine (fructosamine test is used to measure overall blood glucose levels) levels in blood samples of type-2 diabetic patients (Ahangarpour et al., 2014). A comparative study on antidiabetic effect of *B. serrata* gum resin (250 mg and 500 mg) and placebo twice daily for 8 weeks in addition to their routine antidiabetic treatments, revealed that *B. serrata* gum resin (250 mg) treated group significantly reduced fasting blood sugar, glycosylated hemoglobin and triglyceride levels in 56 diabetic patients. *B. serrata* gum resin with a daily dose of 500 mg had no better glucose and lipid-lowering effect than placebo in diabetic patients (Saeed et al., 2018).

6.3.11.8 ANTI-ASTHMATIC STUDIES

Asthma is a chronic inflammatory disorder of the respiratory airways, characterized by increased mucus production and airway hyper-responsiveness resulting in decreased airflow, and marked by recurrent episodes of wheezing, coughing, and shortness of breath (Miller, 2001). *B. serrata* gum resin and its chemical constituent's BAs have been reported to inhibit leukotriene biosynthesis, which are naturally produced eicosanoid lipid mediators, which may be responsible for the effects related to asthma, allergies, and inflammation (Wildfeurer et al., 1998; Ammon, 1996). Gupta et al. (1998) conducted a clinical study on the anti-asthmatic activity of *B. serrata* gum resin in a double-blind, placebo-controlled study. Around 40 asthma patients, 23 males and 17 females in the age range of 18–75 years were selected and test drug *B. serrata* gum resin 300 mg given orally thrice daily for a period of 6 weeks. Around 70% of patients showed improvement of disease as evidenced by the disappearance of physical symptoms and signs such as dyspnoea, rhonchi, number of attacks, increase in forced expiratory volume (FEV), forced vital capacity (FVC) and peak expiratory flow rate (PEFR) as well as decrease in eosinophilic count and ESR.

Alcoholic extract of *B. serrata* gum resin on oral administration strongly inhibited the antibody production and cellular responses to sheep red blood cells in mice. It significantly inhibited the infiltration of polymorphonuclear leucocytes and reduced the volume of pleural exudate in carrageenan-induced pleurisy in rats (Sharma et al., 1988). Leukotrienes and 5-hydroxy-eicosa-tetra-enoic acid (5-HETE) produced from endogenous arachidonic acid in calcium and ionophore stimulated rat peritoneal polymorphonuclear leukocytes (PMNL) elicited with glycogen were strongly suppressed by alcoholic extract of *B. serrata* gum resin. The extract strongly decreased 5-LOx mediated metabolization of exogenously added arachidonic acid to leukotrienes and 5-HETE (Ammon et al., 1991). Higher levels of inflammatory factors transforming growth factor-b1 (TGF-b1) and tumor necrosis factor (TNF), macrophages infiltrate, accumulated eosino and neutrophils in the lung and aggregation of fibroblasts in the fibrotic area, as well as high content of hydroxyproline and lipoxygenase enzyme in bleomycin-induced pulmonary fibrosis in rat models were significantly attenuated by boswellic acid extracts (Ali and Mansour, 2011). Liu et al. (2015) investigated the anti-asthmatic potential of boswellic acid in ovalbumin (OVA) and aerosol allergen challenged BALB/c mice. Boswellic acid significantly suppressed allergic airway inflammation, airway hyperresponsiveness. Further, boswellic acid significantly attenuated OVA-specific IgE and Th2 cells secreted cytokines phosphorylated signal transducer and activator of transcription-6 (pSTAT6) and GATA-binding protein-3 (GATA3) in a dose-dependent manner in mice. Yugandhar et al. (2018) evaluated the anti-asthmatic activity of *B. serrata* gum resin and *Aegle marmelos* fruit extracts in equal ratio in a double-blind and placeo-controlled clinical experiment indicated that the formulation exhibited moderate activity in the subjects.

6.3.11.9 ANTI-CANCER STUDIES

BAs isolated from *B. serrata* and *B. carterii* exhibited potential anticancer effects associated with inhibition of protein kinase B (AKT), NF-κB activation, and down-regulation of anti-apoptotic proteins in *in-vitro* studies. *In vivo* clinical experiments conducted in glioblastoma patients, BAs significantly improved clinical symptoms in terms of edema reduction, but no reduction in tumor size (Efferth and Greten, 2011). BAs exhibited positive effects on DOX-induced tumors of Ehrlich's ascites carcinoma (EAC) grown in mice. BAs significantly reduced doxorubicin-induced tumors by enhancing the antioxidant enzymes reduced GSH, SOD, and catalase activities (Ali et al., 2015). Anti-proliferative effect of acetyl-11-keto-β-boswellic acid (AKBA) and amyrin was investigated for its possible effect on leukemic cell growth, proliferation of HL-60 and CCRF-CEM cells by Hoernlein et al. (1999) and Chashoo et al. (2011). Flow cytometric analysis of propidium iodide-stained cells and DNA relaxation assay with phiX174RF DNA, indicated that, HL-60 and CCRF-CEM cells underwent apoptosis by AKBA may be due to inhibition of topoisomerase I and II in these cells.

Among the tested four BAs (βBA, 3-O-acetyl-β-boswellic acid, 11-keto-β-boswellic acid, and 3-O-acetyl-11-keto-β-boswellic acid) 3-O-acetyl-11-keto-β-boswellic acid strongly suppressed tumor formation in human leukemia HL-60 cells by inhibiting synthesis of DNA, RNA, and proteins. 3-O-acetyl-11-keto-β-boswellic acid significantly inhibited the cellular growth of HL-60 cells, but did not affect cell viability (Shao et al., 1998). Among the proteins involved in protein synthesis, proteasome, 14-3-3 and some ribosomal proteins were most interesting targets in the modulation of the cancer progression. Mass spectrometry-based chemoproteomic study revealed that β-boswellic acid significantly modulated the ribosomal proteins involved in cancer progression (Casapullo et al., 2016). Yuan et al. (2013) evaluated the comparative effect of acetyl-11-keto-β-boswellic acid (AKBA) and aspirin on human colonic adenocarcinoma growth in APC (Min/+) mouse model. AKBA significantly inhibited human colon adenocarcinoma growth, by arresting the cell cycle in G1 phase and induction of apoptosis. It effectively delayed the growth of HT-29 xenografts by modulation of epidermal growth factor receptor (EGFR) and ATM/P53 signaling pathways in the HT-29 xenografts. The activity of AKBA was more potent than that of aspirin.

External application of boswellin (BE), a methanol extract of *B. serrata* gum resin to the backs of mice markedly inhibited 12-O-tetradecanoylphorbol-13-acetate (TPA)-induced skin inflammation, epidermal proliferation, the number of epidermal cell layers and 7,12-dimethylbenz[a] anthracene (DMBA) induced tumors in mice. Oral administration of 0.2% BE in the diet to CF-1 mice for 10–24 weeks reduced the accumulation of parametrial fat under the abdomen, and inhibited azoxymethane (AOM)-induced formation of aberrant crypt foci (ACF) by 46% (Huang et al.,

2000). The molecular mechanism of the anti-tumor effect of acetyl-11-keto-β-boswellic acid (AKBA) in the human prostate tumor xenograft mice was elucidated by Pang et al. (2009). The mechanism involved that, AKBA at 10 mg/kg concentration significantly inhibited blood vessel formation in the Matrigel plug assay in mice and effectively suppressed VEGF-induced micro vessel sprouting in rat aortic ring assay *ex vivo*. It also inhibited VEGF-induced cell proliferation, chemotactic motility, and the formation of capillary-like structures from primary cultured human umbilical vascular endothelial cells in a dose-dependent manner.

Park et al. (2011) conducted *in vitro* and *in vivo* experiments to evaluate the anticancer potential of acetyl-11-keto-β-boswellic acid (AKBA) isolated from *B. serrata* gum resin, in four different pancreatic cancer cell lines and orthotopic nude mouse model of pancreatic cancer. In *in-vitro* AKBA at different concentrations suppressed NF-kB activation and by inducing apoptosis in pancreatic cancer cell lines AsPC-1, PANC-28, and MIA PaCa-2 with K-Ras and p53 mutations, and BxPC-3 with wild-type K-Ras and p53 mutation. In mice models AKBA at 100 mg/kg inhibited the metastasis by decreasing Ki-67, a biomarker of proliferation, and CD31, a biomarker of microvessel density and downregulated the expression of COX-2, MMP-9, CXCR4, and VEGF in the tumor tissue. A similar experiment was conducted by Yadav et al. (2011) on the effect of acetyl-11-keto-β-boswellic acid (AKBA) on growth and metastasis of human colorectal cancer in an orthotopic mouse model. AKBA at 50–200 mg/kg significantly inhibited the growth of CRC tumors in mice, resulting in a decrease in tumor volumes than vehicle-treated mice in a dose-dependent manner. At molecular level, AKBA significantly inhibited markers of tumor proliferation index Ki-67, the microvessel density CD31, NF-κB activation and pro-inflammatory markers like COX2, tumor survival (bcl-2, bcl-xL, IAP-1, survivin), proliferative (cyclin D1), invasive (ICAM-1, MMP-9) and angiogenic (CXCR4 and VEGF) biomarkers. Schneider and Weller (2016) investigated anticancer potential of BAs acetyl-11-keto-β-boswellic acid (AKBA), β-boswellic acid (βBA) and α-boswellic acid (αBA) on nine long-term human Glioma stem-like cells, five glioma-initiating cell lines and examined the mechanism of cell cycle progression and the mode of cell death induction anti-clonogenic properties and assessed the potential synergy with temozolomide (TMZ) or irradiation.

Mazzio et al. (2017) evaluated the anticancer potential of 3-O-acetyl-β-BA (3-OAβBA) and BSE on MDA-MB-231 breast cancer cells by *in vitro* methods. Both BSE and/or 3-OAβBA considerably downregulated the expression of cancer progression genes and upregulated the expression of tumor suppressor genes (TSGs), such as glutathione-depleting ChaC glutathione-specific γ-glutamylcyclotransferase-1 (CHAC1) and the mTOR inhibitors – sestrin 2 (SESN2) tribbles homolog 3 (TRIB3), homocysteine-inducible, endoplasmic reticulum stress-inducible, ubiquitin-like domain member 1 (HERPUD1), and cystathionine gamma-lyase (CTH). *B. serrata* gum resin extract and various analogs of BAs, including AKBA, β-boswellic acid and a novel modified boswellic acid, cyano enone of methyl boswellates (CEMB) exhibited cytostatic and apoptosis-inducing activity against glioma brain cancer cells in *in vitro* studies (Li et al., 2018; Conti et al., 2018).

In *in-vivo* studies, intratumor administration of CEMB significantly inhibited the tumor growth in C6 glioma tumor xenograft mice model (Ravanan et al., 2011). Further, inhibitory effect of 3-α-propionyloxy-β-BA (POBA) on poly (ADP-ribose) polymerase (PARP) cleavage and perturbance in mitochondrial membrane potential in cervical cancer cells was evaluated by Qurishi et al. (2012). The apoptotic and antiproliferative effects of AKBA in colon cancer cells was evaluated (Liu et al., 2006; Hostanska et al., 2002). In *in-vivo* studies, AKBA showed its chemopreventive action in APC (Min/+) mice by inhibiting Wnt/β-catenin and NF-kB/Cox-2 signaling pathways (Wang et al., 2014). While, βBA exhibited its anticancer effects via induction of apoptosis and cell cycle arrest, as well as abrogation of PI3K/Akt signaling pathway in *in vitro* model (Wang et al., 2018). AKBA alone and in combination with curcumin significantly modulated the upregulation of cancer-related genes in colorectal cancer cells (Takahashi et al., 2012; Liu et al., 2002; Shen et al., 2012). Antiproliferative effect of BAs on MCF-7 cells was assessed in *in-vitro* method, revealed that BAs notably suppressed the markers associated with proliferation of breast cancer cell lines (Saraswati & Agarwal, 2012).

Boswellic acid derivatives showed anticancer effects by inducing apoptosis through caspase-mediated pathways in myeloid leukemia cells (Jing et al., 1999; Xia et al., 2005) or by disrupting PI3K/AKT/Hsp-90 cascade, mitochondrial integrity, and other cell survival signaling events in HL-60 cells (Khan et al., 2012). Keto boswellic acid (KBA) and AKBA showed potent anti-proliferative and induced apoptosis through the caspase-8-dependent pathway in liver cancer cells (Liu et al., 2002). Antitumor potential of 11-carbonyl-β-boswellic acid was evaluated in lung cancer cells. The results showed that it activated the JNK signaling pathway, which cause the cleavage of PARP, and downregulated survivin protein expression in lung cancer cells (Huang et al., 2018). *In vitro* and *in vivo* experiment results indicate that prostate cancer cell lines treated with acetyl-boswellic acids strongly inhibited Ikappa B kinase activity and it promoted apoptosis (Syrovets et al., 2005). AKBA showed apoptotic effects in LnCaP and PC-3 prostate cancer cells by modulating caspase-3 and caspase-8 activation, as well as PARP cleavage induction (Lu et al., 2008).

In another study, AKBA strongly suppressed VEGFR2-mediated angiogenesis in prostate cancer cell lines and blocked activation of STAT3 and Akt signaling pathways in docetaxel-resistant prostate cancer cells (Pang et al., 2009; Liu et al., 2019). 3-cinnamoyl-11-keto-β-BA (C-KβBA), a semi-synthetic triterpenoid derivative, showed antiproliferative and proapoptotic activity in cancer cell lines PC-3, LnCaP, and DU-145, as well as in PC-3 prostate cancer xenografts, by downregulating the activation of p70 ribosomal S6 kinase (Glaser et al., 1999). AKBA significantly inhibited cell growth and downregulated the expressions of inflammatory genes such as Ki-67, CD31, Cox-2, MMP-9, CXCR4 and VEGF in human pancreatic tumor tissues (Park et al., 2011). Recently, in combination treatment of the anti-diabetic drug metformin and boswellic acid nanoparticles significantly inhibited the growth of pancreatic cancer cells (Yadav et al., 2011). The anticancer effect of BC-4 is an isomeric compound containing both α- and β-boswellic acid acetate, was evaluated in metastatic melanoma and fibrosarcoma cells (Zhao et al., 2003). In results, BC-4 arrests the cell cycle in the G1 phase during cell growth and attenuated topoisomerase II activity.

6.3.11.10 ANALGESIC EFFECT

The analgesic effect induced by secobarbitone-induced hypnosis in rats was significantly potentiated by non-phenolic fraction of *B. serrata* gum resin. Further, the secondary-conditioned response (SCR) was specifically blocked in the trained rats, whereas the conditioned-avoidance response (CAR) was insignificant. Nalorphene, a standard drug control caused only an insignificant effect to the analgesic effect (Menon and Kar, 1971). The analgesic activity of AKBA at different concentrations by tail flick and acetic acid-induced writhing method in mice was examined (Bishnoi et al., 2005). In acetic acid-induced writhing method, a dose-dependent increase in anti-nociceptive activity of AKBA was shown while in tail flick method, 100 mg/kg of AKBA showed similar response to 200 mg/kg. AKBA was found to be significantly better than nimesulide (positive control). The essential oil (10 ml/kg), gum, resin, oleo-resin, and oleo-gum-resin significantly reduced carrageenan-induced paw edema in rats. In acetic acid-induced writhing assay oleo-gum-resin fraction showed maximum inhibition (60.54) as compared to oil (20.70) and gum fraction (54.88). All fractions at 20 mg/kg dose showed greater effect on hot plate induced analgesic effect than indomethacin at 10 mg/kg dose. Essential oil, gum, oleo-resin, and oleo-gum-resin at 100 mg/kg body wt. strongly inhibited formalin-induced inflammation than indomethacin, while resin part showed feeble effect (Sharma et al., 2010).

6.3.11.11 HEPATOPROTECTIVE STUDIES

Non-alcoholic fatty liver disease (NAFLD) is closely linked to insulin resistance, oxidative stress, and cytokine imbalance. Zaitone et al. (2015) assessed the protective effects of BAs in high fat diet-induced NAFLD in rat models. Rats treated with BAs (125 or 250 mg/kg) or pioglitazone showed improved insulin sensitivity and a reduction in liver index, activities of liver enzymes, serum TNF-α and IL-6 as well as hepatic iNOS expression and 4-hydroxy-2-nonenal compared to control group. At molecular level BAs at 250 mg/kg significantly ameliorated the expression of thermogenesis-related mitochondrial uncoupling protein-1 and carnitine palmitoyl transferase-1 in white adipose tissues of the studied animal model. A comparative study on effect of *B. serrata* gum resin extracts (50 mg/kg/day) and *Salvia miltiorrhiza* extracts (150 mg/kg/day) on dimethyl nitrosamine (DMN) induced hepatic fibrosis was evaluated by Sferra et al. (2012). The immune-histological severity of the hepatic fibrosis showed a marked improvement in samples treated with BSEs and *S. miltiorrhiza* extracts. This was associated with a reduction in the hepatic expression of α-SMA, collagen I-III, CTGF, TGF-β1, Smad3, and Smad7.

Isobolographic analysis was carried out to evaluate the combined effect of *B. serrata* gum resin extract alone and in combination with DOX against HepG2 and Hep3B cell lines. The Isobolographic results showed a combination index (CI) of DOX and BSE of suggesting synergistic behavior against the two cell lines. BSE also caused a dose-dependent increase in caspase-3 activity, TNF-α level, and IL-6 level which was higher with DOX (1 μM) and BSE (20 μg/mL) combination (Khan et al., 2014). Thabet et al. (2020) investigated that BAs significantly ameliorated bisphenol and IR induced hepatotoxicity in animal models.

6.3.11.12 ANTIVIRAL STUDIES

Chikungunya virus is a mosquito-transmitted alphavirus, which causes chikungunya fever in humans. Rhein et al. (2015) evaluated the inhibitory effect of curcumin and *B. serrata* gum resin extract against chikungunya virus and vesicular stomatitis virus using *in vitro* methods. Both the tested herbs significantly inhibited the chikungunya virus and vesicular stomatitis virus infection. HSV, a highly contagious pathogen, is responsible for causing lifelong oral to genital infection in humans. The effect of *B. serrata* gum resin extract and βBA against HSV was studied by Gowsami et al. (2018). *B. serrata* gum resin extract and βBA significantly inhibited the upregulation of pro-inflammatory markers like NF-kB, p38 MAP-kinase activation, TNFα, interleukin (IL)-1β, IL-6. Methanolic extract of *B. serrata* gum resin strongly inhibited hepatitis C-virus (HCV) protease (PR) in *in vitro* studies (Hussein et al., 2000).

6.3.11.13 NEUROPROTECTIVE STUDIES

The nuclear factor erythroid 2-related factor 2 (Nrf2; encoded by the Nfe2l2 gene) is a transcription factor responsible for the regulation of antioxidant genes/proteins, including heme oxygenase-1 (HO-1). HO-1 is an inducible enzyme, its dysregulation or deficiency is associated with a variety of pathologies such as neurodegenerative (Alzheimer's disease (AD), Parkinson's disease (PD)), multiple sclerosis, and neuroinflammatory diseases (Habtemariam, 2019). Treatment with AKBA significantly reduced infarct volumes and apoptotic cells, and also increased neurologic scores by elevating the Nrf2 and HO-1 expression in brain tissues in middle cerebral artery occlusion (MCAO) rats at 48 hours post-reperfusion. Further, AKBA increased the Nrf2 and HO-1 expression in oxygen and glucose deprivation induced oxidative stress in primary cultured neurons (Ding et al., 2014). In another experiment, *B. serrata* at 50 mg/kg, significantly improved memory retrieval and step through latency in streptozotocin induced Alzheimer disease in rat models (Zaker et al., 2015). Treatment with aqueous extract of *B. serrata* gum resin for a period of 42 days, significantly reduced streptozotocin induced Alzheimer disease symptoms, i.e., decreased step-through latency, increased the number of step-through into the dark compartment and increased the time spent in the dark compartment (Beheshti and Aghaie, 2016).

6.3.11.14 ANTI-OBESITY STUDIES

The effect of *B. serrata* gum resin on reduction capacity of obesity was assessed in 45 patients, resulted significant reduction in obesity induced anthropometric indices (Fatima et al., 2017). Similarly, Gomma et al. (2019) studied the effect of *B. serrata* gum resin extract on obesity in Wistar rat models, indicated that extract exhibited a noteworthy reduction in lipid profile and inhibited the activity of pancreatic lipase activity.

6.3.11.15 ANTI-LEISHMANIASIS ACTIVITY

Anti-Leishmaniasis activity of 52 essential oils of commercial importance, which includes *Boswellia* oil, was evaluated against *Leishmania amazonensis.* Among the tested essential oils, three oils, i.e., coriander, frankincense, and wintergreen oils, exhibited beneficial activity against the tested pathogen (Monzote et al., 2019).

6.3.11.16 HYPOCHOLESTEROLEMIC PROPERTY

Pandey et al. (2005) reported that water extract of *B. serrata* possesses strong hypocholesterolemic properties along with an increase in serum HDL. It inhibits the LPS-induced NO production by the activated rat peritoneal macrophages and show hepato-protective and reno-protective property.

6.3.12 *BOSWELLIA SOCOTRANA*

6.3.12.1 ANTIOXIDANT ACTIVITY

Antioxidant activity of essential oil isolated from *B. socotrana* oleo gum resin was evaluated by DPPH method (Ali et al., 2008). In results, oil exhibited strong DPPH quenching activity by expressing lower IC_{50} value 121.4 μg/ml. The oil was more potent than the other two oils tested, i.e., *B. ameero* and *B. elongata*. This may be due to the synergistic effect among the components present in the essential oil. Similarly, DPPH scavenging capacity of essential isolated from *B. socotrana* stem bark was assessed. The oil expressed very weak DPPH quenching activity (Mothana et al., 2011a). Antioxidant activity of methanol and hot water extracts of *B. socotrana* along with 25 other plants collected from Yemen was investigated. Methanol extracts of *B. socotrana* along with three other medicinal plants showed potent antioxidant activity at low concentrations (Mothana et al., 2009).

6.3.12.2 ANTIMICROBIAL ACTIVITY

The antimicrobial activity of methanol and hot water extracts of *B. socotrana* along with 25 other plants collected from Yemen was investigated. The extracts of *B. socotrana* along with other 8 plants exhibited strong antimicrobial activity against the tested pathogens (Mothana et al., 2009). Whereas antimicrobial effects of essential oil isolated from *B. socotrana* stem bark was investigated. The oil showed antimicrobial activity against gram-positive bacteria with MIC-values between 1.8 mg/ml and 17.2 mg/ml (Mothana et al., 2011a).

6.3.12.3 ANTICANCER ACTIVITY

Anticancer activity of methanol and hot water extracts of *B. socotrana* along with 25 other plants collected from Yemen was investigated against three human cancer cell lines (A-427, 5637, and MCF-7). The methanol extracts of *B. socotrana*, along with four medicinal plants exhibited noticeable antiproliferative potency with lowest IC_{50} values (Mothana et al., 2009).

6.3.12.4 ANTICHOLINESTERASE ACTIVITY

Anticholinesterase activity of chloroform and methanol extracts of *B. socotrana*, along with 13 other endemic medicinal plants of Socotra

Island, was investigated. Chloroform extracts of *B. socotrana*, along with three other endemic plants reported to have more than 50% inhibition at 200 μg/ml concentration (Bakthira et al., 2011).

6.4 CONCLUSION

The frankincense is the main ingredient in the formulations of a various of modern perfumes and such as "*Replique*" (Colonia), "*Me!*" (Frances Denney), "*Mennen Millionaire*" (Mennen), "*Nino Cerruti Pour Homme*" (Uniperf), "*Onna*" (Gary Farn), "*Sculptura*" (Jovan), "*Volcan d'Amour I*" (Diane von, Furstenburg), "*Paul Sebastian* V.S.O.P." (Sebastian) and "*Gambler*" by Jovan, Fragnance Foundation (Tucker, 1986). Besides its perfumery applications, it has developed as a prominent medicament for the inflammation and rheumatoid pains, cancer treatment, and many above-mentioned biological activities. The *Boswellia* has many product names and primarily they possess certain secondary metabolites such as pinene, thujene, BE, olibanumol, serratol, caryophellene, and so on. More than 300, which may stand for the rich diversity of terpenes and many interesting bio-effects. Hence, the species of *Boswellia* require more number of scientific studies for their wide range of medicinal applications apart from their conservation and sustainable utilization for the sake of future generations.

KEYWORDS

- **anticancer**
- **antidiabetic and rheumatoid arthritis**
- **antimicrobial**
- **biological activities**
- ***Boswellia* species**
- **frankincense**
- **olibanum**

REFERENCES

Abdallah, E., & Khalid, A., (2012). A preliminary evaluation of the antibacterial effects of *Commiphora molmol* and *Boswellia papyrifera* oleo-gum resins vapor. *Intern. J. Chem. Biochem. Sci., 1*, 1–5.

Abdel-Tawab, M., Werz, O., & Schubert-Zsilavecz, M., (2011). *Boswellia serrata*: An overall assessment of *in vitro*, preclinical, pharmacokinetic and clinical data. *Clin. Pharmacokinet., 50*(6), 349–369.

Abdou-Latif, F. M., Obame, L. C., Bassole, I. M. N., & Dicko, M. H., (2012). Antimicrobial activities of essential oil and methanol extract of *Boswellia sacra* Flueck. and *Boswellia papyrifera* (Del.) Hochst from Djibouti. *Intern. J. Management, Modern Sci. Technologies, 1*(1), 1–10.

Adebisi, I., Abubakar, A., Abubakar, K., & Giaze, T., (2018). Analgesic effect and anti-inflammatory activity of aqueous extract of *Boswellia dalzielii* (Burseraceae) stem bark. *Intern. J. Pharm. Pharmaceut. Sci., 10*. 139–142.

Ahangarpour, A., Hamid, H., Ramezani, A. A. F., Mostafa, P., Hajeye, S., Iraj, A., Zahra, M., & Babadi, H. M., (2014). Effect of *Boswellia serrata* supplementation on blood lipid, hepatic enzymes and fructosamine levels in type 2 diabetic patients. *J. Diabetes & Metabolic Disorders, 13*(1), 29.

Ahangarpour, A., Ramezani, A. A. F., Heidari, H., Pakmehr, M., Shahbazian, H., & Ahmadi, I., (2013). The effect of *Boswellia serrata* on blood glucose, insulin level and insulin resistance in type 2 diabetic patients. *DMed, 20*(6), 11–18.

Ahmed, A., Al-sobarry, M., Rachad, A., Yahia, C., & Katim, A., (2012). Analgesic and anti-inflammatory activities of *Boswellia elongata* Balf. methanolic extracts, as endemic plants in Yemen. *J. Biologically Active Products from Nature, 2*(2), 90–98.

Akihisa, T., Tabata, K., Banno, N., Tokuda, H., Nishimura, R., Nakamura, Y., Kimura, Y., et al., (2006). Cancer chemopreventive effects and cytotoxic activities of the triterpene acids from the resin of *Boswellia carterii*. *Biol. Pharmaceut. Bull., 29*(9), 1976–1979.

Akthar, N., Miller, M. J., & Haqqi, T. M., (2011). Effect of a herbal-leucine mix on the IL-1α-induced cartilage degradation and inflammatory gene expression in human chondrocytes. *BMC Complement. Altern. Med., 11*, 66.

Alam, M., Khan, H., Samiullah, L., & Siddique, K. M., (2012). A review on phytochemical and pharmacological studies of Kundur (*Boswellia serrata* Roxb. ex Colebr.)–a Unani drug. *J. Appl. Pharm. Sci., 2*, 148–156.

Alemika, T. E., Onawunmi, G. O., & Olugbade, T. A., (2006). Antibacterial phenolics from *Boswellia dalzielii*. *J. Nat. Prod. Med., 10*, 108–110.

Ali, E. N., & Mansour, S. Z., (2011). Boswellic acids extract attenuates pulmonary fibrosis induced by bleomycin and oxidative stress from gamma irradiation in rats. *Chin Med., 6*, 36.

Ali, M., Hosseini-Sharifabad, A., Monsef-Esfahani, H. R., Yazdinejad, A. R., Khanavi, M., Roghani, A., Beyer, C., & Sharifzadeh, M., (2011). Evaluation of systemic administration of *Boswellia papyrifera* extracts on spatial memory retention in male rats. *J. Nat. Med., 65*(3, 4), 519–525.

Ali, N. A. A., Samaha, H. A., Arnold, N., Alhaj, M., & Wessjohann, L., (2014). Screening of extracts from endemic Socotraen medicinal plants for antiradical and anti-fungal activity. *World J. Pharmaceut. Res., 3*(8), 823–830.

Ali, N. A. A., Wurster, M., Arnold, N., Teichert, A., Schmidt, J., Lindequist, U., & Wessjohann, L., (2008). Chemical composition and biological activities of essential oils from the oleogum resins of three endemic Soqotraen *Boswellia* species. *Rec. Nat. Prod., 2*(1), 6–12.

Ali, S. A., Zaitone, S. A., & Moustafa, Y. M., (2015). Boswellic acids synergize antitumor activity and protect against the cardiotoxicity of doxorubicin in mice bearing Ehrlich's carcinoma. *Can. J. Physiol. Pharmacol., 93*(8), 695–708.

Ali, W. A., & Hashim, H. R., (2020). Inhibit of the virus infections by using aromatic oils, *Saccharomyces cerevisiae, Boswellia carterii* and Zinc. *Sys. Rev. Pharm., 11*(4), 561–569.

Aliyu, R., Gatsing, D., & Jaryum, K. H., (2007). The effect of *Boswellia dalzielii* (Burseraceae) aqueous bark extract on rat liver function. *Asian J. Biochem., 2*(5), 359–363.

Al-kuraishy, H. M., Al-gareeb, A. I., Ashoor, A. W., & Al-windy, S. A., (2012). Modifying effects of *Boswellia carterii* on clarithromycin action: *In vitro* antibacterial study against common sensitive bacterial strains. *Intern. J. Drug Dev. Res., 4*(3), 155–162.

Al-Saidi, S., Rameshkumar, K. B., Hisham, A., Sivakumar, N., & Al-Kindy, S., (2012). Composition and antibacterial activity of the essential oils of four commercial grades of Omani Luban, the oleo-gum resin of *Boswellia sacra* Flueck. *Chemistry & Biodiversity, 9*, 615–624.

Amir, F., Golbarg, G., Farshchi, S., & Khatabi, P., (2010). Effects of *Boswellia papyrifera* gum extract on learning and memory in mice and rats. *Iranian J. Basic Med. Sci., 13*(2), 9–15.

Ammon, H. P., (1996). Salai guggal – *Boswellia serrata*: From a herbal medicine to a non-redox inhibitor of leukotriene biosynthesis. *Eur. J. Med. Res., 1*(8), 369, 370.

Ammon, H. P., (2002). Boswellic acids (components of frankincense) as the active principle in treatment of chronic inflammatory diseases. *Wien Med Wochenschr., 152*, 337–378.

Ammon, H. P., (2006). Boswellic acids in chronic inflammatory diseases. *Planta Med., 72*, 1100–1116.

Ammon, H. P., (2010). Modulation of the immune system by *Boswellia serrata* extracts and boswellic acids. *Phytomedicine., 17*(11), 862–867. doi: 10.1016/j.phymed.2010.03.003.

Ammon, H. P., Mack, T., Singh, G. B., & Safayhi, H., (1991). Inhibition of leukotriene B4 formation in rat peritoneal neutrophils by an ethanolic extract of the gum resin exudate of *Boswellia serrata*. *Planta Med., 57*, 203–207.

Ammon, H. P., Safayhi, H., Mack, T., & Sabieraj, J., (1993). Mechanism of anti-inflammatory actions of curcumine and boswellic acids. *J. Ethnopharmacol., 38*, 113–119.

Ammon, H., (2016). Boswellic acids and their role in chronic inflammatory diseases. In: *Anti-Inflammatory Nutraceuticals and Chronic Diseases* (pp. 291–327). Springer International Publishing AG, Switzerland.

Anitha, G., & Sudarsanam, G., (2012). Studies on antimicrobial activity of *Boswellia ovalifoliolata* against *Xanthomonas citri* and *Salmonella typhymurium*. *Intern. J. Appl. Biol. Pharmaceut. Technol., 4*(1), 243–247.

Atal, C. K., Gupta, O. P., & Singh, G. B., (1981). Salai guggal an anti-arthritic and anti-hyperlipidemic agent. *Br. J. Pharm., 74*, 203.

Atal, C. K., Singh, G. B., Batra, S., Sharma, S., & Gupta, O. P., (1980). Salai guggal ex *Boswellia serrata* a promising anti-hyperlipidemic and antiarthritic agent. *Ind. J. Pharm., 12*, 56.

Awadi, F., Fatania, H., & Shamte, U., (1991). The effect of plant mixture extract on liver gluconeogenesis in streptozotocin induced diabetic rats. *Diabetes Res., 18*, 163–168.

Ayub, M. A., Hanif, M. A., Sarfraz, R. A., & Shahid, M., (2018). Biological activity of *Boswellia serrata* Roxb. oleo gum resin essential oil: Effects of extraction by supercritical carbon dioxide and traditional methods. *Intern. J. Food Properties, 21*(1), 808–820.

Azemi, M. E., Namjoyan, F., Khodayar, M. J., Ahmadpour, F., Darvish, P. A., & Panahi, M., (2012). The antioxidant capacity and anti-diabetic effect of *Boswellia serrata* Triana and Planch aqueous extract in fertile female diabetic rats and the possible effects on reproduction and histological changes in the liver and kidneys. *Jundishapur J. Nat. Pharm. Prod., 7*(4), 168–175.

Badria, F. A., Abu-Karam, M., Mikhaeil, B. R., Maatooq, G. T., & Amer, M. M., (2003a). Anti-herpes activity of isolated compounds from frankincense. *Biosci. Biotechnol. Res. Asia, 1*(1), 1–10.

Badria, F. A., Mikhaeil, B. R., Maatooq, G. T., & Amer, M. M. A., (2003b). Immunomodulatory triterpenoids from the oleogum resin of *Boswellia carterii* Birdwood. *Z. Naturforsch, 58c*, 505–516.

Bakthira, H., Awadh, A. N. A., Arnold, N., Teichert, A., & Wessjohann, L., (2011). Anticholinesterase activity of endemic plant extracts from Socotra. *Afr. J. Tradit. Complement. Altern. Med., 8*(3), 296–299.

Bannuru, R. R., Osani, M. C., Al-Eid, F., & Wang, C., (2018). Efficacy of curcumin and *Boswellia* for knee osteoarthritis: Systematic review and meta-analysis. *Semin Arthritis Rheum, 48*(3), 416–429.

Barre, S. M., Mirghani, E. L. M., Abdulkarim, I. S. M. B., & Olorunnisola, K. S., (2014). Anti-bacterial activities of crude extracts of some East African oleo gum resins (*Burseraceae*) and their respective extraction yield. *International Conference on Chemical, Environment & Biological Sciences (CEBS-2014)*. Kuala Lumpur (Malaysia). Conference paper.

Basar, S., (2005). *Phytochemical Investigations on Boswellia Species: Comparative Studies on the Essential Oils, Pyrolysates and Boswellic Acids of Boswellia Carterii Birdw., Boswellia Serrata Roxb., Boswellia Frereana Birdw., Boswellia Neglecta S. Moore and Boswellia Rivae Engl.* PhD Thesis, Institute of Organic Chemistry, University of Hamburg.

Beheshti, S., & Aghaie, R., (2016). Therapeutic effect of frankincense in a rat model of Alzheimer's disease. *Avicenna J. Phytomed., 6*(4), 468–475.

Benelli, G., Rajeswary, M., Vijayan, P., Senthilmurugan, S., Alharbi, N., Km, S., Khaled, J., & Govindarajan, M., (2018). *Boswellia ovalifoliolata* (Burseraceae) essential oil as an eco-friendly larvicide? Toxicity against six mosquito vectors of public health importance, non-target mosquito fishes, backswimmers, and water bugs. *Environ. Sci. Poll. Res., 25*, 10264–10271.

Bhutada, S., Farhan, M. M., & Dahikar, S., (2017). Preliminary phytochemical screening and antibacterial activity of resins of *Boswellia serrata* Roxb. *J. Pharmacogn. Phytochem., 6*(4), 182–185.

Bishnoi, M., Patil, C. S., Kumar, A., & Kulkani, S. K., (2005). Analgesic activity of acetyl-11-keto-b-boswellic acid, a lipooxygenase – enzyme inhibitor. *Indian J. Pharmacol., 37*, 255, 256.

Blain, E. J., Ali, A. Y., & Duance, V. C., (2010). *Boswellia frereana* (frankincense) suppresses cytokine-induced matrix metalloproteinase expression and production of pro-inflammatory molecules in articular cartilage. *Phytother. Res., 24*(6), 905–912.

Boonsri, R., & Eswaraiah, M. C., (2014). Preliminary phytochemical analysis and analgesic activity of root extracts of *Boswellia ovalifoliolata* Bal. & Henry. *Intern. Res. J. Pharmaceut. Appl. Sci., 4* (6), 18–21.

Bothon, F. T. D., Abou, L., Atindehou, M., Yarou, K. A., Ewedje, E. B. K., Agbangnan, P. S. D., Latifou, L., et al., (2019). Evaluation of antibacterial, free radical scavenging activities and phytochemical composition of *Boswellia dalzielii* hutch extracts. *J. Pharmacogn. Phytochem., 8*(4), 414–418.

Burkill, H. M., (1985). *Useful plants of west tropical Africa* (Vol. I, p. 300). Royal Botanical Gardens Kew.

Casapullo, A., Cassiano, C., Capolupo, A., Del Gaudio, F., Esposito, R., & Tosco, A., (2016). β-Boswellic acid, a bioactive substance used in food supplements, inhibits protein synthesis by targeting the ribosomal machinery. *J. Mass Spectrom., 51*(9), 821–827.

Chashoo, G., Singh, S. K., Sharma, P. R., Mondhe, D. M., Hamid, A., Saxena, A., Andotra, S. S., et al., (2011). A propionyloxy derivative of 11-keto-beta-boswellic acid induces apoptosis in HL-60 cells mediated through topoisomerase I & II inhibition. *Chem.-Biol. Interact., 189*, 60–71.

Chatterjee, G. K., & Pal, S. D., (1984). Anti-inflammatory agents from Indian medicinal plants. *Indian Drugs, 21*, 431.

Chaurasia, A., & Gharia, A., (2017). Antifungal activity of medicinal plant *Boswellia serrata*. *J. Ultra Chem., 13*, 88–90.

Chevrier, M. R., Ryan, A. E., Lee, D. Y. W., Zhongze, M., Wu-Yan, Z., & Via, C. S., (2005). *Boswellia carterii* extract inhibits TH1 cytokines and promotes TH2 cytokines *in vitro*. *Clin. Diagnostic Lab. Immunol., 12*(5), 575–580.

Chopra, A., Saluja, M., Tillu, G., Sarmukkaddam, S., Venugopalan, A., Narsimulu, G., Handa, R., et al., (2013). Ayurvedic medicine offers a good alternative to glucosamine and celocoxib in the treatment of symptomatic knee osteoarthritis: A randomized, double-blind, controlled equivalence drug trail. *Rheumatol (Oxford), 52*(8), 1408–1417.

Chopra, R. N., Nayer, S. L., & Chopra, I. C., (1956). *Glossary of Indian Medicinal Plants*. Council of Scientific and Industrial Research, New Delhi.

Committee of the Chinese Pharmacopeia, (2015). *Chinese Pharmacopeia* (Vol. 1, p. 223). Chinese Medical Science Press.

Conti, S., Vexler, A., Edry-Botzer, L., Kalich-Philosoph, L., Corn, B. W., Shtraus, N., Meir, Y., et al., (2018). Combined acetyl-11-keto-beta-boswellic acid and radiation treatment inhibited glioblastoma tumor cells. *PLoS One, 13*, e0198627.

Cuaz-Perolin, C., Billie, L., Bauge, E., Copin, C., Scott-Algara, D., Genze, F., Buchele, B., et al., (2008). Anti-inflammatory and antiatherogenic effects of the NF-kappaB inhibitor acetyl-11-ketobeta – boswellic acid in LPS-challenged ApoE-/-mice. *Arterioscler. Thromb. Vasc. Biol., 28*, 272–277.

Danlami, U., Gbubele, J. D., Bwai, M. D., & Kabir, M. G., (2015). Phytochemical, nutritional and antimicrobial screening of hexane, ethyl acetate and ethanolic extracts of *Boswellia dalzielii* leaves and bark. *J. Biosci. Bioeng., 3*(8), 76–79.

Dhanalakshmi, M., & Maddilety, B., (2012). Antiulcer activity of methanolic extract of leaves of *Boswellia ovalifoliolata* Bal. & Henry. *Pharmatutor.* Art-1370.

Dinesh, T., Manish, K. J., Shweta, S., Aparna, A., Veera, G. V., Samyuktha, V., Prashanth, K., Naidu, V. G. M., Ashutosh, K., & Ramakrishna, S., (2014). *Boswellia ovalifoliolata* abrogates ROS mediated NF-κB activation causes apoptosis and chemosensitization in triple negative breast cancer cells. *Environ. Toxicol. Pharmacol., 38*, 58–70.

Ding, Y., Chen, M., Wang, M., Wang, M., Zhang, T., & Park, J., (2014). Neuroprotection by acetyl-11-keto-[bgr]-boswellic acid, in ischemic brain injury involves the Nrf2/HO-1 defense pathway. *Sci. Rep., 4*.

Efferth, T., & Greten, H. J., (2011). Anti-inflammatory and anti-cancer activity of boswellic acids from frankincense (*Boswellia serrata* Roxb. et Colebr, *B. carterii* Birdw.). *For. Immunopathol. Dis. Ther., 2*, 303–313.

El Ghazali, G. E., (1997). *The Promising Medicinal Plants of the Sudan* (1st edn., p. 63). Khartoum: University Press.

El Kichaoui, A., Abdelmoneim, A., Elbaba, H., & El Hindi, M., (2017). The antimicrobial effects of *Boswellia carterii*, *Glycyrrhiza glabra* and *Rosmarinus officinalis* some pathogenic microorganisms. *IUG J. Nat. Studies, 25*(2), 208–213.

Etuk, E. U., Agaie, B. M., Onyeyili, P. A., & Ottah, C. U., (2006a). Anti-diarrhoea effect of *Boswellia dalzielii* stem bark extract in albino rats. *J. Pharmacol. Toxicol., 1*(6), 591–596.

Etuk, E. U., Agaie, B. M., Onyeyili, P. A., & Ottah, C. U., (2006b). Toxicological studies of aqueous stem bark extract of *Boswellia dalzielii* in albino rats, *Indian J. Pharmacol., 38*(5), 359, 360.

Etzel, R., (1996). Special extract of *Boswellia serrata* (H15) in the treatment of rheumatoid arthritis. *Phytomedicine, 3*, 91–94.

Fatima, S., Ahmad, T., Shahid, M., & Sofi, G., (2017). Comparative study of Kurdur (*Boswellia serrata*) and Tareeq (Diaphoresis) in the management of samne mufrit (Obesity): A randomized clinical trial. *Int. J. Health Sci. Res., 7*(10), 186–196.

Fatimah, A. O., Albasher, G., Alharbi, R. I., & Alsaggabi, N. S., (2019). Antifungal potential of aqueous extract of *Boswellia carterii*. *J. Pure Appl. Microbiol., 13*(4), 2375–2381.

Fitchl, R., & Admasu, A., (1994). *Honeybee Flora of Ethiopia* (p. 510). Margraf Verlag, Weikersheim, Germany.

Frank, M. B., Yang, Q., Osban, J., Azzarello, J. T., Saban, M. R., Saban, R., Ashley, R. A., et al., (2009). Frankincense oil derived from *Boswellia carterii* induces tumor cell specific cytotoxicity. *BMC Comple. Altern. Med., 9*, 6.

Garg, S. C., (1974). Antifungal activity of some essential oils. *Ind. J. Pharmacol., 36*, 47–47.

Gaurea, S. H., & Bapat, U. C., (2016). Study of antibacterial activity of resins of *Boswellia serrata* Roxb. ex Colebr., *Commiphora mukul* (Hook. ex Stocks) Engl., *Gardenia resinifera* Roth. and *Shorea robusta* Gaertn. *Intern. J. Pharm. Pharmaceut. Sci., 8*(1), 29–31.

Gayathri, B., Manjula, N., Vinay, K. K. S., Lakshmi, B. S., & Balakrishnan, A., (2007). Pure compound from *Boswellia serrata* extract exhibits anti-inflammatory property in human PBMCs and mouse macrophages through inhibition of TNF alpha, IL-1beta, NO and MAP kinases. *Int. Immunopharmacol., 7*, 473–482.

Geetha, K., & Ganapaty, S., (2013a). Antihyperlipidemic activity of *Boswellia ovalifoliolata* Bal. & Henry in atherogenic diet induced rats. *Intern. J. Phytother. Res., 3*(3), 11–17.

Geetha, K., & Ganapaty, S., (2013b). *In vitro* antioxidant studies on ethanolic extracts of *Boswellia ovalifoliolata* and *Saccharum spontaneum* by DPPH, nitric oxide and lipid peroxidation methods. *Intern. J. Res. Pharmacol. Pharmacotherapeut., 2*(1), 324–330.

Geetha, K., Srikanth, K., & Ganapaty, S., (2014). Screening of crude plant extracts for anti-adipogenesis activity in 3T3-L1 cells. *J. Pharm. Res., 8*(1), 81–86.

Getachew, M. F., Asnake, G. E., & Aman, D., (2013). Biocide value characterization of essential oil from *Boswellia neglecta* S. against pathogenic termite, cockroach, ticks, *Escherichia coli* and *Staphylococcus aureus*. *Intern. J. Modern Chem., 5*(3), 145–158.

Ghani, N., (2002). *Khazainul Advia* (pp. 1069, 1070). Idara Kitabus Shifa, New Delhi.

Glaser, T., Winter, S., Groscurth, P., Safayhi, H., Sailer, E. R., Ammon, H. P., Schabet, M., & Weller, M., (1999). Boswellic acids and malignant glioma: Induction of apoptosis but no modulation of drug sensitivity. *Br. J. Cancer., 80*, 756–765.

Gomma, A. A., Farghaly, H. S. M., El-Sers, D. A., Farrag, M. M., & Al-Zokeim, N. I., (2019). Inhibition of adiposity and related metabolic disturbances by polyphenol-rich extract of *Boswellia serrata* gum through alteration of adipo/cytokine profile. *Inflammopharmacol, 27*(3), 549–559.

Gowsami, D., Mahapatra, A. D., Banerjee, S., Kar, A., Ojha, D., Mukherjee, K. P., & Chattopadhyay, D., (2018). *Boswellia serrata* oleo-gum resin and β-boswellic acid inhibits HSV-1 infection *in vitro* through modulation of NF-kB. and p38 MAP kinase signaling. *Phytomedicine, 51*, 94–103.

Groom, N., (1981). *Frankincense and Myrrh: A Study of the Arabian Incense Trade*. Longman, London and New York.

Gupta, I., Gupta, V., Parihar, A., Gupta, S., Ludtke, R., Safayhi, H., & Ammon, H. P., (1998). Effect of *Boswellia serrata* gum resin in bronchial asthma. *Eur. J. Med. Res., 3*, 511–514.

Gupta, P., Samarakoon, S., Chandola, H., & Ravishankar, B., (2011). Clinical evaluation of *Boswellia serrata* (Shallaki) resin in the management of Sandhivata (osteoarthritis). *Ayu, 32*(4), 478–482. 10.4103/0974-8520.96119.

Habtemariam, S., (2019). The Nrf2/HO-1 axis as targets for flavanones: Neuroprotection by Pinocembrin, Naringenin, and Eriodyctiol. *Oxidative Medicine and Cellular Longevity*, 1–15.

Hassan, H. S., Musa, A. M., Usman, M. A., & Abdulaziz, M., (2009). Preliminary phytochemical and anti-plasmodic studies of the stem bark of *Boswellia dalzielii*. *Nigerian J. Pharmaceut. Sci., 8*(1), 1–6.

Hasson, S. S., Al-Balushi, M. S., Sallam, T. A., Idris, M. A., Habbal, O., & Al-Jabri, A., (2011). *In vitro* antibacterial activity of three medicinal plants-*Boswellia* (Luban) species. *Asian Pacific J. Trop. Biomed., 1*(SUPPL. 2). https://doi.org/10.1016/S2221-1691(11)60151-2.

Henderson, C. J., & Panush, R. S., (1999). Diets, dietary supplements, and nutritional therapies in rheumatic diseases. *Rheum. Dis. Clin. North Am., 25*, 937–968.

Henkel, A., Kather, N., Mönch, B., Northoff, H., Jauch, J., & Werz, O., (2012). Boswellic acids from frankincense inhibit lipopolysaccharide functionality through direct molecular interference. *Biochem. Pharmacol., 83*(1), 115–121. 10.1016/j.bcp.2011.09.026.

Henley-Smith, C. J., Botha, F. S., & Lall, N., (2013). The use of plants against oral pathogens. *Formatex., 30*, 1375–1384.

Hoernlein, R. F., Orlikowsky, T., Zehrer, C., Niethammer, D., Sailer, E. R., & Simmet, T., (1999). Acetyl-11-keto-beta-boswellic acid induces apoptosis in HL-60 and CCRF-CEM cells and inhibits topoisomerase I. *J. Pharmacol. Exp. Ther., 288*, 613–619.

Hostanska, K., Daum, G., & Saller, R., (2002). Cytostatic and apoptosis-inducing activity of boswellic acids toward malignant cell lines *in vitro*. *Anticancer Res., 22*, 2853–2862.

Huang, G., Yang, J., Zhang, L., Cao, L., Zhang, M., Niu, X., Zhou, Z., et al., (2018). Inhibitory effect of 11-carbonyl-beta-boswellic acid on non-small cell lung cancer H446 cells. *Biochem. Biophys. Res. Commun., 503*(4), 2202–2205.

Huang, M. T., Bacdmaev, V., Ding, V., Liu, Y., & Xie, J. G., (2000). Anti-tumor and anti-carcinogenic activities of triterpenoid, beta-boswellic acid. *Biofactors, 13*(1–4), 225–230.

Hussein, G., Miyashiro, H., Nakamura, N., Hattori, M., Kakiuchi, N., & Shimotohno, K., (2000). Inhibitory effects of Sudanese medicinal plant extracts on hepatitis C virus (HCV) protease. *Phytother. Res., 14*(7), 510–516.

Ibn B., (2003). *Jamia al Mufradat al Advia al Aghiza* (Vol. 4, pp. 201–205). Central Council for Unani Research, New Delhi.

Ibrahim, J. A., Muazzam, I., Jegede, I. A., Kunle, O. F., & Okogun, J. I., (2007). Ethnomedicinal plants and methods used by Gwandara tribe of Sabo Wuse in Niger state, Nigeria to treat mental illness. *Afr. J. Trad. Complement. Altern. Med., 4*, 211–218.

Jasim, M. A., Ibrahim, F. M., & Khalifa, N. H., (2019). Evaluation of the cytotoxic effects of *Boswellia sacra* and *Origanum majorana* against mice lymphocytes and RD cell lines *in vitro*. *Iraq Med. J., 3*(1), 23–26.

Jing, Y., Nakajo, S., Xia, L., Nakaya, K., Fang, Q., Waxman, S., & Han, R., (1999). Boswellic acid acetate induces differentiation and apoptosis in leukemia cell lines. *Leuk. Res., 23*, 43–50.

Kafuti, Y. S., Ojerinde, O. S., Balogun, O., Alemika, T. E., Taba, K. M., Mpiana, P. T., & Kindombe, N. M., (2019). Antioxidant and antiproliferative activities of the stem bark extract and fractions of *Boswellia dalzielii* Hutch. *Intern. J. Pharmacogn. Phytochem. Res., 11*(3), 177–182.

Kafuti, Y. S., Ojerinde, S. O., Balogun, O., Alemika, T. E., Taba, K. M., Mpiana, P.T., et al., (2017). Antifungal evaluation and phytochemical screening of methanolic extract and fractions of *Boswellia dalzielii* stem bark. *J. Pharm. Bioresources, 14*(1), 64–69.

Kapoor, L. D., (1990). *CRC Handbook of Ayurvedic Medicinal Plants* (p. 83). CRC Press, Florida, USA.

Khan, M. A., Singh, M., Khan, M. S., Najmi, A. K., & Ahmad, S., (2014). Caspase mediated synergistic effect of *Boswellia serrata* extract in combination with doxorubicin against human hepatocellular carcinoma. *Biomed. Res. Int.*, 294143.

Khan, S., Kaur, R., Shah, B. A., Malik, F., Kumar, A., Bhushan, S., Jain, S. K., Taneja, S. C., & Singh, J., (2012). A novel cyano derivative of 11-keto-beta-boswellic acid causes apoptotic death by disrupting PI3K/AKT/Hsp-90 cascade, mitochondrial integrity, and other cell survival signaling events in HL-60 cells. *Mol. Carcinog., 51*, 679–695.

Khare, C. P., (2004). *Encyclopedia of Indian Medicinal Plants: Rational Western Therapy, Ayurvedic and other Traditional Usage*. Springer.

Kimmatkar, N., Thawani, V., Hingorani, L., & Khiyani, R., (2003). Efficacy and tolerability of *Boswellia serrata* extract in treatment of osteoarthritis of knee—A randomized double-blind placebo controlled trial. *Phytomedicine, 10*(1), 3–7.

Kirtikar, K. R., & Basu, B. D., (1935). The anti-inflammatory action of Indian medicinal plants. In*: Indian Med Plants, 1*, 521–529.

Kohoude, M. J., Gbaguidi, F., Agbani, P., Ayedoun, M. A., Cazaux, S., & Bouajila, J., (2017). Chemical composition and biological activities of extracts and essential oil of *Boswellia dalzielii* leaves. *Pharm. Biol., 55*, 33–42.

Kokkiripati, P. K., Bhakshu, Md. L., Swathi, M., Padmasree, K., Anupama, T. R., Agepati, S. R., & Sarada, D. T., (2011). Gum resin of *Boswellia serrata* inhibited human monocytic (THP-1) cell activation and platelet aggregation. *J. Ethnopharmacol., 137*, 890–901.

Kubmarawa, D., Akiniyi, J. A., & Okorie, D. A., (2013). Ethnomedicinal survey of the traditional medicine of Lala people of Nigeria. *Int. J. Med. Plants Altern. Med., 1*(3), 039–057.

Kumar, B. D., Sainath, K., Neetha, M., & Viswateja, (2012). *In vivo* analgesic activity of petroleum ether extract of *Boswellia ovalifoliolata* Linn. leaves. *Intern. J. Res. Pharmacol. Pharmacotherapeut., 1*(1), 2278–2656.

Kumar, D. S., Banji, D., Harani, A., Pavan, K. C., & Ravi, V. J. N., (2014). Role of *Boswellia ovalifoliolata* Bal. & Henry extract of high fat diet-induced hyper cholesterolemia. *Pharmacogn. J., 6*(4), 108–116.

Kumar, R., Singh, S., Saxen, A. K., Pal, R., Jaiswal, R., & Kumar, R., (2019). Effect of *Boswellia serrata* extract on acute inflammatory parameters and tumor necrosis factor-α in complete Freund's adjuvant-induced animal model of rheumatoid arthritis. *Int. J. Appl. Basic Med. Res., 9*(2), 100–106.

Langmead, L., & Rampton, D. S., (2006). Review article: Complementary and alternative therapies for inflammatory bowel disease. *Aliment. Pharmacol. Ther., 23*, 341–349.

Li, W., Liu, J., Fu, W., Zheng, X., Ren, L., Liu, S., Wang, J., et al., (2018). 3-O-acetyl-11-keto-beta-boswellic acid exerts anti-tumor effects in glioblastoma by arresting the cell cycle at G2/M phase. *J. Exp. Clin. Cancer Res., 37*, 132.

Liu, J. J., Huang, B., & Hooi, S. C., (2006). Acetyl-keto-beta-boswellic acid inhibits cellular proliferation through a p21-dependent pathway in colon cancer cells. *Br. J. Pharmacol., 148*, 1099–1107.

Liu, J. J., Nilsson, A., Oredsson, S., Badmaev, V., Zhao, W. Z., & Duan, R. D., (2002). Boswellic acids trigger apoptosis via a pathway dependent on caspase-8 activation but independent on Fas/Fas ligand interaction in colon cancer HT-29 cells. *Carcinogenesis, 23*, 2087–2093.

Liu, Y. Q., Wang, S. K., Xu, Q. Q., Yuan, H. Q., Guo, Y. X., Wang, Q., Kong, F., et al., (2019). Acetyl-11-keto-beta-boswellic acid suppresses docetaxel-resistant prostate cancer cells *in vitro* and *in vivo* by blocking Akt and Stat3 signaling, thus suppressing chemo resistant stem cell-like properties. *Acta Pharmacol. Sin., 40*, 689–698.

Liu, Z., Liu, X., Sang, L., Liu, H., Xu, Q., & Liu, Z., (2015). Boswellic acid attenuates asthma phenotypes by downregulation of GATA3 via pSTAT6 inhibition in a murine model of asthma. *Int. J. Clin. Exp. Pathol., 8*(1), 236–243.

Lu, M., Xia, L., Hua, H., & Jing, Y., (2008). Acetyl-keto-beta-boswellic acid induces apoptosis through a death receptor 5-mediated pathway in prostate cancer cells. *Cancer Res., 68*, 1180–1186.

Mahesh, B. U., Shrivastava, S., Rajeswara, R. P., Naidu, V. G. M., & Ramakrishna, S., (2014). Antioxidant and hepatoprotective effects of *Boswellia ovalifoliolata* bark extracts. *Chinese J. Nat. Med., 12*(9), 663–671.

Majeed, M., Majeed, S., Narayanan, N. K., & Nagabhushanam, K., (2019). A pilot, randomized, double-blind, placebo-controlled trial to assess the safety and efficacy of novel *Boswellia serrata* extract in the management of osteoarthritis of the knee. *Phytother. Res, 33*(5), 1457–1468.

Majeed, M., Nagabhushanam, K., Lawrence, L., Nallathambi, R., Thiyagarajan, V., & Mundkur, L., (2021). *Boswellia serrata* extract containing 30% 3-acetyl-11-keto-boswellic acid attenuates inflammatory mediators and preserves extracellular matrix in collagen-induced arthritis. *Front Physiol., 12*, 735247. doi: 10.3389/fphys.2021.735247. eCollection 2021.

Majeed, M., Vaidyanathan, P., Natarajan, S., Majeed, S., & Kumar, V. K., (2016a). Effect of boswellin® super on knee pain in Japanese adults: A randomized double-blind placebo-controlled trial. *Eur. J. Biomed. Pharmaceu. Sci., 3*(10), 293–298.

Majeed, M., Vaidyanathan, P., Sankaran, N. S. M., & Vuppala, K. K., (2016b). Effect of nilitis® SR on knee pain in Japanese adults: A randomized double-blind placebo-controlled study. *Int. J. Innov. Res. Med. Sci., 1*(6), 285–293.

Martina, B., Gloria, I., Federica, M., Giulia, A., Irvin, T. U., Paola, R., Monica, F., & Chiara, B., (2018). Anti-inflammatory activity of *Boswellia serrata* extracts: An *in vitro* study on porcine aortic endothelial cells. *Oxidative Medicine and Cellular Longevity*, 1–9.

Mazzio, E. A., Lewis, C. A., & Soliman, K. F. A., (2017). Transcriptomic profiling of MDA-MB-231 cells exposed to *Boswellia serrata* and 3-O-acetyl-B-boswellic acid; ER/UPR-mediated programmed cell death. *Cancer Genom. Proteom., 14*, 409–425.

Mbiantcha, M., Almas, J., Atsamo, A. D., Ateufack, G., Shabana, S. U., Bomba, T. D. F., Yousseu, N. W., & Nida, D., (2018). Anti-inflammatory and anti-arthritic effects of methanol extract of the stem bark of *Boswellia dalzielii* hutch (*Burseraceae*) in rats. *Inflammopharmacol., 26*(6), 1383–1398.

Mbiantcha, M., Khalid, R., Atsamo, D. A., Njoku, I. S., Mehreen, A., Ateufack, G., Hamza, D., Nana, W. Y., Ur Naeem, R., & Izhar, A., (2020). Anti-hypernociceptive effects of methanol extract of *Boswellia dalzielii* on STZ-induced diabetic neuropathic pain. *Adv. Tradit. Med., 20*, 405–417.

Medugu, A. N., Yakubu, J., Medugu, U. N., Marte, H. I., Tata, F. Y., & Balami, V. M., (2020). Phytochemical and anti-epileptic studies of ethanol extract of *Boswellia dalzielii* (frankincense Tree) stem bark. *Europ. J. Med. Plants, 31*(8), 94–100.

Menon, M. K., & Kar, A., (1971). Analgesic and psychopharmacological effects of the gum resin of *Boswellia serrata. Planta. Med., 19*(4), 333–341.

Michael, S., Sophia, J. L., Katharina, W., Luay, J. R., Tatiana, S., & Thomas, S., (2019). Comparative analysis of pentacyclic triterpenic acid compositions in oleogum resins of different *Boswellia* species and their *in vitro* cytotoxicity against treatment-resistant human breast cancer cells. *Molecules, 24*, 2153. doi: 10.3390/molecules24112153.

Miller, A. L., (2001). Effects of *Boswellia serrata* on asthma. *Altern. Med. Rev., 6*(1), 20–47.

Mohamed, A. A., Ali, S. I., Kabiel, H. F., Hegazy, A. K., Kord, M. A., & EL-Baz, F. K., (2015). Assessment of antioxidant and antimicrobial activities of essential oil and extracts of *Boswellia carterii* resin. *Intern. J. Pharmacogn. Phytochem. Res., 7*(3), 502–509.

Monzote, L., Herrea, I., Satyal, P., & Setzer, W. N., (2019). *In vitro* evaluation of 52 commercially-available essential oils against *Leishmania amazonensis. Molecules, 24*(7).

Mothana, R. A. A., Sidgi, S. H., Wulf, S., Annette, M., & Ulrike, L., (2011a). Phytochemical composition and *in vitro* antimicrobial and antioxidant activities of essential oils of three endemic Soqotraen *Boswellia* species, *Food Chemistry, 126*(3), 1149–1154.

Mothana, R. A., & Lindequist, U., (2005). Antimicrobial activity of some medicinal plants of the island Soqotra. *J Ethnopharmacol., 96*(1, 2), 177–181.

Mothana, R. A., (2011b). Anti-inflammatory, antinociceptive and antioxidant activities of the endemic Soqotraen *Boswellia elongata* Balf. f. and *Jatropha unicostata* Balf. f. in different experimental models. *Food Chem. Toxicol., 49*(10), 2594–2599.

Mothana, R. A., Lindequist, U., Gruenert, R., & Bednarski, P. J., (2009). Studies of the *in vitro* anticancer, antimicrobial and antioxidant potentials of selected Yemeni medicinal plants from the island Soqotra. *BMC Complement. Altern. Med., 9*, 7.

Mothana, R. A., Mentel, R., Reiss, C., & Lindequist, U., (2006). Phytochemical screening and antiviral activity of some medicinal plants from the island Soqotra. *Phytother Res., 20*(4), 298–302.

Moussaieff, A., Shohami, E., Kashman, Y., Fride, E., Schmitz, M. L., & Renner, F., (2007). Incensole acetate, a novel anti-inflammatory compound isolated from *Boswellia* resin inhibits nuclear factor-kappa B activation. *Mol Pharmacol., 72*, 1657–1664.

Mustafa, G. I., Abdalla, A. A., Ahmed, K. E., Al-magboul, A. Z., & Eltayeb, F. E., (2020). Antimicrobial activity of essential oils and extracts of oleo-gum resins from *Boswellia papyrifera* (Tarak tarak) grown in some parts of the Sudan. *Arabian J. Med. Arom. Plants, 6*(1), 22–35.

Mustafa, I. G., (2012). *Phytochemical Composition and Bioassay of Oleogum Resins of Boswellia papyrifera (Tarak tarak), Grown in Some Parts of the Sudan.* Ph.D. Thesis. University of Khartoum, Sudan.

Nagaraju, N., & Rao, K. N., (1990). A survey of plant crude drugs of Rayalaseema, Andhra Pradesh, India. *J. Ethnopharmacol., 29*, 137–158.

Natarajan, S., & Majeed, M., (2012). To assess the efficacy and safety of NINLIN^{SR} Tablets in the management of osteoarthritis of the knee. *Int. J. Pharm Life Sci., 3*, 2012–1413.

Nazifi, B. A., Danjuma, N., Olurishe, T., & Ya'u, J., (2017). Anticonvulsant activity of methanol stem bark extract of *Boswellia dalzielii* Hutch. (Burseraceae) in mice and chicks. *African J. Pharmacol. Therapeut., 6*(2), 64–71.

Ni, X., Suhail, M. M., Yang, Q., Cao, A., Fung, K., Postier, R., Woolley, C., et al., (2012). Frankincense essential oil prepared from hydrodistillation of *Boswellia sacra* gum resins induces human pancreatic cancer cell death in cultures and in a xenograft murine model. *BMC Complem. Altern. Med., 12*, 253.

Nwinyi, F. C., Binda, L., Ajoku, G. A., Aniagu, S. O., Enwerem, N. M., Orisadipea, et al., (2004). Evaluations of the aqueous extract of *Boswellia dalzielii* stem bark for antimicrobial activities and gastrointestinal effects. *Afr. J. Biotechnol., 3*(5), 284–288.

Nwude, N., & Ibrahim, M. A., (1980). Plants used in traditional veterinary medical practice in Nigeria. *J. Vet. Pharmacol. Ther., 3*, 261–273.

Ohadoma, S. C., Amazu, L. U., & Felix, N., (2016). Osuala, F. N., Iwuji, S. Assessment of antimicrobial activity of ethanol extracts of *Commiphora africana* and *Boswellia dalzielii*. *UK J. Pharmaceut. Biosci., 4*(3), 1–5.

Oleski, A., Lindequist, U., Mothana, R. A., & Melzig, M. F., (2006). Screening of selected Arabian medicinal plant extracts for inhibitory activity against peptidases. *Pharmazie, 61*(4), 359–361.

Olukemi, A. M., Kandakai-Olukemi, Y. T., & Mawak, J. D., (2005). Antibacterial activity of the stem bark of *Boswellia dalzielii*. *J. Pharm. Bioresources, 2*(2), 131–136.

Pandey, G. S., & Chunekar, K. C., (1967). *Bhavaprakash Nighantu* (pp. 139–141). Chaukhambha Vidya Bhawan, Varanasi.

Pandey, R. S., Singh, B. K., & Tripathi, Y. B., (2005). Extract of gum resins of *Boswellia serrata* L. inhibits lipopolysaccharide induced nitric oxide production in rat macrophages along with hypolipidemic property. *Indian J. Exp. Biol., 43*(6), 509–516.

Pang, X., Yi, Z., Zhang, X., Sung, B., Qu, W., & Lian, X., (2009). Acetyl-11-keto-b-boswellic acid inhibits prostate tumor growth by suppressing vascular endothelial growth factor receptor 2–mediated angiogenesis. *Cancer, 69*(14), 5893–5900.

Paranjpe, P., (2001). *Boswellia serrata*. In: *Indian Medicinal Plants – Forgotten Healers: A Guide to Ayurvedic Herbal Medicine* (pp. 233, 234). Chaukhamba Sanskrit. Pratishthan Publishers, Delhi.

Park, B., Prasad, S., Yadav, V., Sung, B., & Aggarwal, B. B., (2011). Boswellic acid suppresses growth and metastasis of human pancreatic tumors in an orthotopic nude mouse model through modulation of multiple targets. *Plos One, 6*, e26943.

Parr, C., & Ali, A. Y., (2018). *Boswellia frereana* suppresses HGF-mediated breast cancer cell invasion and migration through inhibition of c-Met signaling. *J. Transl. Med., 16*, 281.

Patel, N. B., & Patel, K. C., (2015). Antibacterial activity of *Boswellia serrata* Roxb. ex Colebr. ethnomedicinal plant against gram positive UTI pathogens. *Ayudh., 16*, 129–137.

Paul, M. J., Basha, S. K. M., (2015). Phytochemical analysis and antibacterial activities of *Boswellia ovalifoliolata*. *Intern. J. Sci. Res., 6*(7), 589–598.

Poeckel, D., & Werz, O., (2006). Boswellic acids: Biological actions and molecular targets. *Current Medicinal Chemistry, 13*(28), 3359–3369.

Prabhakar, Y. K., Subhanali, M. D., Jyothikumar, M. V., Krishna, T. T., & Apparao, C. H., (2013). Evaluation of antioxidant activities of aqueous extract of stem bark of *Boswellia ovalifoliolata* in streptozotocin induced diabetic rats. *J. Pharm. Chem., 7*(4), 19–24.

Prasanna, A. R. L., Venkata, R. K., Bhakshu, Md. L., Narasimha, R. B., & Veeranjaneya, R. L., (2015). Chemical profile, antioxidant and antimicrobial activity of essential oils from *Boswellia ovalifoliolata* Bal. et. Henry. *Int. J. Pharmace. Clin. Res., 7*(1), 96–101.

Qurishi, Y., Hamid, A., Sharma, P. R., Wani, Z. A., Mondhe, D. M., Singh, S. K., Zargar, M. A., et al., (2012). PARP cleavage and perturbance in mitochondrial membrane potential by 3-alpha-propionyloxy-beta-boswellic acid results in cancer cell death and tumor regression in murine models. *Future Oncol., 8*, 867–881.

Raja, A. F., Ali, F., Inshad, A. K., Abdul, S. S., Daljit, S, A., Bhahwal, A. S., & Subhash, C. T., (2011a). Anti-staphylococcal and biofilm inhibitory activities of acetyl-11-keto-b-boswellic acid from *Boswellia serrata*. *BMC Microbiol., 11*, 54.

Raja, A. F., Ali, F., Khan, I. A., Shawl, A. S., & Arora, D. S., (2011b). Acetyl-11-keto-bboswellic acid (AKBA); targeting oral cavity pathogens. *BMC Res. Notes, 4*, 406.

Ramireddy, K. V., John, P. M., Basha, S. K. M., (2017). Studies on antibacterial activity of *Boswellia ovalifoliolata* and *Syzygium alternifolium* against *Bacillus subtilis* and *Pseudomonas auriginosa*. *Intern. J. Eng. Res. Technol., 6*(6), 326–330.

Rasoul, M., Yadegari, M., Fariborz, M., & Masoumeh, S., (2006). Antifungal activity of *Boswellia serrata*'s essential oil against fluconazole resistant and susceptible isolates of *Candida albicans*. *J. Isfahan Medical School, 24*, 30–36.

Ravanan, P., Singh, S. K., Rao, G. S., & Kondaiah, P., (2011). Growth inhibitory, apoptotic and anti-inflammatory activities displayed by a novel modified triterpenoid, cyano-enone of methyl boswellates. *J. Biosci., 36*, 297–307.

Reddy, V., Ravinder, K., Srinivasulu, M., Goud, T., Reddy, S., Dondapati, S., Rao, T., Murthy, U., & Venkateswarlu, Y., (2003). Two new macrocyclic diaryl ether heptanoids from *Boswellia ovalifoliolata*. *Chem. Pharmaceut. Bull., 51*, 1081–1084.

Ren, J., Wang, Y. G., Wang, A. G., Wu, L. Q., Zhang, H. J., Wang, W. J., Su, Y. L., & Qin, H. L., (2015). Cembranoids from the gum resin of *Boswellia carterii* as potential anti-ulcerative colitis agents. *J. Nat. Prod., 78*, 2322–2331.

Renu, C., Manjeet, K., Masood, R., Simmi, S., Anjali, P., Sarang, P., Samar, S. A., et al., (2014). Anti-inflammatory terpenoids from *Boswellia ovalifoliolata*. *Royal Soc. Chem., 4*, 8632–8637.

Rhein, C., Weidner, T., Hen, L., Martin, J., Weber, C., Sliva, K., & Schnierle, B., (2015). Curcumin and *Boswellia serrata* gum resin extract inhibits chikungunya and vesicular stomatitis virus infection *in vitro*. *Antiviral Res., 125*, 51–57.

Sa'id, A. M., Ibrahim, M. S., Mashi, J. A., & Daha, I. U., (2017). Hepatoprotective effect of aqueous bark extract of *Boswellia dalzielii* against paracetamol induced hepatotoxicity in rabbits. *J. Adv. Med. Pharmaceut. Sci., 12*(3), 1–11.

Sadhasivam, S., Palanivel, S., & Ghosh, S., (2016). Synergistic antimicrobial activity of *Boswellia serrata* Roxb. ex Colebr. (*Burseraceae*) essential oil with various azoles against pathogens associated with skin, scalp and nail infections. *Lett. Appl. Microbiol., 63*(6), 495–501.

Saeed, M., Bahreh, T., Hasan, F. H., Seyed, H. M., & Mojtaba, H., (2018). The effects of *Boswellia serrata* gum resin on the blood glucose and lipid profile of diabetic patients: A double-blind randomized placebo-controlled clinical trial. *J. Evidence-Based Integr. Med., 23*, 1–7.

Safayhi, H., Mack, T., Sabieraj, J., Anazodo, M. I., Subramanian, L. R., & Ammon, H., (1992). Boswellic acids: Novel, specific, nonredox inhibitors of 5-lipoxygenase. *J. Pharmacol. Exp. Therapeut., 261*(3), 1143–1146.

Safayhi, H., Rall, B., Sailer, E. R., Ammon, H. P. T., (1997). Inhibition by boswellic acids of human leukocyte elastase. *J. Pharmacol. Exp. Therapeut., 281*(1), 460–463.

Sander, O., Herborn, G., & Rau, R., (1998). Is H15 (resin extract of *Boswellia serrata*, "incense") a useful supplement to established drug therapy of chronic polyarthritis? Results of a double-blind pilot study. *Zeitschrift für Rheumatologie, 57*(1), 11–16.

Saraswati, S., & Agarwal, S. S., (2012). Antiangiogenic and cytotoxic activity of boswellic acid on breast cancer MCF-7 cells. *Biomed. Prev. Nutr., 2*(1), 31–37.

Saritha, M., Prabhakar, Y. K., Subhan, A. M., Shaik, A. N., Jyothi, K., Venkateshwarulu, M., Rajesh, N., & Appa, R. C., (2014). Antihyperlipidemic activity of the stem bark of *Boswellia ovalifoliolata* in high fat diet fed rats. Intern. *J. Med. Plants, 107*, 572–576.

Savithramma, N., & Sulochana, C., (1998). Endemic medicinal plants from Tirumala hills Andhra Pradesh, India. *Fitoterapia, 69*, 253–254.

Savithramma, N., Linga, R. M., & Suvarnalatha, D. P., (2011). Evaluation of antibacterial efficacy of biologically synthesized silver nanoparticles using stem barks of *Boswellia ovalifoliolata* Bal. & Henry and *Shorea tumbaggaia* Roxb. *J. Biol. Sci., 11*, 39–45.

Savithramma, N., Linga, R. M., Venkateswarlu, P., & Bhumi, G., (2014). Evaluation of biologically synthesized silver nanoparticles from gum extract of *Boswellia ovalifoliolata* Bal. & Henry: An endemic endangered medicinal plant of Tirumala hill range of Andhra Pradesh, India. *Intern. J. Pharmaceut. Res. Scholars, 3*(1), 187–195.

Sayed, A. S., & El Sayed, N. S., (2016). Co-administration of 3-acetyl-11-keto-beta-boswellic acid potentiates the protective effect of celecoxib in lipopolysaccharide-induced cognitive impairment in mice: Possible implication of anti-inflammatory and antiglutamatergic pathways. *J. Mol. Neurosci., 59*(1), 58–67.

Schneider, H., & Weller, M., (2016). Boswellic acid activity against glioblastoma stem-like cells. *Oncol. Lett., 11*(6), 4187–4192.

Sedighi, B., Pardakhty, A., Kamali, H., Shafiee, K., & Hasani, B. N., (2014). Effect of *Boswellia papyrifera* on cognitive impairment in multiple sclerosis. *Iran J. Neurol., 13*(3), 149–153.

Selvan, A. T., Anusha, R., & Suthakaran, R., (2013). Phytochemical evaluation and anti-inflammatory activity of *Boswellia ovalifoliolata. Intern. J. Pharmaceut. Sci. Rev. Res., 22*(1), 148–151.

Sengupta, K., Kolla, J. N., Krishnaraju, A. V., Yalamanchili, N., Rao, C. V., & Golakoti, T., (2011). Cellular and molecular mechanisms of anti-inflammatory effect of Aflapin: A novel *Boswellia serrata* extract. *Mol. Cell Biochem., 354*, 189–197.

Sferra, R., Vetuschi, A., Catitti, V., Ammanniti, S., Pompili, S., & Melideo, D., (2012). *Boswellia serrata* and *Salvia miltiorrhiza* extracts reduce DMN-induced hepatic fibrosis in mice by TGF-beta1 downregulation. *Eur. Rev. Med. Pharmacol. Sci., 16*(11), 1484–1498.

Shao, Y., Ho, C. T., Chin, C. K., Badmaev, V., Ma, W., & Huang, M. T., (1998). Inhibitory activity of boswellic acids from *Boswellia serrata* against human leukemia HL-60 cells in culture. *Planta Med., 64*, 328–331.

Sharma, A., Bhatia, S., Kharya, M. D., Gajbhiye, V., Ganesh, N., & Namdeo, A. G., (2010). Anti-inflammatory and analgesic activity of different fractions of *Boswellia serrata. Int. J. Phytomed., 2*, 94–99.

Sharma, M. L., Bani, S., & Singh, G. B., (1989). Anti-arthritic activity of boswellic acids in bovine serum albumin (BSA)-induced arthritis. *Int. Immunopharmacol., 11*, 647–652.

Sharma, M. L., Khajuria, A., & Kaul, A., (1988). Effect of Salai guggul ex-*Boswellia serrata* on cellular and humoral immune responses and leucocyte migration. *Agents Actions, 24*, 161–164.

Shehata, A. M., Quintanilla-Fend, L., Bettio, S., Jauch, J., Scior, T., & Scherbaum, W. A., (2015). 11-Keto-b-boswellic acids prevent the development of autoimmune reactions, insulitis and reduce hyperglycemia during induction of multiple low-dose streptozotocin (MLD-STZ) diabetes in mice. *Horm. Metab. Res., 47*(6), 463–469.

Shehata, A. M., Quintanilla-Fend, L., Bettio, S., Singh, C. B., & Ammon, H. P., (2011). Prevention of multiple low-dose streptozotocin (MLD-STZ) diabetes in mice by an extract from gum resin of *Boswellia serrata* (BE). *Phytomedicine, 18*, 1037–1044.

Shen, Y., Takahashi, M., Byun, H. M., Link, A., Sharma, N., Balaguer, F., Leung, H. C., et al., (2012). Boswellic acid induces epigenetic alterations by modulating DNA methylation in colorectal cancer cells. *Cancer Biol. Ther., 13*, 542–552.

Shenvi, S., Kiran, K. R., Kumar, K., Diwakar, L., & Reddy, G. C., (2015). Synthesis and biological evaluation of boswellic acids – NSAID hybrid molecules as anti-inflammatory and anti-arthritic agents. *Eur. J. Med. Chem., 98*, 170–178.

Shinkafi, T. S., Bello, L., Hassan, S. W., & Ali, L., (2015). An ethnobotanical survey of antidiabetic plants used by Hausa–Fulani tribes in Sokoto, Northwest Nigeria. *J. Ethnopharmacol., 172*, 91–99.

*Siddiqui, M. Z., (*2011*). Boswellia serrata*, a potential anti-inflammatory agent: An overview. *Indian J. Pharm. Sci., 73*(3), 255–261.

Sina, I., (1998). *Al Qanoon fit Tibb.* [Trans: English] (Vol. 2, pp. 399, 400). Jamia Hamdard, New Delhi.

Soliman, A. M., Majeed, R. R., Himyari, H. F. A., UM, D., & Khan, S. A., (2020). *In vitro* cytotoxic, antioxidant and antimicrobial activities of alcoholic and chloroform extracts of Dhofari frankincense. *Dhaka Univ. J. Pharmaceut. Sci., 19*(2), 105–110.

Sudhakar, A., (1998). Pharmacognosy of some indigenous medicinal plants of Chittoor district, Andhra Pradesh, India. *Fitoterapia, 69*, 390–400.

Suhail, M. M., Wu, W., Cao, A., Mondalek, F. G., Fung, K. M., Shih, P. T., Fang, Y. T., et al., (2011). *Boswellia sacra* essential oil induces tumor cell-specific apoptosis and suppresses tumor aggressiveness in cultured human breast cancer cells. *BMC Complement. Altern. Med., 11*, 129.

Sumantran, V. N., Joshi, A. K., Boddul, S., Koppikar, S. J., Warude, D., Patwardhan, B., Chopra, A., et al., (2011). Antiarthritic activity of a standardized, multiherbal, Ayurvedic formulation containing *Boswellia serrata*: *In vitro* studies on knee cartilage from osteoarthritis patients. *Phytother Res, 25*(9), 1375–1380.

Sun, X., Geng, Y., Wang, X., Qin, D., & Yu, J., (2020). Cembrane-type diterpenoids from the gum resin of *Boswellia carterii* and their biological activities. *RSC Advances, 10*, 746–755.

Syrovets, T., Gschwend, J. E., Buchele, B., Laumonnier, Y., Zugmaier, W., Genze, F., & Simmet, T., (2005). Inhibition of IkappaB kinase activity by acetyl-boswellic acids promotes apoptosis in androgen-independent PC-3 prostate cancer cells *in vitro* and *in vivo*. *J. Biol. Chem., 280*, 6170–6180.

Takahashi, M., Sung, B., Shen, Y., Hur, K., Link, A., Boland, C. R., Aggarwal, B. B., & Goel, A., (2012). Boswellic acid exerts antitumor effects in colorectal cancer cells by modulating the expression of the let-7 and miR-200 microRNA family. *Carcinogenesis, 33*, 2441–2449.

Tausch, L., Henkel, A., Siemoneit, U., Poeckel, D., Kather, N., Franke, L., & Werz, O., (2009). Identification of human cathepsin G as a functional target of boswellic acids from the anti-inflammatory remedy frankincense. *J. Immunol., 183*(5), 3433–3442.

Thabet, N. M., Abdel-Rafei, M. K., & Moustafa, E. M., (2020). Boswellic acids protects against Bisphenol-A and gamma radiation-induced hepatic steatosis and cardiac remodeling in rats: Role of hepatic PPAR-alpha/P38 and cardiac Calcineurin-A/NFATcl /P38 pathway. *Arch Physiol. Biochem.*, 1–19.

Tilahun, G., (1997). *Boswellia papyrifera (Del.) Hochst. From western Tigray: Opportunities, Constraints and Seed Germination Responses* (pp. 1–58). MSc thesis. Report No. 1996: 12. Swedish University of Agricultural Sciences. Faculty of Forestry, Sweden.

Tripathi, Y. B., Birendra, S. K., Ravi, P. S., & Mohan, K., (2005). BHUx: A patent polyherbal formulation to prevent atherosclerosis. *eCAM, 2*(2), 217–221.

Tucker, A., (1986). Frankincense and myrrh. *Economic Botany, 40*(4), 425–433.

Umar, S., Umar, K., Saarwar, A. H., Khan, A., & Ahmad, N., (2014). *Boswellia serrata* extract attenuates inflammatory mediators and oxidative stress in collagen induced arthritis. *Phytomedicine, 21*(6), 847–856.

Uphof, J. C. Th., (1959). *Dictionary of Economic Plants.* Weinheim Publishers.

Ur Rehman, N., Halim, S. A., Al-Azri, M., Khan, M., Khan, A., Rafiq, K., Al-Rawahi, A., et al., (2020). Triterpenic acids as non-competitive α-glucosidase inhibitors from *Boswellia elongata* with structure-activity relationship: *In vitro* and in silico studies. *Biomolecules, 10*(5), 751.

Van, B. G., (1960). Frankincense and myrrh. *Bibli. Archaeol., 23*(3), 7095.

Van Vuuren, S. F., Kamatou, G. P., & Viljoen, A. M., (2010). Volatile composition and antimicrobial activity of twenty commercial frankincense essential oil samples. *S. Afr. J. Bot., 76*, 686–691.

Vedavathy, S., Mrudula, V., & Sudhakar, A., (1995). Herbal medicines for birth control, ante – and post-partum treatments from Chittoor District, Andhra Pradesh, India. *Fitoterapia, 66*(6), 501–506.

Védékoi, J., Selestin, S. D., Abdoulaye, H., Justin, K., Djamilah, Z., & Pierre, K., (2019). Investigation of antioxidant activity of the ethanol extract of the resin exudates of trunk bark of *Boswellia dalzielii* Hutch (Burseraceae). *J. Mater. Environ. Sci., 10*(12), 1413–1419.

Venkata Ratnam, K., & Venkata Raju, R.R. (2008). *In vitro* antimicrobial activity of leaf extracts of *Boswellia ovalifoliolata. J. Trop. Med. Plants, 9*(1), 27–29.

Venkateswarlu, P., Savithramma, N., & Ravisekhar, P., (2012). Anti-ulcer activity of gum extract of *Boswellia ovalifoliolata* Bal. & Henry against aspirin, ethanol and indomethacin induced gastric mucosal injury in albino rats. *Intern. J. Rev. Life Sci., 2*(1), 1–6.

Wang, D., Ge, S., Bai, J., & Song, Y., (2018). Boswellic acid exerts potent anticancer effects in HCT-116 human colon cancer cells mediated via induction of apoptosis, cell cycle arrest, cell migration inhibition and inhibition of PI3K/AKT signaling pathway. *J. BUON, 23*, 340–345.

Wang, R., Wang, Y., Gao, Z., & Qu, X., (2014). The comparative study of acetyl-11-keto-beta-boswellic acid (AKBA) and aspirin in the prevention of intestinal adenomatous polyposis in APC Min/+ mice. *Drug Discoveries & Therapeutics., 8*(1), 25–32.

Wildfeurer, A., Neu, I. S., Safayhi, H., Metzger, G., Wehrmann, M., Voge, U., & Ammon, H. P., (1998). Effects of boswellic acids extracted from a herbal medicine on the biosynthesis of leukotrienes and the course of experimental autoimmune encephalomyelitis. *Arzneimittelforschung, 48*(6), 668–674.

Xia, L., Chen, D., Han, R., Fang, Q., Waxman, S., & Jing, Y., (2005). Boswellic acid acetate induces apoptosis through caspase-mediated pathways in myeloid leukemia cells. *Mol. Cancer Ther., 4*, 381–388.

Yadav, V. R., Prasad, S., Sung, B., Gelovani, J. G., Guha, S., & Krishnan, S., (2011). Boswellic acid inhibits growth and metastasis of human colorectal cancer in orthotopic mouse model by downregulating inflammatory, proliferative, invasive, and angiogenic biomarkers. *Int. J. Cancer, 130*, 2176–2184.

Yoshikawa, M., Morikawa, T., Oominami, H., & Matsuda, H., (2009). Absolute stereostructures of olibanumols A, B, C, H, I, and J from olibanum, gum-resin of *Boswellia carterii*, and inhibitors of nitric oxide production in lipopolysaccharide-activated mouse peritoneal macrophages. *Chem. Pharm. Bull., 57*, 957–964.

Younoussa, L., Nukenine, E. N., & Esimone, C. O., (2016). Toxicity of *Boswellia dalzielii* (Burseraceae) leaf fractions against immature stages of *Anopheles gambiae* (Giles) and *Culex quinquefasciatus* (Say) (Diptera: Culicidae). *Int. J. Insect Sci., 8*, 23–31.

Yu, J. Q., Geng, Y. L., Wang, D. J., Zhao, H. W., Guo, L., & Wang, X., (2018). Terpenes from the gum resin of *Boswellia carterii* and their NO inhibitory activities. *Phytochemistry Letters, 28*, 59–63.

Yu, J., Zhao, L., Sun, X., Geng, Y., & Wang, X., (2020). Bioactive cembrane diterpenoids from the gum resin of *Boswellia carterii. Fitoterapia, 146*, 104699. doi: 10.1016/j.fitote.2020.104699.

Yuan, Y., Cui, S. X., Wang, Y., Ke, H. N., Wang, R. Q., & Lou, H. X., (2013). Acetyl-11-keto-beta-boswellic acid (AKBA) prevents human colonic adenocarcinoma growth through modulation of multiple signaling pathways. *Biochem Biophys Acta, 1830*, 4907–4916.

Yusuf, A., Salami, H., Babakura, M., Sandabe, U., Sambo, N., & Abubakar, A., (2018). Modulation of vascular endothelial growth factor and tissue necrosis factor alpha using ethanol stem bark extract of *Boswellia dalzielii* H. attenuates ethanol-induced gastric ulcer in albino rats. *Eurasian J. Med. Oncol., 2*(2), 97–104.

Zaitone, S. A., Barakat, B. M., Bilasy, S. E., Fawzy, M. S., Abdelaziz, E. Z., & Farag, N. E., (2015). Protective effect of boswellic acids versus pioglitazone in a rat model of diet-induced non-alcoholic fatty liver disease: Influence on insulin resistance and energy expenditure. *Naunyn Schmiedeb Arch Pharmacol., 388*(6), 587–600.

Zaker, S., Beheshti, S., Aghaie, R., & Noorbakhshnia, M., (2015). Effect of olibanum on a rat model of Alzheimer's disease induced by intracerebroventricular injection of streptozotocin. *Physiol. Pharmacol., 18*, 477–489.

Zhao, W., Entschladen, F., Liu, H., Niggemann, B., Fang, Q., Zaenker, K. S., & Han, R., (2003). Boswellic acid acetate induces differentiation and apoptosis in highly metastatic melanoma and fibrosarcoma. *Cancer Detect Prev., 27*, 67–75.

CHAPTER 7

PROPAGATION OF *BOSWELLIA* SPECIES

K. VENKATA RATNAM[1] and LEPAKSHI MD. BHAKSHU[2]

[1]*Department of Botany, Rayalaseema University, Kurnool – 518007, Andhra Pradesh, India*

[2]*Department of Botany, PVKN Government College, Chittoor – 517002, Andhra Pradesh, India*

**Corresponding author. E-mail: drvenkatapkd@gmail.com*

ABSTRACT

The genus *Boswellia* is the chief source of gum resin secreted by its stem portion, which had immense medicinal value in traditional systems. Most of the *Boswellia* species are propagated through seeds. The species growing in arid and terrarium areas act as a barrier, because these areas are poor in nutrients to germinate seeds. To overcome this and to improve *Boswellia* populations, botanists started propagation of *Boswellia* species through branch cuttings and other methods. Plant taxonomists and reproductive biologists observed that the following Socotra *Boswellia* species like *B. ameero, B. elongata, B. nana, B. dioscoridis, B. popoviana, B. bullata, B. socotrana*, and Indian species *B. serrata* and *B. ovalifoliolata* are propagated through seeds. These species produce seeds with limited success due to less viability or harsh environments. The alternate propagation method is through vegetative reproduction. This is also reported in a few species, such as *B. ovalifoliolata, B. serrata, B. papyrifera,* and *B. sacra.* Hence more propagative methods that are cost-effective to be implemented for a

Frankincense – Gum Olibanum: Botany, Chemistry, Extraction, Utilization, Propagation, Biotechnology, and Conservation, T. Pullaiah, K. Venkata Ratnam, Mallappa Kumara Swamy, Lepakshi Md. Bhakshu (Eds.)

large-scale production of saplings. Also, a wide range of plantation plots to be encouraged for both conservation and sustainable utilization of natural olio gum-yielding plant species.

7.1 INTRODUCTION

Most of the oleo gum resin-producing trees of the genus *Boswellia*, with about 24 species, are IUCN-listed as threatened and are under the risk of extinction. The gum resin is important since it is used in the religious ceremonies throughout the World, leaving huge demand for the supply, which is not compensated through the natural resources in the existing woods. Inappropriate tapping methods and exhaustive exploitation leading the trees at their life risk.

Poor seed setting and scanty seed germination restricts the distribution of *B. serrata* to a great extent. Dry seeds of *B. serrata* can be stored in tins for not more than 6–9 months. Moreover, excessive exploitation for tapping oleo-gum-resin with irregular harvesting practices also impedes its natural regeneration and jeopardizing its population (Purohit et al., 1995; Orwa et al., 2009; Ghorpade et al., 2010; Vijay, 2014). Brendler et al. (2018) have suggested for establishing sustainable supply chains and applying sustainable harvesting and management practices to protect *B. serrata* from overexploitation which is jeopardizing its population.

B. papyrifera may have self-incompatibility leading to reduced chances of successful reproduction. Besides, lack of uniform seed germination and seed dormancy along with forest fires, beetle-attacks, and grazing are some of the causes hampering conservation efforts for maintaining viable population of *B. papyrifera* (Rijkers et al., 2006; Eshete et al., 2005; Groenendijk et al., 2012; Gidey et al., 2020). Lemenih and Kassa (2011) have developed a guide with suitable directions for defensible production of frankincense from *B. papyrifera* as excessive tapping of trees for gum olibanum has limited its population in Ethiopia.

In *B. sacra* also, low flower and fruit production and production of non-viable seeds reduces natural regeneration. Moreover, tree cutting, animal grazing and leaf collection for fodder purposes are also some of the causes of *B. sacra* degradation (Raffaelli et al., 2008). Recently, unsustainable harvesting of Frankincense at high level without significantly increasing livelihood outcomes or community development has raised issues regarding sustainable management to break such damaging cycles of *Boswellia* species population in Somaliland (DeCarlo et al., 2020).

Moreover, these species have limited individuals, and even the seed setting and its germination success rate also are very low. Hence the propagative methods, in addition to seeds, have been investigated and practiced (Gonzalez, 2020). Vegetative propagation is recommended, and more emphasis has to develop the nurseries to produce high-quality clones to suit the arid environments existing in the native lands of this genus. The reports on the propagation are very limited and noted with a few species, such as *B. ovalifoliolata, B. serrata, B. papyrifera,* and *B. sacra,* and the other species were yet to be reported as per the authors' knowledge. The propagation methods of the above-mentioned species are discussed herein.

7.2 PROPAGATION METHODS

Al-Harrasi et al. (2019) gave a detailed account on propagation methods of different species of *Boswellia.* Different species of *Boswellia* grow under varied environmental conditions in the tropics. *B. serrata* grows in the humid region of tropical zone, *B. sacra* grows in tropical dry regions while *B. socotrana* prefers the winter. Almost all *Boswellia* species need humidity and heat (Eslamieh, 2017).

Boswellia plants grow well in soil medium that had equal parts of quarter-inch solid granular material, screened perlite, and sterilized compost (Eslamieh, 2017). However, various researchers or growers have replaced perlite with vermiculite. Granular gypsum, decomposed granite, clay balls, and other rock-type bonsai mixes can be used instead of pumice (Al-Harrasi et al., 2019). Proper irrigation for the plants during growth is necessary. Regular irrigation is required under straight compost, and growers should never allow the soil to dry during the active month for humid-tropical species (Eslamieh, 2017).

Different species of *Boswellia* show different rates of germination. Species like *B. nana* show a high germination rate, while *B. sacra* has one of the lowest germination rates (with the highest seed production). Similarly, the viability of *Boswellia* seeds also varies with the conditions (humidity, temperature, and light), and most seeds can grow after 3 or more years (Al-Harrasi et al., 2019).

The amount and frequency of watering depend on various conditions and factors. *Boswellia* species lack an appropriate system for retaining water for longer times during harsh conditions because they are semi-succulents. However, they can develop tubers during possible drought seasons. A *Boswellia* plant can survive in a pot as well as in the field without water for some time, but after a few weeks, it will start defoliating. Supplying sufficient water after defoliation can lead to a quick recovery for the plants (Al-Harrasi et al., 2019).

Basic fertilizers with balanced NPK and chelated metal are sufficient for better plant growth. Some additional elements are needed for better plant growth. Seaweed-based fertilizer is used by some growers for better growth for some humid-tropical species such as *B. serrata, B. ovalifoliolata*, and *B. dalzielii.* For *Boswellia* propagation and flowering, sufficient nutrients are necessary. For increased vegetative growth of cuttings, the fertilizer should be enriched with nitrogen (Al-Harrasi et al., 2019).

Propagation by root suckers is another way to produce *Boswellia* plants (Eslamieh, 2017). Propagation by this method employing root sprouts arise from exposed roots of a mature *B. papyrifera* plant, which can grow into a mature tree under protected conditions. The propagation rate of *B. serrata* in natural conditions is too slow, when older trees are propagated through root suckers (Lemenih and Kassa, 2011).

7.2.1 *BOSWELLIA OVALIFOLIOLATA* BAL. & HENRY

The efficacy of silver nanoparticles prepared from *B. ovalifoliolata* stem bark on germination of seeds and growth of seedlings was assessed using *in vitro* conditions. Nanoparticles-treated seeds showed 90% germination than control seeds (70%). The seeds treated with nanoparticles started sprouting from the 7th day onwards and germination was completed between 7 and 10 days, whereas control seeds started from the 15th day onwards, and germination was completed in a time schedule of 10–20 days. Further, the particle enhanced seedlings growth in a concentration-dependent manner (Savithramma et al., 2012).

7.2.2 *BOSWELLIA PAPYRIFERA* (CAILL. EX DELILE) HOCHST

Study on propagation of *B. papyrifera* using branch cuttings treated with IAA rich *Euphorbia abyssinica* latex using basal dip method. The experiment resulted that, 54% of branch cuttings survived and only 40% of cuttings with roots were observed (Negussie et al., 2009). A comparative propagation study on effect of collection time, length of branch cutting on survival rate of *B. papyrifera* collected from Ethiopia was carried out. Results demonstrated that the survival capacity of branch cuttings was proportional to the length of branch cuttings collected in the months of May, March, and October. The highest survival capacity (80%) of branch cuttings were observed in the branch cuttings with 2 m length collected in May month than in cuttings with 1 m (57%) and 1.5 m (75%). Very poor survival capacity of cuttings was observed in cuttings collected in October month, during the leaf-shedding period (Abiyu et al., 2016).

Lemenish and Kassa (2011) described a method of vegetative propagation through rooted cuttings of *B. papyrifera.* Rooted cuttings should be gathered from mature plants which are healthy and vigorous. Suitable season for collection of root cuttings and for planting is when the donor trees are dormant during the dry season. Cuttings should be 1.0 to 1.5 m long and with 20 cm girth for propagation. To protect them from desiccation/dehydration branch cuttings should be planted immediately. Pits size for planting should be 45–60 cm deep with 20 cm diameter. The thick branches should be sliced at 45° slanting angle and cutting samples inserted into the soil at a leaning angle. To escape dehydration and decay, the tip of the cuttings should be covered with fungicide treated water repellent coatings. Proper care can be taken during the raise of plantlets as proposed by Lemenih and Kassa (2011), such as air tightening in the pits, water accumulation, heaping or mounding around the cuttings and best agricultural practices to protect from pests, weeds, or browsing/grazing animals.

Lemenih and Kassa (2011) observed that large amount of seed germination was observed in natural environment during rainy season, but most of them failed to survive during dry season. They also succeeded propagation of *B. papyrifera* through root cuttings of 1–1.5 m length collected during dry season, when the donor plant in dormant condition. Propagation of *B. papyrifera* through root suckers were also observed in the natural environment (Lemenih and Kassa, 2011). Propagation by this method employing the exposed roots of a mature *B. papyrifera* tree can initiate a root sprout, which can grow into a mature tree under protected conditions (Figure 7.1).

FIGURE 7.1 Seedlings of *B. payrifera* in their natural environment.

Source: Reprinted from Lemenih and Kassa (2011).

The germination study on *B. papyrifera* seeds showed that germination does not exceed 7% under different treatments and the seedlings survival rates ranged from 27.6–31.8% (Adam, 2003). Seed germination test was carried out on *B. papyrifera* using different treatment methods such as cold water, hot water, and different concentrations of sulfuric acid (Adam and El-Tayeb, 2008). The results indicated that seeds treated with cold and hot water showed very feeble germination capacity; less than 8% only. The seeds treated with concentrated sulfuric acid found to be destructive. The authors studied the effect of seed storage on viability, those seeds sown after immediate collection showed 4.7%, while seeds sowed after one year of collection showed 2.5% germination (Adam and El-Tayeb, 2008). In contrast to this, *B. papyrifera* seeds collected from Eritrea exhibited significantly higher germination rates (80%–94%) from untapped trees and 14–16% for seeds collected from tapped populations (Ogbazgahi, 2001).

The effect of collection time and size of branch cuttings on shooting and rooting of *B. papyrifera* was assessed (Haile et al., 2011). It revealed that stem cuttings collected in May showed very poor shooting number, leaf size and number, rooting percentage than the cuttings collected between February–March. Regarding cutting size medium to long size cuttings survived better than short and weak branch cuttings.

7.2.3 *BOSWELLIA SACRA* FLÜCK.

A comparative seed germination study was carried out by using seed samples collected from different localities, at the sea level and 400 m above mean sea level and three methods, i.e., open tray, leaf litter and utilizing terrarium (Swartout and Solowey, 2018). The results, based on seed collection locality, revealed that seeds collected from tree species growing at 400 m sea level showed the highest germination capacity than the seeds collected from trees growing at sea level. Regarding methods results, open tray method showed good germination capacity than leaf litter and terrarium methods (Figure 7.2).

FIGURE 7.2 Seedlings of *B. sacra* growing in the greenhouse after 14 days.

Source: Reprinted from Swartout and Solowey (2018). © 2013 Net Journals. Open access.

Aa per horticultural results carried out by Eslamieh (2017) on propagation of *Boswellia* species, revealed that seeds of *B. sacra* readily germinate and seedlings will grow luxuriantly on any type of soil. After 4–5 months, the seedlings will reach plantable size. Seedlings developed from root cuttings in nurseries showed very poor survival rates when grown in natural conditions (Eslamieh, 2017). Most of the *Boswellia* species can propagate through the development of root tubers/thickened roots. The selection of root tuber/thickened root exposed major portion to the above the soil in *B. sacra* will develop caudex portion, which will eventually have characteristics of a seed-grown plant (Eslamieh, 2017).

Al-Harrasi et al. (2019) gave a detailed account on propagation of *B. sacra. B. sacra* grows in dry tropical regions, and has adapted to the Sonoran Desert with its very short winter season.

7.2.4 *BOSWELLIA SERRATA* ROXB.

This drought-resistant and frost-hardy tree is recommended for soil reclamation, and afforestation on poorer soils and typically grows on rocky ridges; it grows well on neutral soils above limestone, gneiss, quartzite, sandstone, and schist. The species can withstand and grows well in soils poor in nutrients, where most of its associated species remain stunted. The species is suitable for plantation in soils very poor in nutrients and received a mean annual rainfall of 500–1,250 mm. It is valuable tree species for plantation on dry sandstone hills, where soil is markedly ferruginous and interference of biotic factors are more. The species has great assets such as i) to withstand the adverse effects of the environment (for example, forest fire), ii) regeneration capacity after repeated grazing of animals, iii) drought-resistant capacity iv) reproduction capacity through seed and root suckers. Regular practice of coppicing (pruning) to emerge new branches and pollarding is a practice to remove the tree at soil surface level which also initiates new or emerging or spreading branches and this plant reported good response for coppices and pollards even for stem/root cutting propagation. The capacity of the production of vegetative shoots depends on environmental factors. Seeds have poor viability, but they can be stored dry in storage boxes for not more than 6–9 months. There are 13,400–25,600 seeds/kg (Orwa et al., 2009) (Figure 7.3).

FIGURE 7.3 Seedlings of *B. serrata* found on the forest floor at various locations in Madhya Pradesh in India.

Source: Dr. N. Venugopal, with permission.

Vegetative propagation of *B. serrata* through branch cuttings was carried out by Vaishnav and Janghel (2018). Five branch cuttings collected from different *B. serrata* measuring about 25 cm in length and 1–2 cm diameter and one branch cutting from each plant having 1 m length and 4 cm diameter were selected for propagation. All branch cuttings were treated with IAA and IBA at 500 ppm to 2,000 ppm concentrations. The branch cuttings treated with growth hormones were planted in pots mixed with sand, soil, and vermicompost (2:1:1 ratio). After 50 days of plantation, no growth and bud formation were observed in 25 cm branch cuttings. Three out of 5 of 1 m length samples of branch cuttings survived and bud formation, leaf differentiation and shoot formation were observed at 12, 22, and 47 days of plantation, respectively (Figure 7.4) (Vaishnav and Janghel, 2018). Similar experiment was conducted by Bedi et al. (2011) propagation of *B. serrata* with seedling branch cuttings 10–12 week aged, resulted 80% success.

FIGURE 7.4 Propagation of *Boswellia serrata* Roxb. (A) Bottom to the top view of a tree trunk; (B) & (C) thick and longer branch cuttings propagated; (D) & (E) shoot proliferation from large sized cuttings; (F) root proliferation from large size cuttings digging out after 65 days; (G) seeds used for planting; and (H) 25 cm long cuttings propagated.

Source: Reprinted from Vaishnav and Janghel (2018), with permission.

For rooting stem cuttings were treated with different auxins. Indole-3-acetic acid (IAA) Indole-3-butyric acid (IBA), and Naphthalene acetic acid (NAA) were used, and IBA showed a better result. Stem cuttings were treated with 500 ppm IBA for 15 min, which yielded a better result in combination with sand: soil (1:1) rooting media (Vijay, 2014). When stem cuttings were treated with different concentrations of IAA and NAA, the cuttings showed low rooting % and sprouting, and the cuttings survived for nearly 10 days (Vijay, 2014).

The effect of different treatments, which includes mechanical scarification, hot water treatment, sulfuric acid, KNO_3 and thiourea, and growth regulators such as IAA and GA_3 on seed germination of *B. serrata,* was studied by Khan (2015). The results demonstrate that seeds treated with hot water for half an hour showed 63% germination, mechanical scarification showed 68% germination. Seeds pre-treated with 1% thiourea/HNO_3 and sulfuric acid for a period of 30 minutes resulted in 78, 76, and 75%, respectively. Seeds treated with growth hormones GA_3 and IAA at 750 ppm showed the highest germination capacity than the other treatments. In another study, 60–65% germination capacity was noticed in *B. serrata* seeds sowed under shade nets (Rawat et al., 2015). Eslamieh (2017) noted that rooting of stem cuttings of *B. serrata* is a very easy and inexpensive method to propagate the species.

7.3 CONCLUSION

The genus *Boswellia* is one of the commercial olio gum resins-producing taxon, which is distributed over restricted areas of the globe. The tribal people residing with the adjacent areas depend on gum extraction and for its commercial values. They make a livelihood during the drought seasons, and many people rely on these trees. Moreover, with ever-increasing demand, the trees are over-exploited for the gum, which is leading plants to the risk of life and consequently becoming endangered (EN). Hence more action is to be taken to improve the plants, and more plantations need to be taken up using possible propagation measures. It also noted to establish a greater number of *Boswellia* orchards and sustainable gum-yielding protocols. Even the wild trees also need to be taken individual care during the extraction of gum in addition to providing training to dependent people by engaging governmental or non-governmental agencies.

KEYWORDS

- **indole-3-acetic acid**
- **naphthalene acetic acid**
- **olio gum resin**
- **over-exploitation**
- **propagation methods**
- **sustainable utilization *Boswellia* species**

REFERENCES

Abiyu, A., Dejene, T., Eshete, A., & Sisay, K., (2016). Vegetative propagation of *Boswellia papyrifera*: Time of collection and propagule size affect survival and establishment. *J. Arid Environments, 133*, 122–124.

Adam, A. A., & El Tayeb, A. M., (2008). A comparative study of natural regeneration of *B. papyrifera* and other tree species in Jebel Marra Darfur Sudan. *Res. J. Agric. Biol. Sci., 4*(1), 94–102.

Adam, A. A., (2003). *Some Aspects of Ecology and Management of Boswellia papyrifera (Del) Hochst. in Jebel Marra Aream Darfur-Sudan*. Ph.D. Diss. University of Khartoum.

Al-Harrasi, A., Khan, A. L., Asaf, S., & Al-Rawahi, A., (2019). Propagation and conservation of *Boswellia sacra*. In: Al-Harrasi, et al., (eds.), *Biology of Genus Boswellia* (pp. 71–78). Springer.

Bedi, Y., Dutt, H., & Kaur, H., (2011). *Plants of Indian Systems of Medicine: Monographs on Twenty Commercially Important Medicinal Plants: Boswellia serrata Roxb* (Vol. I). Saarbrücken: Lambert.

Brendler, T., Brinckmann, J. A., & Schippmann, U., (2018). Sustainable supply, a foundation for natural product development: The case of Indian frankincense (*Boswellia serrata* Roxb. ex Colebr.). *J. Ethnopharmacol., 225*, 279–286.

DeCarlo, A., Ali, S., & Ceroni, M., (2020). Ecological and economic sustainability of non-timber forest products in post-conflict recovery: A case study of the frankincense (*Boswellia* spp.) resin harvesting in Somaliland (Somalia). *Sustainability, 12*, 3578. doi: 10.3390/su12093578.

Eshete, A., Teketay, D., & Håkan, H., (2005). The socio-economic importance and status of populations of *Boswellia papyrifera* (Del.) Hochst. in Northern Ethiopia: The case of North Gonder zone. *For. For. Trees Livelihoods, 15*, 55–74.

Eslamieh, J., (2017). *The Genus Boswellia: Preservation Through Horticulture* (p. 80527). Fort Collins, CO: A Book's Mind.

Ghorpade, R. P., Chopra, A., & Nikam, T. D., (2011). Influence of biotic and abiotic elicitors on four major isomers of boswellic acid in callus culture of *Boswellia serrata* Roxb. *Plant Omics J., 4*(4), 169–176.

Gidey, T., Hagos, D., Juhar, H. M., Solomon, N., Negussie, A., Crous-Duran, J., Oliveira, T. S., et al., (2020). Population status of *Boswellia papyrifera* woodland and prioritizing its conservation interventions using multi-criteria decision model in northern Ethiopia. *Heliyon, 6*(10), e05139.

Gonzalez, K., (2020). *Impacts of Unsustainable Harvesting of Frankincense-Producing Boswellia Trees* (pp. 1–40). Paper Submitted to Oregon State University, In partial fulfilment of the requirements for the degree of Master of Natural Resources.

Groenendijk, P., Eshete, A., Sterck, F., Zuidema, P., & Bongers, F., (2012). Limitations to sustainable frankincense production: Blocked regeneration, high adult mortality and declining populations. *J. Appl. Ecol., 49*(1), 164–173.

Haile, G., Gebrehiwot, K., Lemenih, M., & Bongers, F., (2011). Time of collection and cutting sizes affect vegetative propagation of *Boswellia papyrifera* (Del.) Hochst through leafless branch cuttings. *J. Arid Environments, 75*(9), 873–877.

http://apps.worldagroforestry.org/treedb/AFTPDFS/Boswellia_serrata.PDF (accessed on 10 October 2022).

Khan, M. R., (2015). Studies on seed germination of a threatened endangered medicinal plant species *Boswellia serrata* Roxb. *Studies, 2321*, 7871.

Lemenih, M., & Kassa, H., (2011). *Management Guide for Sustainable Production of Frankincense.* CIFOR, Bogor, Indonesia.

Negussie, A., Aerts, R., Gebrehiwot, K., Prinsen, E., & Muys, B., (2009). *Euphorbia abyssinica* latex promotes rooting of *Boswellia* cuttings. *New Forests, 37*, 35–42.

Ogbazghi, W., (2001). *The Distribution and Regeneration of Boswellia papyrifera (Del.). Hochst. in Eritrea.* PhD Thesis, Wageningen University and Research Centre, The Netherlands.

Orwa, C., Mutua, A., Kindt, R., Jamnadass, R., & Anthony, S., (2009). *Agroforestree Database: A Tree Reference and Selection Guide Version 4.0.* http://www.worldagroforestry.org/sites/treedbs/treedatabase (accessed on 10 October 2022).

Purohit, S. D., Tak, K., & Kukda, G., (1995). *In vitro* propagation of *Boswellia serrata* Roxb. *Biol. Plant., 37*, 335–340.

Raffaelli, M., Tardelli, M., & Mosti, S., (2008). Preserving and restoring the frankincense tree (*Boswellia sacra*) at Wadi Doka: A work in progress. A port in Arabia between Rome and the Indian Ocean (3rd C BC–5th C. AD). *Khor Rori Report, 2*, 715–723.

Rawat, R. B. S., Singh, S. K., & Singh, R., (2015). Propagation of Sal (*Shorea robusta*) associates in Tarai region of Uttarakhand through improved nursery practices. *Advances in Tree Seed Science and Silviculture, 200*.

Rijkers, T., Ogbazghi, W., Wessel, M., & Bongers, F., (2006). The effect of tapping for frankincense on sexual reproduction in *Boswellia papyrifera*. *J. Appl. Ecol., 43*, 1188–1195. https://doi.org/10.1111/j.1365-2664.2006.01215.x.

Savithramma, N., Ankanna, S., & Bhumi, G., (2012). Effect of nanoparticles on seed germination and seedling growth of *Boswellia ovalifoliolata* – an endemic and endangered medicinal tree taxon. *Nano Vision, 2*, 61–68.

Swartout, B. T., & Solowey, E., (2018). Increasing *Boswellia sacra* seeds' germination viability and genetic variability utilizing various methods. *Net J. Agric. Sci., 6*, 29–34.

Vaishnav, V., & Janghel, U., (2018). A note on the clonal propagation of depleted threatened species *Boswellia serrata* Roxb. through branch cuttings. *Trop. Plant Res., 5*(1), 27, 28.

Vijay, N., (2014). Studies on *in vitro* cloning of mature tree of *Capparis decidua* and *Boswellia serrata*. Ph.D. thesis submitted to Arid Forest Research Institute, Jodhpur. http://hdl.handle.net/10603/43660 (accessed on 10 October 2022).

CHAPTER 8

IN VITRO PROPAGATION AND BIOTECHNOLOGICAL ASPECTS OF *BOSWELLIA* SPECIES

VARTIKA JAIN,[1] MIMOSA GHORAI,[2] DEVENDRA KUMAR PANDEY,[3] ABDEL RAHMAN AL-TAWAHA,[4] ERCAN BURSAL,[5] MAHIPAL S. SHEKHAWAT,[6] GABER EL-SABER BATIHA,[7] BOREGOWDA PURUSHOTHAM,[8] MALLAPPA KUMARA SWAMY,[8] and ABHIJIT DEY[2*]

[1]*Department of Botany, Government Meera Girls' College, Udaipur – 313001, Rajasthan, India*

[2]*Department of Life Sciences, Presidency University, Kolkata, West Bengal, India*

[3]*Department of Biotechnology, Lovely Professional University, Phagwara, Punjab, India*

[4]*Department of Biological Sciences, Al-Hussein Bin Talal University, Ma'an, Jordan*

[5]*Department of Biochemistry, Mus Alparslan University, Turkey*

[6]*Department of Plant Biology and Biotechnology, Kanchi Mamunivar Government Institute for Postgraduate Studies and Research, Pondicherry, Tamil Nadu, India*

[7]*Department of Pharmacology and Therapeutics, Faculty of Veterinary Medicine, Damanhur University, Damanhur, AlBeheira, Egypt*

[8]*Department of Biotechnology, East-West First Grade College, Bangalore, Karnataka, India*

[*]*Corresponding author. E-mail: abhijit.dbs@presiuniv.ac.in*

Frankincense – Gum Olibanum: Botany, Chemistry, Extraction, Utilization, Propagation, Biotechnology, and Conservation, T. Pullaiah, K. Venkata Ratnam, Mallappa Kumara Swamy, Lepakshi Md. Bhakshu (Eds.)

ABSTRACT

The Burseraceae family is popular for resinous trees and shrubs. *Boswellia* is one of the prominent genera of this family, which yields the oleo-gum-resin known as 'Frankincense.' *Boswellia* spp. are widely distributed, and the major frankincense-yielding species are *B. serrata* Roxb. ex Colebr. (India), *B. sacra* Flueck. (Arabia and Somalia), *B. papyrifera* (Del.) Hochst and *B. frereana* Birdw. (Northern Africa). The oleo-gum-resin is rich in terpenes and various phytoconstituents such as boswellic acid and its derivatives and, therefore, exhibits many health-beneficial activities. Traditionally, the gum-resin is not only used as medicine but also used in various socio-cultural events. Excessive tapping of trees and irregular harvesting of gum resin, along with grazing, forest fires, and poor seed setting, are some of the reasons jeopardizing the population of *Boswellia* spp. and raising the need for conservation. The present chapter deals with biotechnological interventions for the propagation of *Boswellia* spp. through micropropagation and the use of elicitors/endophytes and other techniques for *in vitro* synthesis of boswellic acid, along with analysis of genetic diversity carried out in the last three decades.

8.1 INTRODUCTION

Boswellia genus is out of 17 genera described in the family Burseraceae, which is mainly popular for Myrrh and Frankincense. The genus *Boswellia* includes 24 species of trees and shrubs. These species are highly valued for their oleo-gum-resin; a non-timber forest product which possesses immense medicinal properties. India, Ethiopia, and Somalia are major exporting countries of frankincense in the world. There are four main species of *Boswellia* which yield oleo-gum-resins of commercial trade value *viz. B. serrata* Roxb. ex Colebr., *B. sacra* Flueck., *B. frereana* Birdw. and *B. papyrifera* (Caill. ex Delile) Hochst. The oleo-gum-resin is traded by the name 'Frankincense' all over the world (Lemenih and Kassa, 2011; McCutcheon, 2018; Al-Harrasi et al., 2019).

Due to poor regeneration potential, many of the *Boswellia* species are listed in IUCN's Red Data Book. For example, *B. sacra* is kept under category of near threatened, *B. ovalifoliolata, B. elongata, B. ameero, B. pirottae, B. bullata, B. nana, B. popoviana, B. socotrana,* and *B. dioscoridis* are under Vulnerable category and *B. ogadensis* is a critically endangered (EN) species (Website 1: https://www.iucnredlist.org/species/34533/9874201). Therefore, propagation of *Boswellia* spp. through *ex situ* conservation techniques has been the focus of research world over in the last three decades. The present chapter will describe the efforts carried out for *ex situ* conservation of some major *Boswellia* spp., i.e., *B. serrata*, *B. sacra,* and *B. papyrifera*.

Boswellia species are mainly known for the oleo-gum-resins which are also termed as Frankincense having cultural, economic, and medicinal values (Hairfield et al., 1989). It is used in various social and religious celebrations in Arabian culture. It is burnt on hot charcoal and the aromatic smoke is also used for the treatment of psychosomatic disorders. Chemically, resin is rich in mono-, di-, or triterpenes depending on the species. More than 300 components have been identified from the essential oil of different *Boswellia* species (Mertens et al., 2009) and the resin contains more than 100 diterpenes and triterpenes (Al-Harrasi et al., 2021). For the details on phytochemistry and pharmacology reader may refer to Chapters 5 and 6.

In view of high demand of therapeutic chemical compounds obtained from *Boswellia*, its large-scale cultivation is required and due to low rate of natural propagation, *ex situ* conservation tools and techniques should be employed. Plant tissue culture is one of the important means to propagate any plant species within a short period of time. Further sections describe the various measures and biotechnological approaches adopted for the propagation of *Boswellia* species.

8.2 *IN VITRO* PROPAGATION OF *BOSWELLIA* SPP.

As natural regeneration capacity of *Boswellia* species is very low, scientists have worked towards alternative propagation techniques. Plant Tissue culture technique offers rapid multiplication of plant species with minimum quantity of available inoculum and plays a major role in *ex-situ* conservation of rare, EN, and threatened plant species. Micropropagation has been utilized to propagate different *Boswellia* spp. by scientists' world over for its large-scale production. Some of those important findings are presented in subsections.

8.2.1 *B. SERRATA*

Purohit et al. (1995) successfully developed a protocol for *in vitro* multiplication of *B. serrata* using cotyledonary node segments obtained from one-month-old *in vitro* grown seedlings for the first time. Sturdy explants were obtained from the seedlings grown on MS medium rather than on water-agar. Within 22 days, they obtained average of 4 shoots per node on MS medium supplemented with growth regulators 6-benzyl aminopurine (BAP 0.5 mgdm^{-3}) and naphthalene acetic acid (NAA 0.05 mgdm^{-3}). Efficacy of NAA was more than IAA for elongation of shoots and it further helped to induce rooting. After 88 days, 90–100 shoots per node were obtained by doing repeated subculture. No shoots were obtained in cotyledonary nodes cultured on MS medium without growth regulators. For shoot induction, cotyledonary nodes proved as better explants than epicotyledonary nodes. Approximately 80% of shoots developed callus free rooting on MS medium containing ¼ salts and 1% saccharose supplemented with 0.5 mgdm^{-3} IBA and 0.25 mg dm^{-3} IAA within 8–10 days. Browning of cultures was prevented by addition of antioxidants like PVP (50 mg dm^{-3}) and ascorbic acid (100 mg dm^{-3}) in both multiplication and rooting media. Hardening was achieved through shifting of plants from medium to the culture bottles with 1/4 MS salts and without saccharose resulting in 70% survival at pot level containing soil: Soilrite TM (1:1) mixture.

Ghorpade et al. (2010) collected seeds from green mature fruits and cultured excised green zygotic embryos on various media with different concentrations of growth regulators, sucrose, and PVP. Maximum (96%) embryo germination and conversion into seedling was observed on MS medium having 3% sucrose along with 200 mg/l PVP and 91% on MS medium lacking PVP. Out of those germinated embryos, 82% and 64% were converted into seedlings, respectively, and also survived after transferring to the sand soil mixture (1:1). All other media, such as B5 by Gamborg, Woody Plant Medium by McCown and Loyd and Schenk and Hildebrandt did not produce desired results and induced abnormalities in the seedlings along with callus formation. Notably, reduced germination of green mature seeds was observed after the addition of auxins (IAA, IBA, NAA, or 2,4-D) in MS medium along with the formation of white creamy callus embryo surfaces. Moreover, no embryo germination was observed from dry mature untreated and dry mature treated seeds (treatment with wet heat, acid scarification, auxins, cytokinins, and GA_3). This study indicates the importance of employing seeds from green mature fruits of *B. serrata* for better multiplication potential (Figure 8.1).

FIGURE 8.1 *In vitro* zygotic embryo germination in *Boswellia serrata*. (A) Stem branch with green mature fruits (bar 10 mm); (B) excised green mature embryos (bar 10 mm); (C) *in vitro* germinated embryo with well-developed root, hypocotyl, and expanded green cotyledonary leaves on MS + 3% sucrose + 200 mg/l PVP (bar 10 mm); (D) *in vitro* germinated embryos removed from culture tube (bar 10 mm); and (E) plants (3-month-old) in soil under field conditions (bar 25 mm).

Source: Reprinted from Ghorpade et al. (2010), with permission.

Suthar et al. (2011) reported that liquid medium (without agar) or very weakly gelled medium (0.2% agar) produced more than 20 shoots rather than medium having 0.8% agar which produced 11.3 shoots. Moreover, maximum shoot length of 2.83±0.15 cm with the highest number of leaves (126) per cluster was also observed on the liquid medium without any signs of hyperhydricity. The increase in dry and fresh weight and chlorophyll contents of such shoots further suggested the use of reduced agar concentrations for shoot multiplication of *B. serrata*. Similar observations of a maximum of 5.7 roots per shoot with an average length of 4.3 cm were recorded on liquid medium showing the best (75%) rooting response. Overall, the medium gelled with higher concentrations (0.8 and 1.0% w/v) of agar showed reduction in all the growth parameters of shoots and roots indicating that use of liquid media with reduced concentration of agar could be a better alternative for micropropagation of *B. serrata*.

Biological hardening and acclimatization of *in vitro* raised plantlets of *B. serrata* by using endophytic root fungus *Piriformospora indica* has been reported by Suthar and Purohit (2012). Fungus inoculated plantlets showed increased fresh and dry weight (DW), highest chlorophyll content, maximum height, and increased number of leaves after one month of survival in green house conditions, increase in number of roots and root length per plantlet better growth and 7% higher survival percentage in polybags as compared to control. The positive results of this study suggest that biopriming with microbes could be a better alternative for growth and *ex vitro* survival of tissue culture raised plantlets.

For micropropagation of *B. serrata*, nodal segments containing axillary bud which were pre-disinfected with mixture of aqueous solution of fungicide Bavistin (0.2%) and bactericide streptomycin (0.1%) for 10 min and with 0.1% $HgCl_2$ for 7 min were employed. To avoid the leaching of phenolics and browning, use of PVP (100 mg/l) and antioxidants such as ascorbic acid (50 mg/l) and citric acid (30 mg/l) in medium or rapid subculturing in every three days was carried out. This led towards production of 83.33% aseptic explants. Notably, nodal segments collected during months from September to November exhibited maximum bud break response (57.88%) on MS medium supplemented with BAP (8.88 μM) + NAA (0.54 μM) and additives. For macropropagation, stem cutting of *B. serrata* collected during October-November, treated with 500 ppm IBA for 15 min induced rooting in 60% which was better than other auxins used such as IAA and NAA (Vijay, 2014).

8.2.2 *B. SACRA*

Quazi et al. (2014) have developed *in vitro* callus culture using leaf explant of *B. sacra* which was inoculated in MS medium having growth regulators 2,4-D (2,4-dichlorophenoxy acetic acid) and IAA in concentration of 2 mg/L, 5 mg/L, 10 mg/L, 3% sucrose and 1% agar for gelling. Best callus induction (84.6%) was observed on MS medium with 2 mg/L and 10 mg/L concentration of IAA after two weeks. However, callus induction from stem explants with the similar hormone concentration could not be developed.

Khan et al. (2018) have devised a method using near-infrared spectroscopy to assess endogenous phytohormone content from 13 populations of *B. sacra* growing in Oman which showed a trend of higher amount of salicylic acid>indole acetic acid/gibberellic acid>kinetin. This method could be applied to assess plant health and further adopt conservation strategies. However, other techniques to increase the production of frankincense are application of ethrel and methyl jasmonate on debarked wounds on *B. sacra* trees which have shown enhancement in resin production within 7 days of application (Yamomato et al., 2020).

8.2.3 *B. OVALIFOLIOLATA*

Chandrasekhara et al. (2005) utilized cotyledonary nodal explants of *B. ovalifoliolata* which were grown on modified Murashige and Skoog (MS) medium. The highest shoot multiplication (7.1 days 0.2 shoots per node) was observed after 50 days of inoculation on MS medium supplemented with 2.72 μm thidiazuron as compared to medium supplemented with 6-benzyladenine (BA) and kinetin. Rooting was induced in excised shoot cuttings of 3.0 cm using basal MS medium with addition of indole-3-acetic acid (IAA) and indole-3-butyric acid (IBA). Browning of cultures was effectively reduced by adding activated charcoal in a concentration of 100 mg/1 and polyvinylpyrrolidone (PVP) in 40 mg/1 concentration to the medium. Successfully acclimatization was observed in the regenerated plantlets.

8.3 PRODUCTION OF BOSWELLIC ACID BY *IN VITRO* METHODS

In vitro production of secondary metabolites is required, especially in cases where the natural resource is limited or at the verge of extinction. Boswellic acid is a medicinally important constituent present in oleo-gum-resin exuded from *Boswellia* spp. having four major components; β-boswellic acid (βBA), 11-keto-β-boswellic acid (KβBA), acetyl-β-boswellic acid (AβBA), and acetyl-11-keto-β-boswellic acid (AKβBA) besides keto boswellic acid, α-boswellic acid and 3-O-Acetyl-α-boswellic acid (Al-Harassi et al., 2019; Katragunta et al., 2019). Ghorpade et al. (2011) developed callus cultures from embryo explants isolated from mature green fruits of *B. serrata* on MS medium supplemented with 2.5 μM BA, 15 μM IAA and 200 mg/l PVP. Elicitors are utilized to increase the quantity of secondary metabolites in the cultures. In this study, abiotic elicitors namely, sodium chloride, sodium sulfate, ultraviolet light (UV-C) exposure and light intensity and biotic elicitors, *Fusarium oxysporum, Penicillium notatum, Mucor* sp. and yeast extract was used. Both the elicitors inhibited the growth of callus whereas increased the synthesis of boswellic acid. Among the biotic elicitors, yeast extract (500 mg/l) led accumulation of KβBA by 27-fold, AKβBA by 23-fold and βBA by 42-fold whereas *Fusarium oxysporum* (750 mg/l) elicited 37-fold accumulation of AβBA in cultures. Abiotic elicitor Na_2SO_4 (10 mg/l) led towards 10-fold accumulation of AKβBA and 10-fold accumulation of KβBA at 15 mg/l concentration. Interestingly, 10-fold production of AKβBA and 7-fold of βBA was observed after exposure of callus cultures by UV-C light for 5 minutes. This promising study opens future avenues for more production of boswellic acid by doing modifications in type of elicitors, elicitation time, their concentration and type of *in vitro* culture and largescale *in vitro* production of boswellic acid can certainly help in reducing pressure on existing natural population of *B. serrata.*

Nikam et al. (2013) developed a protocol for the production of *in vitro* boswellic acid, the active component present in the oleo-gum resin of *B. serrata* (Figure 8.2). Root, cotyledonary, and leaf node and leaves were used as explants from 5-week-old seedlings which were raised *in vitro*. Explants were grown on MS medium with varying concentrations of cytokinin and/or auxin. Maximum shoots/cotyledonary node and leafy node explants were produced in cultures fortified with BA (2.5 μM) and PVP (200 mg/l). Better shoot regeneration and true-to-type shoots were produced by vertical placement of leafy nodal explants. Maximum rooting (89%) was observed in shoots treated with 2.5 μM IBA. There was survival of 76.5% plantlets after transfer to field conditions. All the developed plants were true-to-type nature after comparing with mother plants in genetic fidelity assessment by RAPD analysis. Auxin:cytokinins ratio>1 in the medium was found optimum for significant callus production from seedling explants (root, stem, cotyledon, and leaf). Levels of different boswellic acid components βBA, KβBA, AβBA, and AKβBA were found to be higher in stem callus (1.2 mgg^{-1} DW) obtained on MS with 15.0 μM IAA, 5.0 μM BA and 200 mg/l PVP. The trend of accumulation of AβBA was higher followed by AKβBA, βBA, KβBA in root, stem, cotyledon, and leaf callus. Though the amount of Boswellic acid was comparatively less than the natural gum (40–50 mg/g DW) obtained directly from plant yet it is important to provide a year-round availability. As the calluses can be regularly sub-cultured and maintained over a period of 24 months, it will be helpful in commercial cultivation and enhancement in boswellic acid contents by applying approaches like elicitation and precursor feeding.

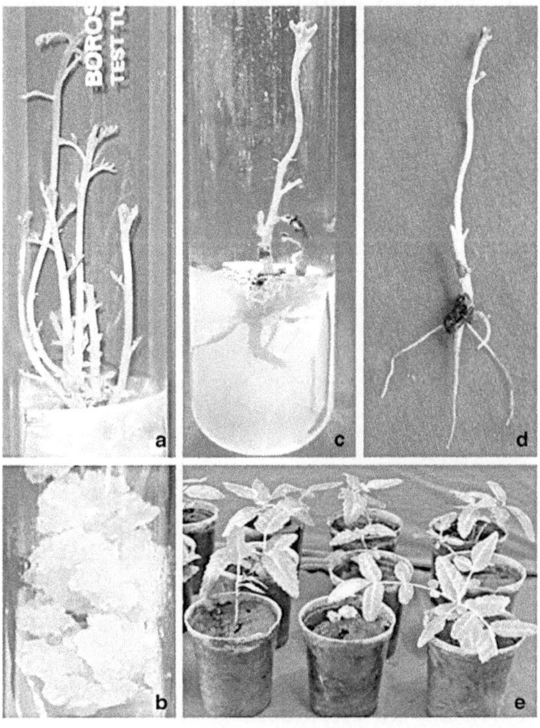

FIGURE 8.2 *In vitro* propagation of *B. serrata.* (A) Multiple shoot induction from leafy node explant on MS with 2.5 μM BA after 4 weeks; (B) callus derived from stem explant on MS with 15.0 μM IAA, 5.0 μM BA, and 200 mgl^{-1} PVP after 4 weeks; (C) & (D) rooting of *in vitro* grown shoot on MS with 1.25 μM IBA; and (E) micro-propagated plant established in plastic pots under *ex vitro* condition after four-week of transfer.

Source: Reprinted from Nikam et al. (2013), with permission.

8.4 CONCLUSION

Boswellia spp. are mainly damaged by excessive and illegal tapping for collection of highly valuable oleo-gum-resin, which is a major threat for population of its trees. Therefore, appropriate intensive management practices, including selective or rotational harvesting of frankincense should be adopted for ensuring sustained supply in future. In this context, propagation of the plants by adopting plant tissue culture techniques and enhancement in production of secondary metabolites through *in vitro* culture during the last three decades keep significance. Adulteration of frankincense is a major issue which is being addressed by identifying biomarkers from various *Boswellia* species to maintain quality control. Researchers have also prepared nanoparticles loaded with phytoconstituents of *Boswellia* spp. with increased bioavailability and medicinal potency, which opens future avenues for the development of such engineered biomolecules to preserve human health.

KEYWORDS

- **arthritis**
- ***Boswellia serrata***
- **boswellic acid**
- **frankincense**
- **Murashige and Skoog**
- **plant tissue culture**

REFERENCES

Abrham, G. G., (2011). *In Vitro Propagation of Boswellia papyrifera (Del.) Hochst.* Haramaya University, Ethiopia.

Addisalem, A., Duminil, J., Wouters, D., Bongers, F., & Smulders, M., (2016). Fine-scale spatial genetic structure in the frankincense tree *Boswellia papyrifera* (Del.) Hochst. and implications for conservation. *Tree Genetics & Genomes, 12*, 86–96.

Al-Harrasi, A., Khan, A. L., Asaf, S., & Al-Rawahi, A., (2019). *Biology of Genus Boswellia.* Springer Nature, Switzerland.

Al-Harrasi, A., Khan, A. L., Rehman, N. U., & Csuk, R., (2021). Biosynthetic diversity in triterpene cyclization within the *Boswellia* genus. *Phytochemistry, 184*(112660).

Brendler, T., Brinckmann, J. A., & Schippmann, U., (2018). Sustainable supply, a foundation for natural product development: The case of Indian frankincense (*Boswellia serrata* Roxb. ex Colebr.). *J. Ethnopharmacol., 225*, 279–286.

Chandrasekhara, T., Hussaina, T. M., & Jayananda, B., (2005). *In vitro* micropropagation of *Boswellia ovalifoliolata. Z. Naturforsch., 60c*, 505–507.

Coppi, A., Cecchi, L., Selvi, F., & Raffaelli, M., (2010). The frankincense tree (*Boswellia sacra*, Burseraceae) from Oman: ITS and ISSR analyses of genetic diversity and implications for conservation. *Genetic Resources and Crop Evolution, 57*(7), 1041–1052.

DeCarlo, A., Ali, S., & Ceroni, M., (2020). Ecological and economic sustainability of non-timber forest products in post-conflict recovery: A case study of the frankincense (*Boswellia* spp.) resin harvesting in Somaliland (Somalia). *Sustainability, 12*, 3578. doi: 10.3390/su12093578.

Eshete, A., Teketay, D., & Håkan, H., (2005). The socio-economic importance and status of populations of *Boswellia papyrifera* (Del.) Hochst. in Northern Ethiopia: The case of North Gonder zone. *For. For. Trees Livelihoods, 15*, 55–74.

Ghorpade, R. P., Chopra, A., & Nikam, T. D., (2010). *In vitro* zygotic embryo germination and propagation of an endangered *Boswellia serrata* Roxb., a source of boswellic acid. *Physiol. Mol. Biol. Plants, 16*(2), 159. doi: 10.1007/s12298-010-0017-7.

Ghorpade, R. P., Chopra, A., & Nikam, T. D., (2011). Influence of biotic and abiotic elicitors on four major isomers of boswellic acid in callus culture of *Boswellia serrata* Roxb. *Plant Omics J., 4*(4), 169–176.

Gidey, T., Hagos, D., Juhar, H. M., Solomon, N., Negussie, A., Crous-Duran, J., Oliveira, T. S., et al., (2020). Population status of *Boswellia papyrifera* woodland and prioritizing its conservation interventions using multi-criteria decision model in northern Ethiopia. *Heliyon, 6*(10), e05139.

Groenendijk, P., Eshete, A., Sterck, F., Zuidema, P., & Bongers, F., (2012). Limitations to sustainable frankincense production: Blocked regeneration, high adult mortality and declining populations. *J. Appl. Ecol., 49*(1), 164–173.

Hairfield, E. M., Hairfield, H. H., & McNair, H. M., (1989). GC, GC/MS, and TLC of β-boswellic acid and O-acetyl-β-boswellic acid from *B. serrata, B. carterii,* and *B. papyrifera. J. Chromatogr. Sci., 27*, 127–133.

Katragunta, K., Siva, B., Kondepudi, N., Rao, V. P. R., Rao, N. R., Tiwari, A. K., & Babu, K. S., (2019). Estimation of boswellic acids in herbal formulations containing *Boswellia serrata* extract and comprehensive characterization of secondary metabolites using UPLC-Q-Tof-MSe. *J. Pharmaceut. Anal., 9*, 414–422.

Khan, A. L., Mabood, F., Akber, F., Ali, A., Shahzad, R., Al-Harrasi, A., et al., (2018). Endogenous phytohormones of frankincense producing *Boswellia sacra* tree populations. *PLoS One, 13*(12), e0207910. https://doi.org/10.1371/journal.pone.0207910.

Lemenih, M., & Kassa, H., (2011). *Management Guide for Sustainable Production of Frankincense: A Manual for Extension Workers and Companies Managing Dry Forest for Resin Production and Marketing*. CIFOR, Bongor, Indonesia.

Maradani, B. S., Gudasalamani, R., Setty, S., & Chandrasekaran, R., (2018). Development of microsatellite markers for the resin-yielding, non-timber forest product species *Boswellia serrata* (Burseraceae). *Appl. Plant Sci., 6*(9), e1180.

Mertens, M., Buettner, A., & Kirchhoff, E., (2009). The volatile constituents of frankincense: A review. *Flavor Fragrance J., 24*(6), 279–300.

Nikam, T. D., Ghorpade, R. P., Nitnaware, K. M., Ahire, M. L., Lokhande, V. H., & Chopra, A., (2013). Micropropagation and non-steroidal anti-inflammatory and anti-arthritic agent boswellic acid production in callus cultures of *Boswellia serrata* Roxb. *Physiol. Mol. Biol. Plants, 19*(1), 105–116.

Purohit, S. D., Tak, K., & Kukda, G., (1995). *In vitro* propagation of *Boswellia serrata* Roxb. *Biol. Plant., 37*, 335–340.

Quazi, M. I. H., Al-Riyami, W. T., Al-Jadeedi, S. K., Ambu-Saidi, H., Ali, A., Al-Busaidi, H., Al-Harassi, A. S., & Hakkim, F. L., (2014). Conservation of Omani frankincense (*Boswellia sacra*, Burseraceae) through plant tissue culture approach. *J. Sci., 4*(10), 632–636.

Raffaelli, M., Tardelli, M., & Mosti, S., (2008). Preserving and restoring the frankincense tree (*Boswellia sacra*) at Wadi Doka: A work in progress. A port in Arabia between Rome and the Indian Ocean (3rd C BC–5th C. AD). *Khor Rori Report, 2*, 715–723.

Rijkers, T., Ogbazghi, W., Wessel, M., & Bongers, F., (2006). The effect of tapping for frankincense on sexual reproduction in *Boswellia papyrifera. J. Appl. Ecol., 43*, 1188–1195. https://doi.org/10.1111/j.1365-2664.2006.01215.x.

Suthar, R. K., & Purohit, S. D., (2012). Biopriming of micropropagated *Boswellia serrata* Roxb. plantlets – role of endophytic root fungus *Piriformospora indica. Indian J. Biotechnol., 11*, 304–308.

Suthar, R. K., Habibi, N., & Purohit, S. D., (2011). Influence of agar concentration and liquid medium on *in vitro* propagation of *Boswellia serrata* Roxb. *Indian J. Biotechnol., 10*(2), 224–227.

Vijay, N., (2014). *Studies on in Vitro Cloning Of mature Tree of Capparis Decidua and Boswellia Serrata*. Ph.D. thesis submitted to Arid Forest Research Institute, Jodhpur. Available online at: http://hdl.handle.net/10603/43660 (accessed on 10 October 2022).

Website 1. https://www.iucnredlist.org/species/34533/9874201 (accessed on 10 October 2022).

Yamamoto, F., Iwanaga, F., Al-Busaidi, A., & Yamanaka, N., (2020). Roles of ethylene, jasmonic acid, and salicylic acid and their interactions in frankincense resin production in *Boswellia sacra* Flueck. trees. *Scientific Reports, 10*, 16760.

CHAPTER 9

ASSESSMENT OF *BOSWELLIA* SPECIES GENETIC DIVERSITY USING MOLECULAR MARKERS

SHREYA NATH[1], SNIGDHA MAJUMDER[1], MIMOSA GHORAI[2], VARTIKA JAIN[3], ABDEL RAHMAN AL-TAWAHA[4], MAHIPAL S SHEKHAWAT[5], DEVENDRA KUMAR PANDEY[6], SUBHABRATA PAUL[1*], MALLAPPA KUMARA SWAMY[7], ABHIJIT DEY[2*]

[1]*School of Biotechnology, Presidency University (2nd Campus), Kolkata – 700156, West Bengal, India*

[2]*Department of Life Sciences, Presidency University, Kolkata, West Bengal, India*

[3]*Department of Botany, Government Meera Girls' College, Udaipur, Rajasthan, India*

[4]*Department of Biological Sciences, Al-Hussein Bin Talal University, Ma'an, Jordan*

[5]*Department of Plant Biology and Biotechnology, Kanchi Mamunivar Government Institute for Postgraduate Studies and Research, Pondicherry, Tamil Nadu, India*

[6]*Department of Biotechnology, Lovely Professional University, Phagwara, Punjab, India*

[7]*Department of Biotechnology, East-West First Grade College, Bangalore, Karnataka, India*

[*]*Corresponding author. E-mail: abhijit.dbs@presiuniv.ac.in*

Frankincense – Gum Olibanum: Botany, Chemistry, Extraction, Utilization, Propagation, Biotechnology, and Conservation, T. Pullaiah, K. Venkata Ratnam, Mallappa Kumara Swamy, Lepakshi Md. Bhakshu (Eds.)

ABSTRACT

Conservation biology deals with the sustainable management and protection of wild species. It is a powerful tool to protect the forest ecosystem. Knowledge of genetic variation within a plant population is immensely important in this regard. For improvement of the conservation status of a plant, its genetic diversity must be maintained. Desirable genetic changes can be introduced by plant breeding. Molecular markers are a promising and helpful tool in the field of genetic diversity study. Being simple and low-cost, molecular markers are a helpful and efficient tool in understanding genetic variation among a population and identify the natural hybrids within. For protecting the forest ecosystem, the genus *Boswellia* is particularly important. *Boswellia* is a deciduous tree with several traditional uses found in India, housing more than 200 phytochemicals. That's why knowledge about the genetic diversity among several *Boswellia* species is very important. This chapter deals with the uses of various molecular markers to access the genetic diversity of *Boswellia* sp.

9.1 INTRODUCTION

Conservation of biodiversity means to protect, scientifically manage, and uplift the biodiversity for a sustainable future (Jaisankar et al., 2018). Conservation of biodiversity is extremely important. Human beings derive benefit from plants, animals, and thus protecting the threatened ecosystem is of considerable value (Maxted, 2001). Improvement of biodiversity means safeguarding ecosystems, genetic diversity, and various species to ensure a better future. Along with immense economic importance, forest trees have ecological importance in many ecosystems. Most of the forest trees are found in natural populations with very little domestication and artificial selection (Savolainen and Pyhäjärvi, 2007). The goal of tree improvement is to maintain the genetic diversity of the population and improve their genetic value. This program improves the genetic value of both exotic and indigenous species. For forest establishment, the tree improvement program provides high-quality seeds from known sources of seed or propagules, which are linked to silviculture systems (Bell et al., 2008). This is labor-intensive and expensive as well. Their major objective is to get maximum gain per unit time and space by selecting and improving genotypes. The main elements of the tree improvement program are as follows, viz., tree improvement strategy, forest tree population, management, research, and development. Two methods are generally used, i.e., provenance studies and tree breeding programs.

Plant breeding is one of the ways to conserve biodiversity. Plant breeding introduces a desirable genetic change in the plant population. It has been successfully used since the 1980s. Plant breeding has added to genetic diversity and increased biodiversity. There are three components to it: producing genetic variation, selecting, and stabilizing them for commercialization. It has been used in crop improvement to allow the accumulation of superior alleles in the gene pool of the population (Borrelli et al., 2009).

B. serrata is a commercially important deciduous tree of India, also called Indian frankincense, produces gum called olibanum. *B. serrata* species has been used as an Ayurvedic herb due to the presence of about 200 bio-active compounds in it. Their roles in anti-inflammation (Ammon, 2016), arthritis (Sumantran et al., 2011), musculoskeletal disorder (Riva et al., 2017), bronchial asthma (Gupta et al., 1998), diabetes (Ammon, 2019), breast cancer (Parr and Ali, 2018) have been studied well. Boswellic acid has been the primary object of study in most researches. Its role has been studied in inflammatory bowels disorder, erythematous eczema, sclerosis, brain tumor-related edema, psoriasis, etc. (Roy et al., 2019; Efferth and Oesch, 2020). This is commercially important, thus the need for its genetic conservation is considerably important. Genetic diversity can occur due to spontaneous mutations, gene flow, inbreeding, natural selection, the Wahlund effect and random genetic drift. Heterogeneity in genetic diversity exists due to the different evolutionary histories of species (Porth and El-Kassaby, 2014).

Molecular markers can be used to decipher the genetic diversity of forest trees, the phylogenetic relationship amongst them, and the identification of natural hybrids. They also provide insight into the origin and domestication of those trees. To detect variations in DNA regions among individuals, molecular markers enable direct detection of the gene of interest, enables linkage analysis by helping in genetic map construction. They are broadly used in DNA probes allowing easy detection or screening (Lidder and Sonnino, 2012). This chapter aims to access the genetic diversity of *Boswellia* species using different molecular markers.

9.2 MOLECULAR MARKERS

Various genetic methods can be used to detect functionally important regions in the genome. Random amplified polymorphic DNA (RAPD) is a widely used polymerase chain reaction (PCR) based technique that involves amplification of short inverted repeats, which are highly polymorphic. No prior information about the sequence is required in this method and can be done with a small amount of DNA is required. Not only these markers are reliable but they are also reusable in studying complex eukaryotic genomes. However, the quality of reaction has to be maintained. They have been used to detect natural hybrids between species (Kloch et al., 2015). Restriction fragment length polymorphisms (RFLP), on the other hand, involve the digestion of DNA by restriction enzymes leading to the generation of DNA fragments varying in size and composition owing to the presence of different restriction sites present on the DNA. This helps to detect polymorphisms between two DNA molecules possessing different numbers of restriction enzyme recognition sites for a particular restriction enzyme. They may also differ due to the difference in the length of fragments generated after restriction digestion. They have high reproducibility and are relatively easy to score which can be transferred easily between laboratories with no prior information about the DNA sequences. However, it needs high quality and a good quantity of DNA. It is also laborious, expensive, and time-consuming. They have been used to detect interpopulation differentiation in forest trees (Neale et al., 1992; Kloch et al., 2015). Cleaved amplified polymorphic sequences (CAPs) is also called restriction fragment length polymorphism (RFLP) PCR. Unlike RFLP, it requires prior information about the sequence. It detects variation in one locus in one particular PCR and can be applied to organism-specific sequences (Mahajan and Gupta, 2012).

Differential folding of single-stranded DNA can produce polymorphisms. Single-strand conformation polymorphism (SSCP) can be utilized for detecting these subtle mutations. It is easy, rapid, and sensitive but quite laborious and expensive at the same time (Dong and Zhu, 2005). Amplified fragment length polymorphism (AFLP) technique blends in the advantages of both PCR and RFLP. The whole-genome fragments are amplified and then separated using the electrophoresis technique. They have been used in the study of biodiversity, analysis of germplasm, constructing gene maps. It requires a very small amount of DNA (Kloch et al., 2015). While simple sequence length polymorphism (SSLP) can be used to map individual loci and also for fingerprinting in plants. There can be mono, di, tri, and tetrameric motifs of repetitive sequences (Huang et al., 2017). The polymorphisms in the DNA can occur due to the alteration of a single nucleotide. Single nucleotide polymorphisms (SNPs) can be used to identify those variations and defining haplotypes (Mahajan and Gupta, 2012).

9.3 MOLECULAR MARKERS AND GENETIC DIVERSITY STUDIES IN *BOSWELLIA* SPECIES

Several molecular markers, like SSR, ISSR, AFLP, RAPD, etc., have been identified not only to classify different species of *Boswellia* but also to pinpoint the genetic diversity among the population of the particular species of *Boswellia.* Through literature study as such, three species of *Boswellia* (*B. sacra, B. papyrifera* and *B. serrata*) have been found in which extensive studies on genetic diversity through different molecular markers.

9.3.1 MARKER STUDIES IN B. SACRA

B. sacra is a member of the Burseraceae family and a primary member of *Boswellia.* It is extensively harvested for frankincense and native plant taxa of the Arabian Peninsula and Somalia. To understand the genetic diversity among and within the populations of *B. sacra* growing in the west coast, east coast, and internal belt of Dhofar, Oman, two molecular markers were used. They were nucleotide sequences of internal transcribed spacer (ITS) of genomic DNA and inter-simple sequence repeat (ISSR) markers. The chosen 511 bp long ITS sequence showed low variation (nearly 6.4%) among the nine populations of *B. sacra* growing with geographical diversity due to the slow rate of evolution in the nuclear genome. Using four selected ISSR primers (ISSR1–$(GT)_7$-YR, ISSR4–CAA-$(GA)_5$, CA6–$(CA)_6$-RG, CT8–$(CT)_8$-RG; where Y = C/T, R = A/G), 93 scorable bands (400 bp–3,500 bp) were found of which 97.1% were polymorphic in 97 individual profiles for seven populations. The total amount of heterozygosity (H_T) among and within the populations (H_S) of *B. sacra* was 0.22 and 0.136, respectively (Coppi et al., 2010). It indicates a long-lived perennial habitat, wide-spread distribution, and appearance in the late-successional stage for the respective taxa (Nybom, 2004). This may be an additive result of outcross and seed dispersal by both wind and water. But interestingly, this value of heterozygosity of other species with similar characteristics was found to be higher than that of *Boswellia.* Hence, it can be inferred that the intrapopulation diversity among *B. sacra* in Dhofar was lower than that to be expected. The reasons include anthropogenic activity, global warming which causes variation in the amount and activity of animal pollinators. However, a molecular variance study revealed that the interpopulation variation among *B. sacra* was very high with clear differentiation between the population of the east coast and west coast. Such drastic differences in two types of diversity could be a result of either inbreeding or pollinations by territorial pollinators within a short geographic distance (Coppi et al., 2010). As an intention of this study, another group of workers selected four more molecular markers (AFLP, SSR, RAPD, and SDS-PAGE) to understand the genetic diversity among 13 populations of *B. sacra* collected from the previous regions of Dhofar. Among the bands obtained by using four sets of SSR primers 88% of them were polymorphic and the band produced from three such markers were 100% polymorphic. Based on SSR markers analysis F_{ST} (mean fixation index) value was 0.567. The pairwise F_{ST} value analysis corresponds to a certain amount of gene flow among the four populations of *B. sacra* namely Dowkha (cultivated), Dowkha (wild), Sadah-b, and Kabout-lower. But, otherwise, geneflow for the rest of the population was low. The total amount of heterozygosity (H_T = 0.413) using SSR markers were higher than that of ISSR markers that were previously studied (Coppi et al., 2010). The most genetic variation was observed between the eastern (Mughsil and Sadah) and western populations (Mughsil and Hasik) because of the geographic barrier based on SSR markers. The other 8 AFLP markers produced 1,431 bands. Among them, the bands produced by marker combination of *Eco*RI-AAG+ MseI-CAT and *Eco*RI-AGT + *Mse*I-CAT were 95% polymorphic (Al-Harrasi et al., 2019). The result showing the genetic variance obtained from AFLP markers was somewhat similar to that of SSR markers. The bands from 5 RAPD markers showed 85% polymorphism and the protein analysis by SDS-PAGE markers showed lower variance. After RAPD analysis, a huge genetic gap was observed between the populations grown in the eastern region (Sadah, Kabout, and Hasik) and the populations grown in the central region (Thumrait) of Dhofar. The genetic distance also might be a result of geographic isolation. The allelic frequencies in the different populations of *B. sacra* were different in correspondence to the markers (Table 9.1).

TABLE 9.1 The Range of Different Allelic Frequencies for Markers Used

Markers	Allelic Frequency Range
SSR	0.127–1.194
AFLP	0.083–1.148
RAPD	0.312–1.82
SDS-PAGE	–

Keeping a similarity with previous reports a cumulative analysis of all the data collected from different populations showed that the interpopulation molecular variance was higher than of intrapopulation variance. The geographic separation was observed between the three central populations (Thumrait, Dowkha-a, and Dowkha-b), one eastern population (Sadah) and one western population (Mughsil) because of their different latitudinal habitat. Among them, the Mughsil population grown in the western region had the most genetic and geographic isolation from the other populations. Interpopular genetic variation was most between Mughsil and Sadah grown in the eastern region and Mugshil and Hasik grown in the eastern region. The cause could be validated by the distance model where geographic distance limits the quantity of gene flow by cross-pollination has given rise to genetic isolation (Schaal et al., 1998). However, a low genetic diversity was observed within the central population as they were grown in national parks where *Boswellia* was well conserved. Accordingly, outcross became capable and successive which caused quality gene flow (Al-Harrasi et al., 2019).

9.3.2 MARKER STUDIES IN *B. PAPYRIFERA*

B. papyrifera plants are broadly found in the dry tropical woods of Ethiopia, Eritrea, and Sudan and are the significant hotspot for frankincense.

Frankincense is a monetarily significant aromatic gum resin gathered from the bark of the tree.

FSSG or Fine-scale spatial hereditary construction is non-arbitrary spatial dissemination of genotype in a populace that relies upon various ecological and evolutionary variables. These include selection by micro-environment (Zhou and Chen, 2010), plant density (Zeng et al., 2012) and limitation in both pollen and seed dispersal (Torres et al., 2003). However, an increase in both pollen and seed dispersal can reduce FSSG of a certain population over a generation.

Adult plants from 12 populations of *B. papyrifera* grown across Ethiopia were collected to understand their genetic diversity and FSSG between the two populations. Around 10 microsatellite loci or SSR markers were selected as molecular markers for the study. However, for two populations AF-BIR and JRG were analyzed using only five SSR loci due to poor quality of DNA. In all cases, the expected amount of heterozygosity (Omondi et al., 2010) was similar to the observed one ($H_0 = 0.681$) with a low inbreeding coefficient ($F_{IS} = 0.01$). Even in KTD population which was represented by only scattering of trees showed the highest amount of heterozygosity ($H_0 = 0.701$). Such a result indicated that though there was a great amount of population decrease in *B. papyrifera* due to anthropogenic activity, there was no proof of inbreeding among the adult population. The presence of a certain amount of intraspecific variation in such a declining plant population may be a result of their long lifespan. F_{ST} values for 10 populations using all 10 markers were 0.084. Through Nei's unbiased pair-wise genetic distance analysis, most genetic distances were observed between B-SHK population of the western region and the AB population of the northeastern region due to geographic separation. After analyzing the results from the data analysis of SSR markers study, it can be hypothesized that geographic isolation develops the spatial structure of the population. Genetic variation in the population is an additive result of the difference in spatial distribution as well as genetic drift. The individual-based structure analysis categorized the 10 populations into 4 geographic regions based on their genetic clusters. The genetic changes present in the populations represented by four genetic clusters were due to adaptation in certain micro-environment (Addisalem et al., 2016a) (Table 9.2).

TABLE 9.2 The Microclimate of Different Genetic Clusters Corresponds to the Micro-Environment

SL. No.	Name of the Population	Genetic Clusters That Correspond to Geographic Region	Micro-Environment of the Respective Region				
			Altitude (Meter Above Sea Level)	Temperature (°C)	Precipitation	Soil Type	References
1.	TRG	North-East (NE)	Higher altitude (1,000 m a.s.l)	Cooler temperature (average maximum temperature 29.3°C)	Modest rainfall (mean annual precipitation 850 mm)	Less fertile, shallow, clayey	Eshete et al. (2011)
2.	AB	–					
3.	KTD	–					
4.	B-KUR	West (W)	Lower altitude (below 900 m a.s.l)	Hot temperature (maximum temperature of 39–42°C)	1,350 mm annual precipitation	Deep red, sandy-loam, and some-times clayey	Teshome et al. (2017)
5.	B-SHK	–					
6.	MQ	North-West (NW)	Lowland (below 900 m a.s.l.)	Hot (average temperature around 36°C)	800–900 mm annual precipitation	Deeper, more clayey, and fertile	Eshete et al. (2011)
7.	MAW	–					
8.	TA-SJ	North (N)	–	Very hot (maximum temperature 41°C)	Dry areas with less than 650 mm precipitation	–	Alemu et al. (2015)
9.	KHB	–					
10.	KHW	–					

Though the adult population of *B. papyrifera* has declined greatly because of anthropogenic influence and geographic isolation, it still manages to retain its genetic diversity. Hence the *circa situm* control of them should not be overlooked, especially for the population at Afar which has been most disturbed.

Addisalem et al. (2015) carried out the sequencing of *B. papyrifera* genomic DNA using Illumina paired-end sequencing in an effort of species conservation. The study revealed the 705 Mb sized haploid genome. Also, the reads consisted a single microsatellite repeat per 5.7 kb. Considering the subset of these repeated sequences, authors have established 46 polymorphic SSR markers that could amplify 2 to 12 alleles in 10 genotypes. This set consisted of 30 tri-nucleotide repeat markers, 4 tetra-nucleotide repeat markers, 6 penta-nucleotide markers and 6 hexanucleotide repeat markers. Interestingly, they noticed the cross-transferable ability of some of these markers between *B. pirottae* and *B. popoviana*. Moreover, the study identified retrotransposons, and the reads when assembled found to have several contigs that were having similarity to genes of the terpenoid and terpene backbone biosynthetic pathways, which form the major constituents of the bark resin.

In another study to analyze the inter popular genetic variance and FSSG, two populations were selected from the western Ethiopia and classified into two cohorts – adult and seedling. Six polymorphic microsatellite loci were utilized called Bp17, Bp20, Bp21, Bp22, Bp23, and Bp39 to genotype the plants. In both cohorts, in both populations, the genetic variance is well maintained. The observed heterozygosity (HO = 0.689) was similar to the expected heterozygosity (HE = 0.693) of tropical plants measured with microsatellites like *Acacia senegal* (Omondi et al., 2010). It suggested a well establish intrapopulation variance among the population. The seedling cohorts had a higher amount of heterozygosity than the adult ones. The acquiescence of genetic diversity in the adult cohorts was also established in their seedlings. However, a low F_{ST} value ($F_{ST} = 0.031$) signified the presence of low genetic variance between the two populations and also the cohorts of each population. As there was no data that explicitly describe the mating system of *B. papyrifera*, the authors obtained the analysis from studying the reproductive process of sister species like *B. serrata* (Sunnichan et al., 2005), *B. ovalifoliolata* (Raju et al., 2012). It was hypothesized that due to pre-zygotic self-incompatibility plants were obliged to outcross. It resulted in extensive pollen dispersal and gene flow among the population (Addisalem et al., 2016b).

9.3.3 MARKER STUDIES IN B. SACRA

Akbar et al. (2019) have studied genetic diversity among 13 populations of *B. sacra* growing in central, eastern, and western Dhofar regions in Oman using RAPD analysis and total leaves protein analysis using SDS-PAGE and found high levels of genetic diversity is still present in *B. sacra* trees.

9.3.4 MARKER STUDIES IN B. SERRATA

B. serrata is known for the production of Indian frankincense. It is generally found in the Indian sub-continent, especially in the Northwestern parts of India and Pakistan, and is endemic. Using magnetic bead enrichment method 16 microsatellite markers were developed from three populations of *B. serrata* collected from Cauvery Wildlife Sanctuary (CWS), Biligiri Ranganathaswamy Temple Tiger Reserve and Male Mahadeshwara Hills Wildlife Sanctuary. Among them, 11 markers were found to be polymorphic, and the average number of alleles per locus for these markers ranges from 3 to 10 as per studies in 60 individuals. The allelic cross-over and intrageneric enhancement of these microsatellite markers demonstrate a close relationship between *B. serrata* and *B. ovalifoliolata*, which needs support from the additional examination. These microsatellite markers can also analyze the genetic variation and population structure of *B. serrata* and to give essential data on natural issues for the protection of the species (Maradani et al., 2018).

Tiwari et al. (2017) evaluated genetic variation among *B. serrata* growing in different agroclimatic zones of Madhya Pradesh with the help of RAPD markers. A total of 56 reproducible bands revealed a high degree of genetic diversity in Shivpuri (Gird Region). This also helped in the production of elite genotype from genetically diversified populations.

The population of *B. serrata* is only restricted to the central part of India. In another study, *Boswellia* samples were collected from Chhattisgarh and genotyped using five of each RAPD and ISSR marker for 56 and 42 loci, respectively. The three populations collected from Chhattisgarh were named according to their geographic locations which were Dhamtari (DT), Narayanpur (NP), Sarguja (SG). The WFL (wood fiber length) varies between 0.803 mm and 1.397 mm with cv (coefficient of variation) of 11.36%. Also, in the case of WFW (wood fiber width), the cv was 8,075%. Both of them were higher than that of *Ailanthus altissima* and *Eucalyptus grandis*. Such huge variation indicates better adaptability to local habitat which corresponds to improved fitness. Both Bayesian and non-Bayesian statistics were used to eliminate both over-estimation and under-estimation and to avoid null hypothesis. According to non-Bayesian statistical analysis, DT population was found to be more diverse and poly-morphic. The GD or Nei's gene diversity and Hs or Panmictic heterozygosity (measured by Hickory's method with Bayesian statistics) for DT population were 0.20 ±0.20 and 0.20 ±0.01, respectively which was comparatively higher than the other two populations (Kosman, 2003; Qiao et al., 2010). However, no huge variety in panmictic heterozygosity was found in the three topographical sub-populations because of the dominance of inbreeding. Even though *B. serrata* is an out-crossing tree species with different mechanisms of self-compatibility, this inbreeding dominance might be an outcome of habitat fragmentation due to anthropogenic activity and geographic seclusion. Though intra-population genetic diversity was higher than the inter-population genetic differences it was comparatively lower than other out-crossing tree species (Ansari et al., 2012; Nybom, 2004). The authors hypothesized the reason might be restricted gene flow as a result of morphological anomaly (early shedding of unripe fruits and ripen fruit with no seeds). Thereafter, a cumulative effect of genetic bottleneck and inbreeding depression had given rise to such genetic evaluation. However, genetic diversity among the three populations or the inter-population could be a result of geographical separation leading to isolation. Hence, as a conservation strategy, different genetic programs should be incorporated to maintain and increase the number of plants with elite genotypes.

Vaishnav et al. (2019) assessed the genetic diversity of *B. serrata* population growing in DT, NP, and Sarguja regions of Chhattisgarh by using band-based and allele frequency-based approach with non-Bayesian and Bayesian statistics. There were 56 RAPD + 42 ISSR loci used for the genetic characterization. This study indicated that the genetic resource of *B. serrata* in central Indian region maintains high variation in wood fiber dimension and exhibits higher genetic variability within and among its sub-populations.

9.4 CONCLUSIONS AND FUTURE PROSPECTS

The inter-population of *B. sacra* genetic diversity is greater than the intra-population genetic diversity as the result of extensive outcrossing. But, in the case of *B. serrata* habitat-separation caused some sort of inbreeding dominance. Hence, intra-population diversity was influenced. In *B. papyrifera* both intra-population and inter-population diversity was maintained due to the long-life span of the plant and geographic distance, respectively. As a conservation strategy, *in-situ* conservation is a better choice for the *Boswellia* population as most of them gained adaptive characteristics in their respective habitat. But, due to uneven genetic variance in the *Boswellia* population in Dhofar regions, strategies for *ex-situ* conservations should be taken. However, the intensive method for rapid pest control and inhibition of fungal infections should be administered to preserve the old and elite genotype of the Frankincense trees in the middle-east.

Unfortunately, most of the marker-based studies were only conducted on the *Boswellia* population of middle-east and Ethiopia. Similar attention should also be given to the Indian *Boswellia* population. So that, conservation strategies could be taken accordingly, as it is an economically profitable product and has recently faced rapid decline due to manmade ventures.

Also, to clearly understand the polymorphism and genetic status of the *Boswellia* population all over the world, more work should be conducted using both dominant and co-dominant markers. At the same time, comparative studies about the *Boswellia* population in different continents using molecular markers will give a holistic understanding of their genetic diversity, amount of population declination, and its reason.

KEYWORDS

- ***B. papyrifera***
- ***B. sacra***
- ***B. serrata***
- ***Boswellia***
- **genetic diversity**
- **molecular markers**

REFERENCES

Addisalem, A. B., Bongers, F., Kassahun, T., Smulders, M. J. M., (2016a). Genetic diversity and differentiation of the frankincense tree (*Boswellia papyrifera* (Del.) Hochst) across Ethiopia and implications for its conservation. *For. Ecol. Manage., 360*, 253–260.

Addisalem, A. B., Duminil, J., Wouters, D., Bongers, F., Smulders, M. J. M., (2016b). Fine-scale spatial genetic structure in the frankincense tree *Boswellia papyrifera* (Del.) Hochst. and implications for conservation. *Tree Genet. Genomes, 12*(5), 86–96. doi: 10.1007/s11295-016-1039-2.

Addisalem, A. B., Esselink, G. D., Bongers, F., & Smulders, M. J. M., (2015). Genomic sequencing and microsatellite marker development for *Boswellia papyrifera*, an economically important but threatened tree native to dry tropical forests. *AoB Plants, 7*, 86.

Akbar, F., Khan, A. L., Gilani, S. A., Al-Harrasi, A., Al-Sadi, A. M., & Shinwari, Z. K., (2019). Genetic differentiation in different endemic *Boswellia sacra* (Burseraceae) populations from Oman. *Pak. J. Bot.* doi: 10.30848/Pjb2019-1(18).

Alemu, B., Garedew, E., Eshetu, Z., & Kassa, H., (2015). Land use and land cover changes and associated driving forces in northwestern lowlands of Ethiopia. *Int. Res. J. Agric. Sci. Soil Sci., 5*(1), 28–44.

Al-Harrasi, A., Khan, A. L., Asaf, S., & Al-Rawahi, A., (2019). Genetic diversity and differentiation among species and populations of *Boswellia*. In: *Biology of Genus Boswellia* (pp. 85–101).

Ammon, H. P., (2016). Boswellic acids and their role in chronic inflammatory diseases. *Adv. Exp. Med. Biol., 928*, 291–327.

Ammon, H. P., (2019). Boswellic extracts and 11-keto-ß-boswellic acids prevent type 1 and type 2 diabetes mellitus by suppressing the expression of proinflammatory cytokines. *Phytomedicine, 63*, 153002. doi: 10.1016/j.phymed.2019.153002.

Ansari, S. A., Narayanan, C., Wall, S. A., Kumar, R., Shukla, N., & Rahangdale, S. K., (2012). ISSR markers for analysis of molecular diversity and genetic structure of Indian teak (*Tectona grandis* L. f.) populations. *Ann. For. Res., 55*(1), 11–23.

Bell, F. W., Parton, J., Stocker, N., Joyce, D., Reid, D., Wester, M., Stinson, A., et al., (2008). Developing a silvicultural framework and definitions for use in forest management planning and practice. *For. Chronicles, 84*(5), 678–693.

Borrelli, G. M., Orru, L., Vita, P. D., Barabaschi, D., Mastrangelo, A. M., & Cattivelli, L., (2009). Integrated views in plant breeding: Modern approaches for an old topic. In: Sadras, V., & Calderini, D. F., (eds.), *Crop Physiology*. Elsevier Inc., USA.

Coppi, A., Cecchi, L., Selvi, F., & Raffaelli, M., (2010). The frankincense tree (*Boswellia sacra*, Burseraceae) from Oman: ITS and ISSR analyses of genetic diversity and implications for conservation. *Genet. Resour. Crop Evol., 57*(7), 1041–1052.

Dong, Y., & Zhu, H., (2005). Single-strand conformational polymorphism analysis: Basic principles and routine practice. *Methods Mol. Med., 108*, 149–157.

Efferth, T., & Oesch, F., (2020). Anti-inflammatory and anti-cancer activities of frankincense: Targets, treatments and toxicities. *Semin. Cancer Biol.* doi: 10.1016/j.semcancer.2020.01.015.

Eshete, A., Sterck, F., & Bongers, F., (2011). Diversity and production of Ethiopian dry woodlands explained by climate- and soil-stress gradients. *For. Ecol. Manage., 261*(9), 1499–1509.

Gupta, I., Gupta, V., Parihar, A., G, G., Lüdtke, R., Safayhi, H., & Ammon, H., (1998). Effects of *Boswellia serrata* gum resin in patients with bronchial asthma: Results of a double-blind, placebo-controlled, 6-week clinical study. *Eur. J. Med. Res., 3*(11), 511–514.

Huang, X. Y., Fan, K., Ye, Y. F., Wang, B., Wu, W. R., & Lan, T., (2017). Teaching practice and experiences of verifying the three laws of genetics based on the SSLP marker analysis. *Yi Chuan = Hered., 39*(9), 856–862.

Jaisankar, I., Velmurugan, A., & Sivaperuman, C., (2018). Chapter 19 – Biodiversity conservation: Issues and strategies for the tropical islands. In: Sivaperuman, C., Velmurugan, A., Singh, A. K., & Jaisankar, I., (eds.), *Biodiversity and Climate Change Adaptation in Tropical Islands* (pp. 525–552). Academic Press.

Kloch, M. O., Hawliczek-Strulak, A. N., & Sekrecka-Bielak, A. N., (2015). Molecular markers in forest management and tree breeding: A review. *Annals of Warsaw University of Life Sciences-SGGW: Forestry and Wood Technology, 92*, 193–199.

Kosman, E., (2003). Nei's gene diversity and the index of average differences are identical measures of diversity within populations. *Plant Pathol., 52*(5), 533–535.

Lidder, P., & Sonnino, A., (2012). Chapter 1 – Biotechnologies for the management of genetic resources for food and agriculture. In: Goodwin, S. F., Friedmann, T., & Dunlap, J. C., (eds.), *Advances in Genetics* (Vol. 78, pp. 1–167). Academic Press.

Mahajan, R., & Gupta, P., (2012). Molecular markers: Their use in tree improvement. *J. Forest Sci., 58*(3), 137–144.

Maradani, B. S., Gudasalamani, R., Setty, S., & Chandrasekaran, R., (2018). Development of microsatellite markers for the resin-yielding, non-timber forest product species *Boswellia serrata* (*Burseraceae*). *Appl. Plant Sci., 6*(9), 4–7.

Maxted, N., (2001). Ex situ, in situ conservation. In: *Encyclopedia of Biodiversity* (pp. 683–695).

Neale, D. B., Devey, M. E., Jermstad, I. D., Ahuja, M. R., Alosi, M. C., & Marshall, K. A., (1992). Use of DNA markers in forest tree improvement research. *New For., 6*, 391–407.

Nybom, H., (2004). Comparison of different nuclear DNA markers for estimating intraspecific genetic diversity in plants. *Mol. Ecol., 13*(5), 1143–1155.

Omondi, S., Kireger, E., Dangasuk, O., Chikamai, B., Odee, D., Cavers, S., & Khasa, D., (2010). Genetic diversity and population structure of *Acacia senegal* (L) Willd. in Kenya. *Trop. Plant Biol., 3*, 59–70.

Parr, C., & Ali, A., (2018). *Boswellia frereana* suppresses HGF-mediated breast cancer cell invasion and migration through inhibition of c-Met signaling. *J. Transl. Med., 16*(1), 281.

Porth, I., & El-Kassaby, Y. A., (2014). Assessment of the genetic diversity in forest tree populations using molecular markers. *Diversity, 6*(2), 283–295.

Qiao, Q., Zhang, C. Q., & Milne, R. I., (2010). Population genetics and breeding system of *Tupistra pingbianensis* (Liliaceae), a naturally rare plant endemic to SW China. *J. Syst. Evol., 48*(1), 47–57.

Raju, A. J. S., Lakshmi, P. V., Ramana, K. V., & Chandra, P. H., (2012). Entomophily, ornithophily and anemochory in the self-incompatible *Boswellia ovalifoliolata* Bal. &

Henry (Burseraceae), an endemic and endangered medicinally important tree species. *J. Threat. Taxa, 4*(7), 2673–2684.

Riva, A., Allegrini, P., Franceschi, F., Togni, S., Giacomelli, L., & Eggenhoffner, R., (2017). A novel boswellic acids delivery form (Casperome®) in the management of musculoskeletal disorders: A review. *Eur. Rev. Med. Pharmacol. Sci., 21*(22), 5258–5263.

Roy, N. K., Dey, P., Banik, K., Bordoloi, D., Devi, A., Thakur, K., Padmavathi, G., et al., (2019). An update on pharmacological potential of boswellic acids against chronic diseases. *Int. J. Mol. Sci., 20*(17), 4101. doi: 10.3390/ijms20174101.

Savolainen, O., & Pyhäjärvi, T., (2007). Genomic diversity in forest trees. *Curr. Opin. Plant Biol., 10*(2), 162–167.

Schaal, B. A., Hayworth, D. A., Olsen, K. M., Rauscher, J. T., & Smith, W. A., (1998). Phylogeographic studies in plants. *Mol. Ecol., 7*, 465–474.

Sumantran, V., Joshi, A., Boddul, S., Koppikar, S., Warude, D., Patwardhan, B., Chopra, A., et al., (2011). Antiarthritic activity of a standardized, multiherbal, ayurvedic formulation containing *Boswellia serrata*: *In vitro* studies on knee cartilage from osteoarthritis patients. *Phytother. Res., 25*(9), 1375–1380.

Sunnichan, V. G., Mohan, R. H. Y., & Shivanna, K. R., (2005). Reproductive biology of *Boswellia serrata*, the source of Salai guggul, an important gum-resin. *Bot. J. Linn. Soc., 147*(1), 73–82.

Teshome, M., Eshete, A., & Bongers, F., (2017). Uniquely regenerating frankincense tree populations in western Ethiopia. *For. Ecol. Manage., 389*, 127–135.

Tiwari, S. K., Yadav, S. S., Krishnamurthy, G., & Pandey, A., (2017). Evaluation of genetic variation in endangered tree species of *Boswellia serrata* with the help of genetic markers. *Research & Reviews: A Journal of Life Sciences, 7*(1), 22–26.

Torres, E., Iriondo, J. M., Escudero, A., & Pérez, C., (2003). Analysis of within-population spatial genetic structure in *Antirrhinum microphyllum* (Scrophulariaceae). *Am. J. Bot., 90*(12), 1688–1695.

Vaishnav, V., Mahesh, S., & Kumar, P., (2019). Assessment of genetic structures of the endangered forest species *Boswellia serrata* Roxb. population in Central India. *J. Trop. Plant Sci., 31*(2), 200–210.

Zeng, X., Michalski, S. G., Fischer, M., & Durka, W., (2012). Species diversity and population density affect genetic structure and gene dispersal in a subtropical understory shrub. *J. Plant Ecol., 5*(3), 270–278.

Zhou, H. P., & Chen, J., (2010). Spatial genetic structure in an understorey dioecious fig species: The roles of seed rain, seed and pollen-mediated gene flow, and local selection. *J. Ecol., 98*(5), 1168–1177.

CHAPTER 10

ENDOPHYTES ASSOCIATED WITH *BOSWELLIA* SPECIES: DEVELOPMENTS AND PROSPECTS

CHAMPA KEYA TUDU,[1] ABDEL RAHMAN AL-TAWAHA,[2] SAMAPIKA NANDY,[1] MAHIPAL S. SHEKHAWAT,[3] ERCAN BURSAL,[4] MIMOSA GHORAI,[1] VARTIKA JAIN,[5] POTSHANGBAM NONGDAM,[6] GABER EL-SABER BATIHA,[7] PADMANABH DWIVEDI,[8] TABARAK MALIK,[9] DEVENDRA KUMAR PANDEY,[10] MALLAPPA KUMARA SWAMY,[11] RUPA SANYAL,[12] and ABHIJIT DEY[1]

[1]*Department of Life Sciences, Presidency University, 86/1 College Street, Kolkata – 700073, West Bengal, India*

[2]*Department of Biological Sciences, Al-Hussein Bin Talal University, Ma'an, Jordan*

[3]*Kanchi Mamunivar Government Institute for Postgraduate Studies and Research, Pondicherry, Tamil Nadu, India*

[4]*Department of Biochemistry, Mus Alparslan University, Turkey*

[5]*Department of Botany, Government Meera Girls' College, Udaipur, Rajasthan, India*

[6]*Department of Biotechnology, Manipur University, Canchipur, Imphal – 795003, Manipur, India*

[7]*Department of Pharmacology and Therapeutics, Faculty of Veterinary Medicine, Damanhur University, Damanhur – 22511, AlBeheira, Egypt*

Frankincense – Gum Olibanum: Botany, Chemistry, Extraction, Utilization, Propagation, Biotechnology, and Conservation, T. Pullaiah, K. Venkata Ratnam, Mallappa Kumara Swamy, Lepakshi Md. Bhakshu (Eds.)

[8]*Department of Plant Physiology, Banaras Hindu University, Banaras, Uttar Pradesh, India*

[9]*Department of Biochemistry, College of Medicine and Health Science, School of Medicine, University of Gondar, Ethiopia*

[10]*Department of Biotechnology, Lovely Professional University, Phagwara, Punjab, India*

[11]*Department of Biotechnology, East-West First Grade College, Bangalore, Karnataka, India*

[12]*Department of Botany, Bhairab Ganguly College, Kolkata, West Bengal, India*

**Corresponding author. E-mail: abhijit.dbs@presiuniv.ac.in*

ABSTRACT

The mutual aid between endophytes and their host floras allows changes in plant physiological and metabolic functions. Endophytic fungi interact with their host plants by means of diverse modes of action that mainly involve controlling various physiological and mechanical mechanisms for their better endurance. They establish a vital niche within the host plants by secreting chemical components. The association of endophytes with *Boswellia* species has been well discovered and researched. Several fungi associated with twigs, leaves, and stems of *Boswellia* species have been reported. Maximum research studies described endophytic microbes associated with *B. sacra* and *B. serrata*. The microbial endophyte community of *Boswellia* species is kind of less. Endophytic fungal strains, such as *Penicillium citrinum, P. spinulosum, Alternaria alternata, Fusarium oxysporum, Aspergillus caespitosus, Preussia* sp., *Phoma medicaginis, Thielavia arenaria, Dothideomycetes* sp., *Aureobasidium* sp., etc., have been recognized using genomic DNA extraction, polymerase chain reaction (PCR) amplification, DNA sequencing, and phylogenetic investigation. Several of them were identified to produce secondary metabolites, and few produced different kinds of extracellular enzymes like phosphatase, cellulases, zylanase, etc. Likewise, different phytohormones, such as auxins secreted by endophytes associated with *Boswellia* species improved the plant growth dynamics and physiology. In addition, they have significantly enhanced the development of phyllospheric parts of *Boswellia* species, and improved tolerance against pathogenic attacks. This chapter aims to elucidate the diversity of endophytic microbe in *Boswellia* species and their beneficial role in the plant growth and development.

10.1 INTRODUCTION

Plants and their allied microbes, also called endophytes (fungi, bacteria, and actinomycetes) live symbiotically and enjoy mutual benefits from the association. Endophytes, the microbial population forms a symbiotic connection with herbs, and live exclusively within the plant tissues and complete their life cycle at least for a short span of time and attain benefits from their host plants without affecting any obvious damages (Hardoim et al., 2015; Biswas et al., 2020). The term "endophyte" was evolved by the combination of two Greek words, i.e., 'endo' referring to inside, and 'phyte' denoting to plant (Maehara et al., 2019). Various investigations have documented that the endophytic microbes are widespread, and found occurring in all plant species, however, in most instances, plant-endophyte associations are yet to be understood in detail. Firstly, in 1809, Heinrich Friedrich Link, a German botanist described endophytic fungi, which were earlier commonly recognized to be a distinct class of plant-parasitic fungi. A French Scientist, Béchamp coined the word "microzymas" for these microorganisms. In 1887, Victor Galippe revealed that bacteria commonly occurring within the plant tissues play a useful role regardless of pathogenesis. Later, scientists understood that bacteria can invade and colonize within the interior parts of plants also (Biswas et al., 2020). Several research studies on endophytic microbes have recognized the fact that they carry out several kinds of beneficial functions, particularly to facilitate in protecting the host plants from the attacks of herbivores and insects. They secrete volatile compounds that deter predators. Also, collaboration amongst host plants and their associated endophytes alters the plant physiology and metabolic functions to induce advantageous role. Likewise, the host plants exhibit more resistance to environmental stress conditions, including drought and cold conditions (Maehara et al., 2019; He et al., 2020). In addition, endophytes offer a great potential in controlling environmental contamination by improving the degradation of pollutants or phytoextraction at the contaminated places (He et al., 2020). Various researches have exposed the ecological distribution of host plants and their allied endophytic community, and determined the plant-endophytes distribution patterns and their interactions (Correia et al., 2018). Plant-endophytes mutualistic co-existence patterns can be balanced precisely by understanding the role of secondary metabolites produced from the interactions and regulating the signaling pathways in plants. Based on the host dependency, nutrition specificity and modification in survival tactics have classified them as 'facultative' or 'obligate' endophytes (Su et al., 2010; Nandy et al., 2020).

Boswellia includes nearly 24 different species that are distributed majorly in the regions of India, the Arabian Peninsula, and Eastern parts of Africa (Camarda et al., 2007; Mothana et al., 2011). *B. serrata* is internationally known as Indian frankincense or olibanum. It is a yellowish-brown oleogum resin, which is obtained from the barks of *Boswellia serrata* and has been employed to treat various diseases (Camarda et al., 2007; Mahdian et al., 2020). *Boswellia* has a number of the prominent species consist of *B. serrata, B. frereana, B. sacra, B. ovalifoliolata, B. neglecta, B. papyrifera, B. rivae, B. dioscoridis, B. elongata, B. socotrana*, and so forth (Camarda et al., 2007; Mothana et al., 2011; Mahdian et al., 2020). Customarily, the olibanum of selected *Boswellia* species, such as *B. serrata* and *B. sacra* (Syn.: B. *carteri*) has been used currently in several nations for treating rheumatic and other related inflammatory diseases, including ulcerative colitis, Crohn's disease. Also, the essential oils and extracts of *Boswellia* species tree parts are being utilized as antiseptic agents and used for preparations of a mouthwash in addition to treating cough and asthma conditions (Mothana et al., 2011). Numerous researches have evidenced the therapeutical potential of several *Boswellia* species, including anticancer, immunomodulatory, anti-inflammatory, and antimicrobial activities. *Boswellia* species have been reported for their use in the treatment of rheumatoid arthritis, inflammatory diseases, and for their anti-carcinogenic, antibacterial, anti-fungal properties, and also specifically help in the treatment of breast cancer patients (Suhail et al., 2011). Various animal research studies demonstrated that this plant species has impacts on the antihyperglycemic and antihyperlipidemic activities in streptozotocin-induced diabetic rats (Mahdian et al., 2020).

The endophytes associated with *Boswellia* species secrete several kinds of secondary metabolites, for example, phenolics and hormone-like compounds. Besides the production of plant growth-promoting hormones, endophytes isolated from *Boswellia* species are found to produce several kinds of extracellular enzymes like phosphatase, cellulases, xylanase, and many more (Khan et al., 2016). Some research papers have documented the secretion of gibberellins (GA_3, GA_4, and GA_7), indole acetic acid (auxin) by various endophytic fungi associated with *Boswellia* species. These enzymes together with secondary metabolites secreted from endophytes have a significant role in imparting larger benefits to the host plants during both regular and distress environmental conditions. A wide-ranging endophytic microbes may occur in each plant species, for example, in a tropical tree, about 90 species may occur within a leaf. Likewise, nearly 50 genera can be noticed in the roots of dry land species, while desert plant species involve below 20 species in a leaf (Bayman and Otero, 2006; Khan et al., 2017b).

10.2 ENDOPHYTIC MICROORGANISMS EFFECT ON TREE GROWTH

For the plant growth and development, the phytohormone-like chemical compounds are very necessary. Some plant growth promoting regulators (PGPRs), such as IAA and gibberellic acid have been sequestered from many endophytic fungi, such as *Acremonium coenophialum, Colletotrichum* sp., *Epicoccum purpurascens*, *Aureobasidium pullulans*, etc. (Al-Harrasi et al., 2019). Several biologically active volatile compounds attained from endophytic fungi are being used as effective medications against numerous diseases. A large number of endophytes occur in the rhizospheric vicinity, and they are described to perform key roles in the growth and development of host plants. They alter the physical and biochemical characteristics of the terrains, improve the mineral content, enhance the biodegradation of organic contaminants, and improve water uptake capability of the lands (Nandy et al., 2020). The existence of xenobiotic compounds- or heavy metal-resistant endophytic microbes, and the consequential phytoremediation certainly influence on the plant's growth and development, survival chances, overcoming of stress conditions, including drought and salinity stresses and other metabolic events (Rajkumar et al., 2010). In addition to the secretion of PGPRs and other beneficial secondary metabolites, endophytic fungi produce a wide-ranging exozymes, i.e., extracellular enzymes, such as glucosidase, esterase, phosphatase, cellulase, xylanase, etc. (Corrêa et al., 2014; Esteves et al., 2014). The secretion of these kinds of exozymes is important to the host plants, and reduces the requirement of additional metabolism under the stressed situations, and thus they perform as plant energy savers. Analyzing symbiotic endophytic microorganisms allied with commercially valued and distinctive medicinal plants is indispensable for understanding their plant life and exploring possible alternative resources for obtaining several bioactive phytometabolites (Soussi et al., 2016).

10.3 VARIOUS ENDOPHYTIC MICROBES FROM *BOSWELLIA* SPECIES

Boswellia species can grow in exceptionally subtropical and dry environmental situations, where nutrients and water accessibility are at their dearth. In such conditions, the diversity of endophytic fungi associated with trees helps to understand and recognize the tree's lifespan and their symbiotic co-evolution. Further, the assessment of their bioactive perspective, i.e., secretion of biomolecules by endophytes can be utilized as inoculants, and can be a novel way of eco-friendly application to improve the plant's growth and development through improving soil quality or fertility (Khan et al., 2016; Al-Harrasi et al., 2019). In general, the propagation of *B. sacra* is very challenging in its natural habitat, because of its harsh nature. However, the utilization of endemic microflora, i.e., as PGPRs helps to increase the declining population of this tree species (Hibbett et al., 2011). *B. sacra* constitutes a wide range of phytochemical compounds with the therapeutic significance. Overall, the microbial community profile of *B. sacra* is reported to be very less. Nevertheless, considering at the tree lifespan and its living habitations, *Boswellia* species are very unique, and they often experience varied stresses, such as UV incidences, heat or drought and strong wind. The existence of endophytic fungal association in different organs, especially leaf and stem parts of *Boswellia* species are being documented by several researchers. The collection of sample and processing approaches for isolating fungal endophytes from the *B. sacra* tree is depicted in Figure 10.1.

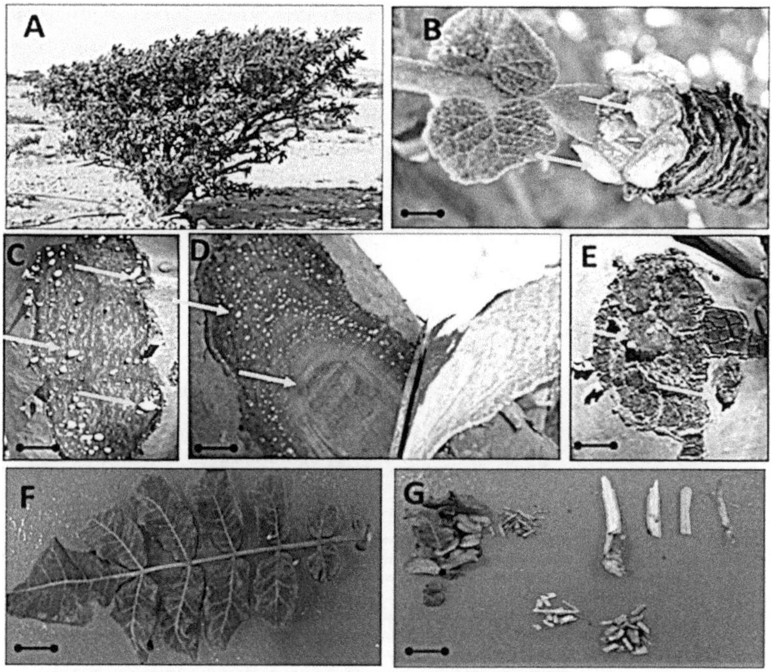

FIGURE 10.1 Sample collection and processing for the isolation of fungal endophytes from the frankincense (*Boswellia sacra*) tree.

Note: (A) Shows the hot and dry habitat of the frankincense tree; (B) resin emerging from the mature leaf of tree branches; (C) the milk-like resin oozing out from the lower epidermal layer of the tree; (D) the inner bark or the cortex region of the tree did not contain resinous ducts and was also evaluated for the presence of endophytes; (E) dried wounded outer bark in which the white milk was converted into crystalline gum; (F) leaf pattern of the frankincense tree; (G) smaller parts of the tree (leaf, stem, and bark parts) were used for isolation. A total of 208 10-mm tissue samples from the leaf and stem (bark) were used to isolate endophytic fungi. The black line is equivalent to 1 inch.

Source: Reprinted from Khan et al. (2016).

A study was carried out to isolate endophytes, and it involved the screening of 30 *B. sacra* trees and around 120 stems and 185 leaves were sterilized and inoculated to grow endophytic fungi. These tissue segments yielded about 77 endophytic fungal isolates. Amongst them, 17 fungi, namely *Chaetomium* sp., *Phoma medicaginis, Penicillium citrinum, Preussia* sp., *Thielavia arenaria, Aureobasidium* sp., *Dothideomycetes* sp., *Phoma* sp., *Aureobasidium pullulans, Sordariomycetes* sp., *Fusarium proliferatum*, etc., were selected and identified on the basis of morphological traits observation (Khan et al., 2016). Further, these endophytic fungi were identified by molecular methods, sequencing, and statistical analysis. The nuclear ribosomal internal transcribed region (ITS) sequences of the rRNA gene from the 17 fungal isolates showed the fragment lengths that ranged from 518 to 630 bp. The average ratio of C, A, T(U), and G was observed to be 26.5, 23.6, 24.5, and 25.3, correspondingly (Figure 10.2).

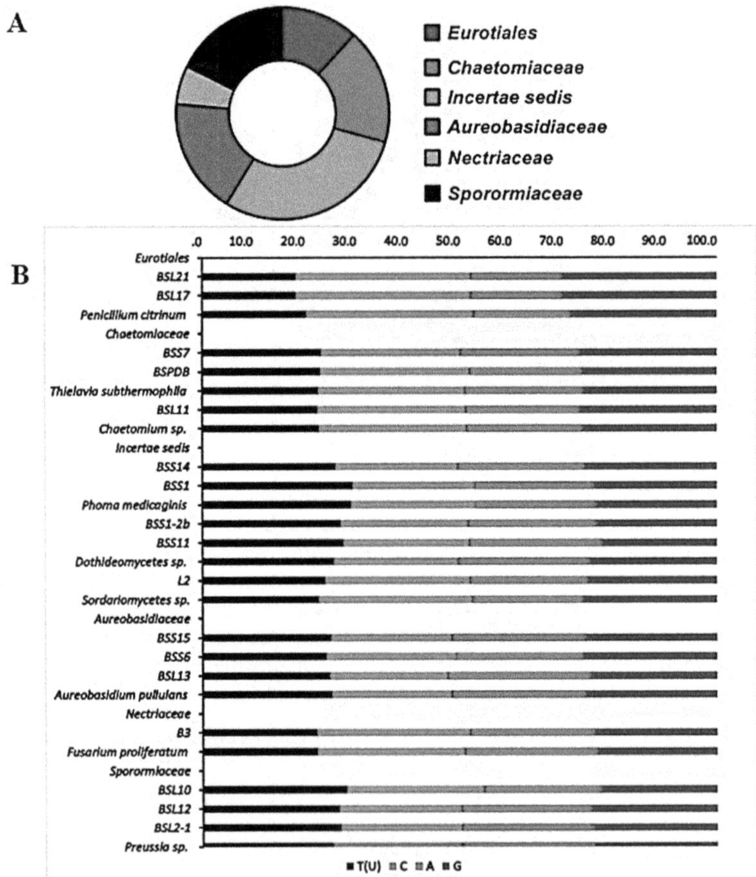

FIGURE 10.2 Endophytic distribution and nucleotide homology.

Note: Endophyte isolated from *B. sacra*, family-wise distribution (A); and comparison of nucleotide composition in the sequence with those from related strains in GenBank (B). A comparative assessment was made with the help of MEGA 6.0.

Source: Reprinted from Khan et al. (2016).

The isolated fungi belonged to Chaetomiaceae (17.7%), Eurotiales (11.8%), Aureobasidiaceae (17.7%), Incertae sedis (29.4%), Nectriaceae (5.9%), and Sporomiaceae (17.7%) from the tree bark and leaf parts (Figure 10.2). The colonization rates, colonization density and isolation rates of fungal diversity were measured as the percentage of segments that are infected by at least one isolate from the total number of segments of each inoculated tissue samples. Also, Shannon Wiener and Simpson diversity indices were recognized for fungal isolates from *B. sacra* using the online system. It was witnessed that the colonization configuration was significantly greater in leaves in comparison to stems (Kumar and Hyde, 2004; Khan et al., 2016; Al-Harrasi et al., 2019). The diversity indices, i.e., the Simpson diversity index and the Shannon-Weiner diversity index was recorded to be 0.583 and 0.8729 for stem and leaf parts, respectively (Al-Harrasi et al., 2019). In another study, the authors have isolated about 530 endophytic fungi isolates from 3,000 tissue segments, including twig, inner bark, bark, and leaf samples of *B. serrata*. All the recovered 530 fungal isolates were observed for their morphological characteristics. Amongst them, 522 isolates were noticed to sporulate, and belonged to 17 genera. Among the 17 genera, 16 belonged to mitosporic fungi (70.58%), 4 Ascomycetes (23.52%), 1 Coelomycetes (5.88%), and other 8 isolates (47.05%) did not sporulate (Sunayana and Prakash, 2012). These endophytic fungi obtained from *B. serrata* were *Acremonium strictum, Botrytis* sp., *Fusarium pallidoroseum, Memnoniella* sp., *Myrothecium verrucaria, Nigrospora oryzae, Pestolotiopsis* spp., *Phoma exigue, Phoma lingam, Phoma sorghina, Phomopsis* sp., *Pithomyces* sp., *Trichothecium roseum, Aspergillus flavus, A. fumigatus, A. nidulans, A. niger* and *Mycelia sterilia.* The endophytes from twig of *B. serrata* showed the highest diversity index (Shannon diversity: 2.16, Simpsons' diversity: 0.137, Species richness: 12), while leaf had the lowest diversity index (Sunayana and Prakash, 2012). Later, endophytic fungal strains like *Penicillium citrinum, P. spinulosum, F. oxysporum, Alternaria alternata,* and *A. caespitosus* were identified from the leaves and stems of *B. sacra* via extracting genomic DNA, amplification using polymerase chain reaction (PCR), sequencing, and phylogenetic investigation. These strains were ascertained to produce the secondary metabolites with enzyme inhibitory activity (Ali et al., 2017).

Khan et al. (2017a) have investigated the occurrence of endophytic fungi and bacteria in the rhizospheric regions of the wild and cultivated *B. sacra* tree populations via the involvement of genomic DNA extraction and NGS (next generation sequencing). The analysis of genomic sequences has disclosed the existence of 60.6±3.1 and 1,006±8.9 OTUs (operational taxonomic units) for fungal and bacterial species, respectively (Khan et al., 2017a). The endophytic diversity of fungal species (*Preussia* sp., *Phoma medicaginis, Thielavia arenaria, Dothideomycetes* sp., and *Aureobasidium* sp.) were relatively found more in the leaf tissues compared to stem parts of *Boswellia* species (Khan et al., 2016). *Dothideomycetes* sp., mostly occurs within the tree bark as an epiphyte or endophyte, and is isolated from both the stem and leaf tissues. It is well-known to play a vital role in the degradation of leaf litter and cellulose (Del Prado et al., 2006). Likewise, *Phoma* sp. produces pectinases, xylanases, cellulases, and proteases, and play a major role in biodegradation (Robl et al., 2013).

10.4 MOLECULAR CHARACTERIZATION OF ENDOPHYTES-ASSOCIATED WITH *BOSWELLIA* SPECIES

Mainly, the endophytic microorganisms are identified based on the morphological and molecular characteristics. The endophytes will be grouped into various morphological forms on the basis of colony formation and its shape, surface texture, margin characteristics, thickness, aerial hyphae color, growth rate and pattern, and growth depth into medium (Arnold et al., 2007). To identify endophytes, the extraction of genomic DNA followed by the PCR amplification of internal transcribed spacer (ITS) or 18S rRNA region, DNA sequencing and phylogenetic analysis are performed (Khan et al., 2012, 2016; Al-Harrasi et al., 2019). The obtained PCR products are sequenced using a Sanger's sequencer instrument. This approach is one of the directly a systematized methods utilized for broad diversities of fungi sequestered from different environments (Rosconi et al., 2013; Deshpande et al., 2016; Raja et al., 2017). The online bioinformatics tool, the basic local alignment search tool (BLASTn) search pipeline was performed to distinguish the homologous conserved ITS regions similarities among the fungal isolates and related fungi sequences found in the databank. The resultant sequence is cleaned from primer adopters and chimera sequences, followed by phylogenetic analysis using the molecular evolutionary genetics analysis (MEGA) software (version 6) of GenBank (Tamura et al., 2013; Al-Harrasi et al., 2019). The obtained sequences were aligned by using pair-wise CLUSTALW alignment tool (Tamura et al., 2013). The missing/ambiguous sequences were deleted or coded as N (Nilsson et al., 2018). By using the neighbor-joining, maximum parsimony, and maximum likelihood trees, the phylogenetic tree of all the aligned sequences were prepared by employing MEGA 6 software. The phylogenetic tree was made by using 2,000 bootstrap replications (Tamura et al., 2013; Han et al., 2016). The aligned sequences data was submitted to obtain GenBank accession numbers.

10.5 GROWTH OF *BOSWELLIA* SPECIES BY ENDOPHYTE INOCULATION

The endophytic inoculation after 3 weeks helped to record higher plant growth promoting characteristics, i.e., increased numbers of leaves, internodes, and length of shoots. Also, improved chlorophyll content (a, b, and total carotenoids) was noticed (Khan and Lee, 2013). In a recent study, it has been reported that inoculating 4-week-old *Boswellia sacra* saplings with the endophyte, *Preussia* sp. BSL10 showed the higher potential of producing enzymes and IAA production (Khan et al., 2016). The inoculated *Preussia* sp. BSL10 to the host saplings' root zone considerably improved the growth dynamics of *B. sacra* when compared to controls. Also, the endophytic fungi significantly augmented the internodes (1%), shoot length (20%), and leaves (23%) as compared to controls, in addition to increase of photosynthetic pigments (Khan et al., 2016). Endophytic fungi, such as *P. formosus* LHL10, *P. janthinellum* LK5, and *P. citrinum* inoculated to the host plants (*Boswellia* species) have shown to improve the growth and stress tolerance (Khan et al., 2012; Limtong et al., 2014; Khan et al., 2015b; Castillo et al., 2015). The endophytic inoculation to the rhizospheric zone improves plant physiology and development of leaves, stem, and seeds (phyllospheric parts) of the host plants, and also increase the tolerance against microbial pathogenic damages (Pedraza et al., 2009; Ritpitakphong et al., 2016).

10.6 PRODUCTION OF PHYTOHORMONES FROM ENDOPHYTES IN *BOSWELLIA* SPECIES

The various classes of endophytic fungi produce phytohormones that can range from several types of phenolics to hormone-like compounds, including gibberellins and auxins (Bömke and Tudzynski, 2009; Khan et al., 2016; Al-Harrasi et al., 2019). These phytohormones formed by various kinds of endophytic microbes help to improve the plant root zone and enhance plant growth and development under normal as well as stress conditions (Khan et al., 2017a). In addition to the initiation of the host-plant symbiotic process, some of the phytohormones and exozymes produced counteract plant pathogenic microbial infections and extend abiotic stress tolerance. However, such interactions have been very less known in trees growing in subtropical deserts like *Boswellia sacra* (Knief et al., 2012; Khan et al., 2017a). Few studies have recorded that the endophytic strains (*Preussia* sp. BSL10 and *Aureobasidium* sp.) associate with *Boswellia* species and produce different types of phytohormones (Khan et al., 2016; Al-Hosni et al., 2018). The endophytic fungal strains have the potential of producing physiologically bioactive plant growth regulators, such as IAA and gibberellins, and ameliorating effects of endophytes are comparable with those of commercially available plant growth regulators under stressed situations (Kusari and Spiteller, 2012; Khan et al., 2015a; Ludwig-Müller, 2015). *B. sarca* tree-associated endophyte, *Aureobasidium* sp. (BSS6) was shown to enhance IAA synthesis significantly by both L-tryptophan-dependent and -independent pathways. Also, *Preussia* sp. BSL10 and *Aureobasidium pollulan* BSS15 exhibits higher synthesis of IAA, considerably (Khan et al., 2016). Several endophytes producing bioactive compounds has been regarded as a resilient feature of endophytes in supporting the growth of the host plants.

10.7 CONCLUSION

The genus, *Boswellia* was reported to have an impactful association of endophytic microbes. Endophytic fungi can produce new secondary metabolites with useful biological activities. The endophytes provide an alternative source of bioactive secondary metabolites. Various fungal endophytes are widely acknowledged for their beneficial role, and they secrete a diverse range of chemicals that can represent as an untapped source of chemical reservoir that finds use in agriculture and therapeutics. The enzymes formed by fungal endophytes inside the host tissues could help supplement the direct uptake of nutrients by microorganisms in the roots. Endophytic fungal strains, such as *P. citrinum, P. spinulosum, F. oxysporum, A. alternata, A. caespitosus, Preussia* sp., *T. arenaria, P.*

medicaginis, Aureobasidium sp., *Dothideomycetes* sp., etc., were identified in *Boswellia* species through genomic DNA extraction, PCR amplification, sequencing, and phylogenetic analysis. Endophytes have the potential to secrete extracellular enzymes (cellulases, phosphatases, and glucosidases) and plant-based hormones, such as auxins, and IAA, which can help the host in harsh environmental conditions and also strengthen the symbiotic bond between the endophyte and the host. The present explanation of endophytic diversity from *B. sacra* grown in extreme drought conditions indicates that the symbiotic association contributed to factors that stimulated the host plants' growth and development. Also, endophytic diversity has been documented to enhance mutualism and protect the host against insect and pathogenic microbes attacks. In the future, more research should be encouraged to explore the occurrence of endophytic microbes and the associated mechanisms involved during symbiosis with *Boswellia* species.

KEYWORDS

- ***Boswellia* species**
- **colonizing**
- **endophytes**
- **fungi**
- **microbe**
- **phyllosphere**
- ***roots***

REFERENCES

Al-Harrasi, A., Khan, A. L., Asaf, S., & Al-Rawahi, A., (2019). *Biology of Genus Boswellia.* Springer, Cham.

Al-Harrasi, A., Rehman, N. U., Khan, A. L., Al-Broumi, M., Al-Amri, I., Hussain, J., Hussain, H., & Csuk, R., (2018). Chemical, molecular and structural studies of *Boswellia* species: β-boswellic aldehyde and 3-epi-11β-dihydroxy BA as precursors in biosynthesis of boswellic acids. *PLoS One, 13*(6), e0198666. https://doi.org/10.1371/journal.pone.0198666.

Al-Hosni, K., Shahzad, R., Latif, K. A., Muhammad, I. Q., Al Harrasi, A., Al Rawahi, A., Asaf, S., et al., (2018). *Preussia* sp. BSL-10 producing nitric oxide, gibberellins, and indole acetic acid and improving rice plant growth. *J. Plant Interactions, 13*(1), 112–118.

Ali, S., Khan, A. L., Ali, L., Rizvi, T. S., Khan, S. A., Hussain, J., Hamayun, M., & Al-Harrasi, A., (2017). Enzyme inhibitory metabolites from endophytic *Penicillium citrinum* isolated from *Boswellia sacra. Arch. Microbiol., 199*(5), 691–700.

Arnold, A. E., Henk, D. A., Eells, R. L., Lutzoni, F., & Vilgalys, R., (2007). Diversity and phylogenetic affinities of foliar fungal endophytes in loblolly pine inferred by culturing and environmental PCR. *Mycologia, 99*(2), 185–206.

Bayman, P., & Otero, J. T., (2006). Microbial endophytes of orchid roots. In: Bayman, P., & Otero, J. T., (eds.), *Microbial Root Endophytes* (pp. 153–177). Springer, Berlin, Heidelberg.

Biswas, D., Biswas, P., Nandy, S., Mukherjee, A., Pandey, D. K., & Dey, A., (2020). Endophytes producing podophyllotoxin from *Podophyllum* sp. and other plants: A review on isolation, extraction and bottlenecks. *South African J. Bot., 134*, 303–313.

Bömke, C., & Tudzynski, B., (2009). Diversity, regulation, and evolution of the gibberellin biosynthetic pathway in fungi compared to plants and bacteria. *Phytochemistry, 70*(15, 16), 1876–1893.

Camarda, L., Dayton, T., Di Stefano, V., Pitonzo, R., & Schillaci, D., (2007). Chemical composition and antimicrobial activity of some oleogum resin essential oils from *Boswellia* spp. (Burseraceae). *Annali di Chimica, 97*(9), 837–844.

Castillo, P., Molina, R., Andrade, A., Vigliocco, A., Alemano, S., & Cassán, F. D., (2015). Phytohormones and other plant growth regulators produced by PGPR: The genus *Azospirillum*. In: Cassán, F. D., Okon, Y., & Creus, C. M., (eds.), *Handbook for Azospirillum* (pp. 115–138). Springer, Cham.

Corrêa, R. C. G., Rhoden, S. A., Mota, T. R., Azevedo, J. L., Pamphile, J. A., De Souza, C. G. M., Polizeli, M. D. L. T. D. M., et al., (2014). Endophytic fungi: Expanding the arsenal of industrial enzyme producers. *J. Industrial Microbiol. Biotechnol., 41*(10), 1467–1478.

Correia, A. M., Lira, S. P., Assis, M. A., & Rodrigues, A., (2018). Fungal endophyte communities in *Begonia* species from the Brazilian Atlantic rainforest. *Current Microbiol., 75*(4), 441–449.

Del Prado, R., Schmitt, I., Kautz, S., Palice, Z., Lücking, R., & Lumbsch, H. T., (2006). Molecular data place trypetheliaceae in dothideomycetes. *Mycol. Res., 110*(5), 511–520.

Deshpande, V., Wang, Q., Greenfield, P., Charleston, M., Porras-Alfaro, A., Kuske, C. R., Cole, J. R., et al., (2016). Fungal identification using a Bayesian classifier and the Warcup training set of internal transcribed spacer sequences. *Mycologia, 108*(1), 1–5.

Esteves, A. C., Saraiva, M., Correia, A., & Alves, A., (2014). Botryosphaeriales fungi produce extracellular enzymes with biotechnological potential. *Can. J. Microbiol., 60*(5), 332–342.

Han, M. L., Chen, Y. Y., Shen, L. L., Song, J., Vlasák, J., Dai, Y. C., & Cui, B. K., (2016). Taxonomy and phylogeny of the brown-rot fungi: *Fomitopsis* and its related genera. *Fungal Diversity, 80*(1), 343–373.

Hardoim, P. R., Van, O. L. S., Berg, G., Pirttilä, A. M., Compant, S., Campisano, A., Döring, M., & Sessitsch, A., (2015). The hidden world within plants: Ecological and evolutionary considerations for defining functioning of microbial endophytes. *Microbiol. Mol. Biol. Rev., 79*(3), 293–320.

He, W., Megharaj, M., Wu, C. Y., Subashchandrabose, S. R., & Dai, C. C., (2020). Endophyte-assisted phytoremediation: Mechanisms and current application strategies for soil mixed pollutants. *Critical Rev. Biotechnol., 40*(1), 31–45.

Hibbett, D. S., Ohman, A., Glotzer, D., Nuhn, M., Kirk, P., & Nilsson, R. H., (2011). Progress in molecular and morphological taxon discovery in fungi and options for formal classification of environmental sequences. *Fungal Biol. Rev., 25*(1), 38–47.

Khan, A. L., & Lee, I. J., (2013). Endophytic *Penicillium funiculosum* LHL06 secretes gibberellin that reprograms *Glycine max* L. growth during copper stress. *BMC Plant Biol., 13*(1), 1–14.

Khan, A. L., Al-Harrasi, A., Al-Rawahi, A., Al-Farsi, Z., Al-Mamari, A., Waqas, M., Asaf, S., et al., (2016). Endophytic fungi from frankincense tree improves host growth and produces extracellular enzymes and indole acetic acid. *PloS One, 11*(6), e0158207.

Khan, A. L., Asaf, S., Al-Rawahi, A., Lee, I. J., & Al-Harrasi, A., (2017a). Rhizospheric microbial communities associated with wild and cultivated frankincense producing *Boswellia sacra* tree. *PloS One, 12*(10), e0186939.

Khan, A. L., Hamayun, M., Kang, S. M., Kim, Y. H., Jung, H. Y., Lee, J. H., & Lee, I. J., (2012). Endophytic fungal association via gibberellins and indole acetic acid can improve plant growth under abiotic stress: An example of *Paecilomyces formosus* LHL10. *BMC Microbiol., 12*(1), 1–14.

Khan, A. L., Hussain, J., Al-Harrasi, A., Al-Rawahi, A., & Lee, I. J., (2015a). Endophytic fungi: Resource for gibberellins and crop abiotic stress resistance. *Critical Rev. Biotechnol., 35*(1), 62–74.

Khan, A. L., Waqas, M., Asaf, S., Kamran, M., Shahzad, R., Bilal, S., Khan, M. A., et al., (2017b). Plant growth-promoting endophyte *Sphingomonas* sp. LK11 alleviates salinity stress in *Solanum pimpinellifolium. Environ. Exp. Bot., 133*, 58–69.

Khan, A. L., Waqas, M., Hussain, J., Al-Harrasi, A., Hamayun, M., & Lee, I. J., (2015b). Phytohormones enabled endophytic fungal symbiosis improve aluminium phytoextraction in tolerant *Solanum lycopersicum:* An example of *Penicillium janthinellum* LK5 and comparison with exogenous GA3. *J. Hazardous Materials, 295*, 70–78.

Knief, C., Delmotte, N., Chaffron, S., Stark, M., Innerebner, G., Wassmann, R., Von, M. C., & Vorholt, J. A., (2012). Metaproteogenomic analysis of microbial communities in the phyllosphere and rhizosphere of rice. *The ISME Journal, 6*(7), 1378–1390.

Kumar, D. S. S., & Hyde, K. D., (2004). Biodiversity and tissue-recurrence of endophytic fungi in *Tripterygium wilfordii. Fungal Diversity, 17*, 69–90.

Kusari, S., & Spiteller, M., (2012). Metabolomics of endophytic fungi producing associated plant secondary metabolites: Progress, challenges and opportunities. *Metabolomics, 10*, 241–266.

Limtong, S., Kaewwichian, R., Yongmanitchai, W., & Kawasaki, H., (2014). Diversity of culturable yeasts in phylloplane of sugarcane in Thailand and their capability to produce indole-3-acetic acid. *World J. Microbiol. Biotechnol., 30*(6), 1785–1796.

Ludwig-Müller, J., (2015). Plants and endophytes: Equal partners in secondary metabolite production? *Biotechnol. Letters, 37*(7), 1325–1334.

Maehara, S., Agusta, A., Tokunaga, Y., Shibuya, H., & Hata, T., (2019). Endophyte composition and cinchona alkaloid production abilities of *Cinchona ledgeriana* cultivated in Japan. *J. Nat. Med., 73*(2), 431–438.

Mahdian, D., Abbaszadeh-Goudarzi, K., Raoofi, A., Dadashizadeh, G., Abroudi, M., Zarepour, E., & Hosseinzadeh, H., (2020). Effect of *Boswellia* species on the metabolic syndrome: A review. *Iranian J. Basic Med. Sci., 23*(11), 1374.

Mothana, R. A., Hasson, S. S., Schultze, W., Mowitz, A., & Lindequist, U., (2011). Phytochemical composition and *in vitro* antimicrobial and antioxidant activities of essential oils of three endemic Soqotraen *Boswellia* species. *Food Chem., 126*(3), 1149–1154.

Nandy, S., Das, T., Tudu, C. K., Pandey, D. K., Dey, A., & Ray, P., (2020). Fungal endophytes: Futuristic tool in recent research area of phytoremediation. *South African J. Bot., 134*, 285–295.

Nilsson, R. H., Kristiansson, E., Ryberg, M., Hallenberg, N., & Larsson, K. H., (2008). Intraspecific ITS variability in the kingdom fungi as expressed in the international sequence databases and its implications for molecular species identification. *Evolutionary Bioinformatics, 4*, EBO-S653.

Pedraza, R. O., Bellone, C. H., De Bellone, S. C., Sorte, P. M. F. B., & Dos, S. T. K. R., (2009). *Azospirillum* inoculation and nitrogen fertilization effect on grain yield and on the diversity of endophytic bacteria in the phyllosphere of rice rainfed crop. *European J. Soil Biol., 45*(1), 36–43.

Raja, H. A., Miller, A. N., Pearce, C. J., & Oberlies, N. H., (2017). Fungal identification using molecular tools: A primer for the natural products research community. *J. Nat. Prod., 80*(3), 756–770.

Rajkumar, M., Ae, N., Prasad, M. N. V., & Freitas, H., (2010). Potential of siderophore-producing bacteria for improving heavy metal phytoextraction. *Trends Biotechnol., 28*(3), 142–149.

Ritpitakphong, U., Falquet, L., Vimoltust, A., Berger, A., Métraux, J. P., & L'Haridon, F., (2016). The microbiome of the leaf surface of *Arabidopsis* protects against a fungal pathogen. *New Phytol., 210*(3), 1033–1043.

Robl, D., Da Silva, D. P., Mergel, C. M., Rojas, J. D., Dos Santos, C. P., Pimentel, I. C., Vicente, V. A., et al., (2013). The capability of endophytic fungi for production of hemicellulases and related enzymes. *BMC Biotechnol., 13*(1), 1–12.

Rosconi, F., Davyt, D., Martínez, V., Martínez, M., Abin-Carriquiry, J. A., Zane, H., Butler, A., et al., (2013). Identification and structural characterization of serobactins, a suite of lipopeptide siderophores produced by the grass endophyte *Herbaspirillum seropedicae. Environ. Microbiol., 15*(3), 916–927.

Soussi, A., Ferjani, R., Marasco, R., Guesmi, A., Cherif, H., Rolli, E., Mapelli, F., et al., (2016). Plant-associated microbiomes in arid lands: Diversity, ecology and biotechnological potential. *Plant and Soil, 405*(1), 357–370.

Su, Y. Y., Guo, L. D., & Hyde, K. D., (2010). Response of endophytic fungi of *Stipa grandis* to experimental plant function group removal in inner *Mongolia steppe*, China. *Fungal Diversity, 43*(1), 93–101.

Suhail, M. M., Wu, W., Cao, A., Mondalek, F. G., Fung, K. M., Shih, P. T., Fang, Y. T., et al., (2011). *Boswellia sacra* essential oil induces tumor cell-specific apoptosis and suppresses tumor aggressiveness in cultured human breast cancer cells. *BMC Complem. Altern. Med., 11*(1), 1–14.

Sunayana, N., & Prakash, H. S., (2012). Fungal endophytes of *Boswellia serrata* Roxb. (Burseraceae), a medicinal tree species. *IOSR J. Pharm. Biol. Sci., 1*(6), 1–5.

Tamura, K., Stecher, G., Peterson, D., Filipski, A., & Kumar, S., (2013). MEGA6: Molecular evolutionary genetics analysis version 6.0. *Mol. Biol. Evol., 30*(12), 2725–2729.

CHAPTER 11

TRADE AND COMMERCE OF FRANKINCENSE

T. PULLAIAH

Department of Botany, Sri Krishnadevaraya University, Anantapur – 515003, Andhra Pradesh, India,
E-mail: pullaiah.thammineni@gmail.com

ABSTRACT

The frankincense is obtained from the oleo gum resin of *Boswellia* tree species, and the gum is used for many applications, which makes it as an economically potential source for the livelihood of tribal people and a rich arena for foreign exchange. The frankincense market has turnover through raw and essential oils of *Boswellia* raised up to US$ 193.4 Mn in 2018, and is estimated to up to US$ 406.1 Mn by 2028. In this chapter, a detailed account of the production, trade, and commerce of frankincense/olibanum is given.

11.1 INTRODUCTION

Boswellia species yield the aromatic resin known as frankincense or olibanum, which is incorporated into a variety of healthcare, home care, aromatherapy, cosmetics, and toiletries, as well as dietary supplements. Tapping the trees generally involves making a repeated incision in the bark over a series of months (Abiyu et al., 2010; Gebrehiwot et al., 2003). The semi-solid, gummy, yellow-brown exudate seeps from the bark and hardens upon exposure to air, often in the shape of a teardrop (Abdoul-Latif

Frankincense – Gum Olibanum: Botany, Chemistry, Extraction, Utilization, Propagation, Biotechnology, and Conservation, T. Pullaiah, K. Venkata Ratnam, Mallappa Kumara Swamy, Lepakshi Md. Bhakshu (Eds.)

et al., 2012; Coppen, 1995; Ogbazghi, 2001). Harvest is mainly seasonal, generally with periods where no harvest occurs during the year (Coppen, 1995; Hassan, 2012).

Frankincense is one of the oldest known globally traded products and has been a valued ingredient in Asian traditional medicine systems for millennia. *B. papyrifera* from Ethiopia has, for the last few decades, been the most traded resin, both internally within Ethiopia and exported for religious and other purposes. Especially in the last decade, there has been a surge in the use of essential oils distilled from traded *Boswellia* resins for aromatherapy and other uses. Most frankincense commercial oil is extracted from *B. sacra* (Syn.: *B. carteri*) resin from Somalia/Somaliland and Oman, with lesser amounts from *Boswellia serrata, B. papyrifera, B. rivae, B. neglecta*, and *B. dalzielii*. So far, four species, *B. serrata, B. sacra, B. papyrifera*, and *B. dalzielii,* have also shown high levels of four main bioactive Boswellia acids (BAs) in the resin (not the essential oils or hydrosol) (Global Frankincense Alliance (GFA), 2021).

11.2 GLOBAL SCENARIO OF THE FRANKINCENSE MARKET AND ITS ECONOMIC POTENTIAL

The frankincense is obtained from the oleo gum resin of *Boswellia* tree species, and the gum is used for many applications, which makes it as an economically potential source for the livelihood of tribal people and a rich arena for foreign exchange. In a study by Prudour (2019), the frankincense market has turnover through raw and essential oils of *Boswellia* raised up to US$ 193.4 Mn in 2018 and estimated to up to US$ 406.1 Mn by 2028 (Figure 11.1), at a CAGR (Compound Annual Growth Rate) of 7.7% and became a valuable commodity. The important frankincense trade leaders are Ambe Phytoextracts Private Limited; Aromaaz International; Katyani Exports; Lala Jagdish Prasad and Company; Kanta Enterprises Private Limited; BO International; Shaanxi Orient Industrial Co. Ltd. and doTERRA International (Prudour, 2019). In a study by Berhanu et al. (2021) reported during the market analysis and explained about the trading chain and the value of profit at each level (urban areas are termed as *Gulit* where pretty traders sell the frankincense; Figure 11.2), the gum-tapper at the site getting additional of 35% of his total income which might be helpful to the livelihood.

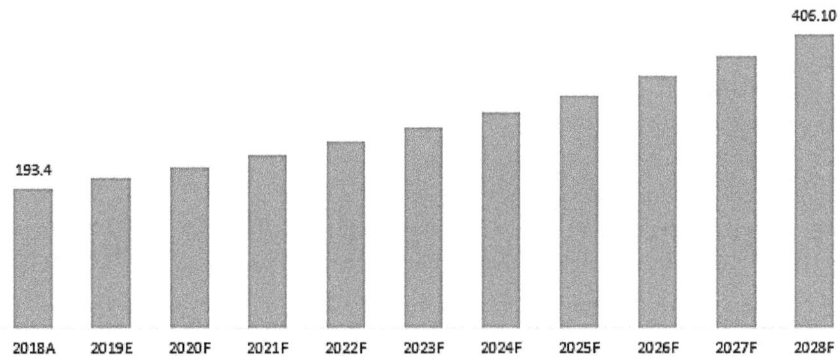

FIGURE 11.1 Analysis of the trend of frankincense market demand.

Source: Prudour (2019); Open source.

FIGURE 11.2 Micro-entrepreneurs in guilt selling the pockets of frankincense.

Source: Reprinted from Berhanu et al. (2021). © 2020 The Authors. Published by Elsevier Ltd. https://creativecommons.org/licenses/by/4.0/

India is the main supplier of frankincense derived from *B. serrata* (Brendler et al., 2018). Most frankincense is harvested in the Sheopur district in Madhya Pradesh, with an estimated 5,302 quintals (530 tons) of frankincense resin harvested annually (Bhattacharya and Hayat, 2004), whereas only 99.8 and 27 tons from Gujarat and Andhra Pradesh, respectively, were harvested in 2008–2013 (Yogi et al., 2014). Brendler et al. (2018) provide a synthesis of frankincense production in India.

Mr. Kamalji Gondwala from Indore, an exclusive gums trader since 1950, says that the Gum-resin of Salai from the Seopur area goes mainly to the extraction market while Indore material goes to Dhoop Industry. According to his estimates, in the year 2020, the Shivpuri area market was 2,000 tons and the Indore market 1,000 tons, Tikamgarh, Chhatarpur, Damoh, and Katni market 500 tons every year. Overall, 3,500 tons is the established market only from Madhya Pradesh. That is 35,00,000 kg @ Rs. 200/- per kg an average Raw Oleoresin price accrues to 70 Cr INR in the market of Salai alone. Not adding value addition generating revenue for the country. We need the export and domestic consumption details to arrive correct figures (Dr. N. Venugopal, personal communication).

Domestic uses of Frankincense abound. In Djibouti, Eritrea, Ethiopia, Somalia, and Sudan, the resin from *B. papyrifera* and *B. sacra* is chewed and used in rituals (Abdoul-Latif et al., 2012). In Ethiopia, *B. papyrifera* paste is used as an antifungal and in perfume and incense (Abdoul-Latif et al., 2012). In Kenya, the smoke of burning *B. neglecta* resin is believed to smoke to repel snakes and flies (Mugah et al., 1997). Similarly, in Oman, *B. sacra* are used as an insect repellent (Hassan, 2012). In India, wood pulp is used for paper; and it makes a good fuel and charcoal that is favored in iron smelting (Orwa et al., 2009; Saha et al., 2015). In India and Ethiopia, *Boswellia* wood is also used for fencing, agricultural implements, inexpensive furniture, packing crates, matches, plywood, and veneers (Eshete, 2002; Orwa et al., 2009). In parts of Africa, the gum and resin may be used as adhesives, dyes, and lithographic ink (Abiyu et al., 2010). There are few recent estimates of domestic consumption; most information is either subregional or must be gleaned from other data. In Oman, the annual domestic consumption of *B. sacra* products for local celebrations is estimated to be 40–50 tons (Hassan, 2012). In 1997, the Roman Catholic and Orthodox Churches in Ethiopia used an estimated about 2 million kg (2,000 tons) of frankincense (Gebrehiwot et al., 2003). Eshete (2002) reported that more than 8,100 tons of natural gums were sold domestically from 1992–2001 – about 80% of which was *B. papyrifera.* Thus, an estimated 6,500 tons were consumed domestically over a 10-year period.

Trade in the Mediterranean region has occurred since at least 1700 BC (Hassan, 2012). Frankincense trade was recorded as early as the 4th century BC in Oman (Al-Gasani, 2000). In Ethiopia, large-scale frankincense export did not begin until the 1930s (Ogbazghi, 2001). The resin has long been used as incense for religious ceremonies around the world (Hassan,

2012; Mugah et al., 1997). There is also a long history of using *Boswellia* species in Ayurvedic, Unani, and Chinese medicine (Al-Harrasi et al., 2018; Brendler et al., 2018; Iram et al., 2017; Jayatissa, 2012; McCutcheon, 2018; Zhu, 1998). *B. serrata* is included in the Ayurvedic Pharmacopoeia of India, the European Pharmacopoeia, and the United States Pharmacopeia (ABC-AHP-NCNPR, 2016; Brendler et al., 2018). The resin and derivatives are incorporated into various dietary supplements (Meins et al., 2016) and often processed into essential oils to be used variously in fragrances, cosmetics, and aromatherapy (ITC, 2014; Sommerlatte, n.d.). With some exceptions, most frankincense is exported from a range of countries as gum or resin (Bhattacharya and Hayat, 2004; Brendler et al., 2018; Eshete, 2002; McCutcheon, 2018; Nour, 2008). Some countries export value-added botanicals (such as essential oils or alcohol extracts derived from resin or bark) and products. Products from Oman include hair care products, aromatherapy aids, and skin cosmetics; fragrance, embalming, insect repellent, and food industry products (Hassan, 2012). Kenya produces commercial incense, essential oils, and shower gel (Mugah et al., 1997; Sommerlatte, n.d.). As with many other commodities based on natural resources, the market value increases as it moves up the value chain, and the essential oil is valued at 10 times that of the resin. There are some exceptions to that trend.

Contemporary international trade patterns are incomplete but are clearly multi-lateral, with a range of countries exporting and importing resin, gum, derivatives, and finished products derived from a variety of *Boswellia* species. *B. frereana* produces a high-quality resin known as "maidi" that is exported by Somalia where it is endemic (Eshete, 2002); *B. neglecta* from Kenya (Mugah et al., 1997; Sommerlatte, n.d.); *B. sacra* from Oman and Somalia (the latter especially in a form locally called "beyo") (Eshete, 2002); *B. papyrifera* from Ethiopia and Sudan (Eshete, 2002); and *B. rivae* from Nigeria (McCutcheon, 2018). According to Brendler et al. (2018), India is the only producer of *B. serrata* oleo-resin for the international market. However, India also imports *B. sacra* and *B. frereana* from the Gulf and North Africa (McCutcheon, 2018). Oman is likely the source of *B. sacra*, but there are no *Boswellia* species native to Northern Africa. So, while *B. frereana* originates from Somalia (where it is endemic), it is unclear what path the raw material takes through the international market. High market demand was reported for *B. ovalifoliolata* (endemic to India) (Saha et al., 2015). The major *B. serrata* importers are Trinidad and Tobago, Germany, Guatemala, Mexico, and the United States of America. Europe is considered the top market for essential oils, especially in France for perfumery (ITC, 2014, 2016).

It is difficult to assess trade levels in parts and derivatives of *Boswellia* species, as the data can be vague or incomplete. Applicable Customs Codes can be broad and may include other species. However, there are a few codes specific to *Boswellia*: Harmonized system Code 1302 1919 pertaining to *B. serrata* extract (BSE); HS Code 1301 2990 Olibanum oil from India, and HS Code 1301 9020 Olibanum, myrrh, and dragon's blood (China only). Brendler et al. (2018) point to sources of trade information from the Indian subcontinent, which is mentioned below. Otherwise, information must be gleaned from industry market reports, i.e., dietary supplements and essential oil channels (McCutcheon, 2018; MarketWatch, 2018).

There are few estimates of overall trade volume. Coppen (1995) compiled some trade statistics for the FAO. International exports of incense gum averaged 252.5 tons/year from Somalia (1975–1980) and 757 tons/year from Ethiopia (1981–1983). World olibanum trade in 1987 was reportedly 3,200 tons (Coulter, 1987). From 1987–1993, India exported an average of 87.5 tons/year of frankincense resin (*B. serrata*). In 1997, the Eritrea Ministry of Agriculture statistics reported an export volume of 543 tons of olibanum resin (FAO, 2001). Annual resin production from *B. sacra* in Oman is ~70–100 tons/year (Hassan, 2012) and 702 tons/year from Ethiopia's North Gandar Zone (in the Amhara Region; Eshete, 2002) (Table 11.1).

TABLE 11.1 Main Sources and Volumes of Olibanum in World Trade in 1987

Species	Commercial Name	Producing Country	Quantity (Tons)
B. papyrifera	Etritrean type	Sudan, Ethiopia	2,000
B. frereana	Maidi	Somalia	800
B. sacra	Beyo	Somalia	200
B. serrata	Indian type	India	200

Source: Adapted from: Coppen (1995).

Coppen (1995) also characterized the international market in terms of main consumers (the Middle East and China; Germany to some extent as well as other parts of Europe and Latin America) and suppliers (mostly Ethiopia, India, Somalia, and Sudan, as well as Kenya). In 1995, olibanum was sourced mainly from Somalia (from the species *B. sacra* (syn. *B. carteri*) and *B. frereana*); from "southern Arabia" (Yemen or Oman) (*B. sacra*); with minor amounts from Somalia (*B. bhau-dajiana* (syn. *B. sacra*) and *B. neglecta*) and Ethiopia (*B. papyrifera*). In 1997, it was reported that "true frankincense is obtained from *B. sacra* (syn. *B. carteri*); and some other species growing in northern Somalia, Dhofar [Oman] and Hadhramaut [Yemen];" and the main species from tropical East Africa were *B. papyrifera* and *B. neglecta* (Mugah et al., 1997, p. 21). In 1998, the most important *Boswellia* species for gum production in Ethiopia were *B. papyrifera, B. ogadensis,* and *B. rivae* (Deffar, 1998). The FAO (2001) reported that Sudan and Ethiopia were the main olibanum exporters from sub-Saharan Africa. Mathe et al. (2004) reported the most important species to be *B. serrata, B. sacra, B. frereana* and *B. sacra* (syn. *B. carteri*); with inferior forms of frankincense from *B. papyrifera.* More recently, Brendler et al. (2018) characterized harvest and trade in *B. serrata* from India. Brendler et al. (2018) estimated that more than 177 metric tons of *B. serrata* was exported from India 2015–2017. Information on global market value is also scant. The *Boswellia* market in the United States is mainly for dietary supplements (ABC-AHP-NCNPR, 2016). There is an increasing international demand for gum olibanum.

Demand for *Boswellia* raw materials is expected to increase as established uses in new appreciation in the personal care products industry and as new pharmaceutical applications are identified. Gesmalla et al. (2015) notes that the cosmetic and pharmaceutical industries hold great potential for development of *Boswellia* commodities. *Boswellia* resin is also used as a substitute for closely-related *Commiphora* species (commonly known as myrrh) (Saha et al., 2015) and for balsam (Orwa et al., 2009). According to one market report, the global essential oil trade is expected to have a market value of US $11.19 billion by 2022, driven by increasing disposable incomes and research and development for food, beverages, cosmetics, and personal care products (MarketWatch, 2018). The U.S. sales of dietary supplements containing *Boswellia* increased from US $1.5 million in 2013 to US $2.1 million in 2016 (McCutcheon, 2016).

With burgeoning studies on the medical applications of BAs, pharmaceutical demand for *Boswellia* raw materials is likely to increase. Recent decades have seen a plethora of research on the therapeutic applications of BAs, including in the treatment of cancer, arthritis, liver, and kidney ailments, diarrhea, cholesterol, asthma, microbial, fungal, analgesic uses (Abdoul-Latif et al., 2012; Iram et al., 2017; Moreillon et al., 2013; Oliff, 2018; Zhang et al., 2016) (Figure 11.3). *Boswellia ovalifoliolata* has also been investigated for veterinary applications (Al-Yasiry et al., 2016) and as an eco-friendly larvicide (Benelli et al., 2017).

In addition, it appears that nurseries in the United States are taking an interest in growing *Boswellia* commercially for home gardeners. *B. sacra* plants are grown as bonsai (Todd's Tropicals 2017; https://www.youtube.com/watch?v=li5i7hvJnE4) and for xeric landscaping (Tortorello 2011; Enlightenment Garden, 2016; https://www.youtube.com/watch?v=_p0lqa-uHrZQ), though frankincense seed is reportedly hard to find and, by some accounts, hard to grow (Tortorello, 2011).

FIGURE 11.3 Some products of *Boswellia* in the market.

KEYWORDS

- **aromatherapy**
- ***Boswellia***
- **essential oil**
- **frankincense**
- **gum resin**
- **olibanum**

REFERENCES

ABC-AHP-NCNPR Botanical Adulterants Program, (2016). Evaluation of the authenticity and quality of top-selling *Boswellia* products by HPLC-triple quadrupole mass spectrometry. *Botanical Adulterants Monitor*, (8).

Abdoul-Latif, F. M., Obame, L. C., Bassole, I. H. N., & Dicko, M. H., (2012). Antimicrobial activities of essential oil and methanol extract of *Boswellia sacra* Flueck. and *Boswellia papyrifera* (Del.) Hochst from Djibouti. *Intern. J. Management Modern Sci. Technologies, 1*(1), 1–10.

Abiyu, A., Bongers, F., Eshete, A., Gebrehiwot, K., Kindu, M., Lemenih, M., Moges, Y., et al., (2010). Chapter 7 – Incense woodlands in Ethiopia and Eritrea: Regeneration problems and restoration possibilities. In: Bongers, F., & Tennigkeit, T., (eds.), *Degraded forests in Eastern Africa: Management and Restoration* (pp. 133–152). Oxford: Earthscan.

Al-Gasani, A., (2000). *Land of Frankincense*. UNESCO World Heritage List website: https://whc.unesco.org/en/list/1010 (accessed on 10 October 2022).

Al-Harrasi, A., Khan, A. L., Asaf, S., & Al-Rawahi, A., (2019). Taxonomy, distribution and ecology of *Boswellia*. In: Al-Harrasi, et al., (eds.), *Biology of Genus Boswellia* (pp. 11–34). Springer.

Al-Yasiry, R. M. A., Jawad, S. A. H., Menati, K. J., Naji, S. A., & Lokman, I. H., (2016). Effects of *Boswellia carterii* and *Boswellia serrata* in drinking water on the growth performance, hematology traits and immune response of broiler chicken. *Research & Reviews: Journal of Food and Dairy Technology, 4*(4), 27–37.

Berhanu, Y., Pal, V., Ayana, A., & Jens, B. A., (2021). The contribution of frankincense to the agro-pastoral household economy and its potential for commercialization: A case from Borana, southern Ethiopia. *J. Arid Environments, 186*, 104423, https://doi.org/10.1016/j.jaridenv.2020.104423.

Bhattacharya, P., & Hayat, S. F., (2004). Sustainable NTFP management for fural development: A case from Madhya Pradesh, India. *Intern. Forestry Review, 6*(2), 161–168.

Brendler, T., Brinckmann, J. A., & Schippmann, U., (2018). Sustainable supply, a foundation for natural product development: The case of Indian frankincense (*Boswellia serrata* roxb. ex colebr.). *J. Ethnopharmacol., 225*, 279–286.

Coppen, J. J. W., (1995). *Flavors and Fragrances of Plant Origin*. Rome, Food and Agriculture Organization of the United Nations. http://www.fao.org/docrep/V5350e/V5350e00.htm#Contents (accessed on 10 October 2022).

Coulter, J., (1987). *Market Study for Frankincense and Myrrh from Somalia*. Tropical Development and Research Institute, London.

Deffar, G., (1998). *Non-Wood Forest Products in Ethiopia*. Rome, Italy: EC-FAO Partnership Program. http://www.fao.org/docrep/003/x6690e/X6690E01.htm#P226_14410 (accessed on 10 October 2022).

Eshete, A., (2002). *Regeneration Status, Soil Seed Bank and Socio-Economic Importance of Boswellia papyrifera (Del.) Hoschst. in Two Woredas of North Gondar Zone, Northern Ethiopia*. Master's Thesis, Swedish University of Agricultural Sciences and Wondo Genet College of Forestry.

FAO (Food and Agriculture Organization), (2001). *Global Forest Resources Assessment 2000*. FAO Forestry Paper No. 140. UN Food and Agriculture Organization, Rome. http://www.fao.org/docrep/004/y1997e/y1997e0f.htm (accessed on 10 October 2022).

Gebrehiwot, K., Muys, B., Haile, M., & Mitloehner, R., (2003). Introducing *Boswellia papyrifera* (Del) Hochst and its non-timber forest product, frankincense. *Int. For. Rev., 5*(4), 348–353.

Gessmalla, A. F., Raddad, E. Y. A., & Ibrahim, G. E. A., (2015). Effects of tapping date, tapping direction and elevation on resin yield from *Boswellia papyrifera* in the Blue Nile State, Sudan. *Natural Resources, Agricultural Development and Food Security – International Working Paper Series, 15*(12), 14. http://economia.unipv.it/naf/ (accessed on 10 October 2022).

Global Frankincense Alliance, (2021). *The Main Outcomes and Road Map*. The Future of Online workshop.

Hassan, A. M. M., (2012). Distribution *Boswellia sacra* in Dhofar Mountains, Sultanate of Oman: Economic value and environmental role. *J. Life Sci., 6*, 632–636.

Iram, F., Khan, S. A., & Husain, A., (2017). Phytochemistry and potential therapeutic actions of Boswellic acids: A mini-review. *Asian Pacific J. Trop. Biomed., 7*(6), 513–523. https://doi.org/10.1016/j.apjtb.2017.05.001.

ITC (International Trade Centre), (2014). *Market Insider: Essential Oils & Oleoresins*. Geneva: International Trade Centre. https://studylib.net/doc/18120304/essential-oils-and-oleoresins-market-insider (accessed on 10 October 2022).

ITC, (2016). *Market Insider: Essential Oils & Oleoresins*. Geneva: International Trade Centre. Geneva: International Trade Centre. https://bit.ly/2BCXdGW (accessed on 10 October 2022).

Jayatissa, L. P., (2012). *Ayurvedic Medicinal Plants of Sri Lanka: Boswellia serrata*. IAAM (Institute of Ayurveda and Alternative Medicine). http://www.instituteofayurveda.org/plants/index.php (accessed on 10 October 2022).

Kandari, L. S., Negi, T., Thakur, A. K., et al., (2015). Ethnobotanical and indigenous knowledge of important plants in East Hararghe, Eastern Ethiopia. *J. Mountain Sci., 12*(6), 1521–1533.

MarketWatch, (2018). *Essential Oil Market to Attain a Value of US $27.49 Bn by 2022; Booming Beauty and Personal Care Industry to Augment Sales of Essential Oils, Says TMR (Transparency Market Research).* Press Release. Online at: https://on.mktw.net/2Ri20Da (accessed on 10 October 2022).

Mathe, C., Culioli, G., Archier, P., & Vieillescazes, C., (2004). High-performance liquid chromatographic analysis of triterpenoids in commercial frankincense. *Chromatographia, 60*(9, 10), 493–499.

McCutcheon, A., (2018). *Adulteration of Boswellia serrata.* Austin, TX: Botanical adulterants prevention program. Botanical Adulterants Bulletin.

Meins, J., Artaria, C., Riva, A., Morazzoni, P., Schubert-Zsilavecz, M., & Abdel-Tawab, M., (2016). Survey on the quality of the top-selling European and American botanical dietary supplements containing Boswellia acids. *Planta Med., 82*(6), 573–579.

Moreillon, J. J., Bowden, R. G., Deike, E., Griggs, J., Wilson, R., Shelmadine, B., Cooke, M., & Beaujean, A., (2013). The use of an anti-inflammatory supplement in patients with chronic kidney disease. *J. Complement. Integr. Med., 10*(1), 1–10.

Mugah, J. O., Chikamai, B. N., Mbiru, S. S., & Casadei, E., (1997). Conservation, management and utilization of plant gums, resins, & essential oils. *Proceedings of a Regional Conference for Africa.* Held in Nairobi, Kenya. www.fao.org/3/a-x0098e.pd (accessed on 10 October 2022).

Nour, L. A. M., (2008). *Production and Productivity of Boswellia papyrifera in Jebel Elgarrie Area (Blue Nile State).* Master's Thesis, University of Khartoum, Sudan.

Ogbazghi, W., (2001). *The Distribution and Regeneration of Boswellia papyrifera (Del.). Hochst. in Eritrea.* PhD Thesis, Wageningen University and Research Centre, The Netherlands.

Oliff, H. C., (2018). *Boswellia*, ginger, & yarrow combination reduces symptoms of irritable bowel syndrome. *HerbClip,* 021854-599.

Orwa, C., Jamnadass, R. H., Kindt, R., Mutua, A., & Simons, A., (2009). *Boswellia serrata. Agroforestree Database: A Tree Reference and Selection Guide, Version 4.0.* World Agroforestry Centre, Nairobi, Kenya.

Prudour, (2019). https://market.us/report/frankincense-essential-oil-market/#major-market-players (accessed on 10 October 2022).

Saha, D., Ved, D., Ravikumar, K., & Haridasan, K., (2015). *Boswellia ovalifoliolata: The IUCN Red List of Threatened Species.* e.T50126567A50131280. http://dx.doi.org/10.2305/IUCN.UK.2015-2.RLTS.T50126567A50131280.en.

Sommerlatte, H., (n.d.). *Arbor Oils of Africa.* Naro Maru, Kenya. http://www.oilsafrica.com/arbor-collectors-arid-land-resources.php (accessed on 10 October 2022).

Zhang, J., Biggs, I., Sirdaarta, J., White, A., & Cock, I. E., (2016). Antibacterial and anticancer properties of *Boswellia carteri* Birdw. and *Commiphora molmol* Engl. oleo-resin solvent extractions. *Pharmacogn. Commn., 6*(3), 120–136.

Zhu, Y. P., (1998). *Chinese Materia Medica: Chemistry Pharmacology and Applications.* Amsterdam: Harwood Academic Publishers.

CHAPTER 12

BOSWELLIA SPECIES: THREATS AND CONSERVATION

LEPAKSHI MD. BHAKSHU,[1] K. VENKATA RATNAM,[2] and T. PULLAIAH[3]

[1]*Department of Botany, PVKN Government College, Chittoor – 517002, Andhra Pradesh, India*

[2]*Department of Botany, Rayalaseema University, Kurnool – 518007, Andhra Pradesh, India*

[3]*Department of Botany, Sri Krishnadevaraya University, Ananthapuramu – 515003, Andhra Pradesh, India*

[*]*Corresponding author. E-mail: lmbakshu@gmail.com*

ABSTRACT

Boswellia species are ecologically and economically medicinally valuable and hence gained commercial exploitation. The over-tapping of this tree species for extracting the commercially valued oleo gum resin has led to their natural population loss severely. Therefore, an alarming situation has raised to make stringent conservation efforts. Many scientific reports and forest management plans have suggested action plans requiring the involvement of people in the vicinity. This chapter enlightens about the conservation progress among the species of *Boswellia* genus, and focuses on implementation of various conservation action plans in the respective countries using *in situ* as well as *ex situ* or *in vitro* methods. Since, *Boswellia* saplings facing major problems, such as grazing, fires cutting for wood, etc., are important constrains to expand its population and can

Frankincense – Gum Olibanum: Botany, Chemistry, Extraction, Utilization, Propagation, Biotechnology, and Conservation, T. Pullaiah, K. Venkata Ratnam, Mallappa Kumara Swamy, Lepakshi Md. Bhakshu (Eds.)

be rectified by utilizing management plans in support of the life of future generations to re-vitalize its golden era.

12.1 INTRODUCTION

For at least 3,000 years, olibanum, also known as frankincense, had been an important trade material for the civilizations located in the Arabian Peninsula, North Africa, and India. Because of its specific fragrance, old kings and queens like Queen of Saba 700 BC (Yemen) got interested and started usage in their cultural aspects. Hence, it gained importance as a precious commercial material in the market. Olibanum is a natural oleo-gum-resin, obtained from stem barks of the genus *Boswellia* (Family Burseraceae).

The genus *Boswellia* consists of about 24 taxa of trees and shrubs, distributed in the dry localities of the Arabian countries, India, and East Africa. Socotra (Yemen), a small Island of the Indian Ocean, consists of 11 unique endemic *Boswellia* species that occur as a result of its prolonged geological isolation and the existing hot and dry climate conditions.

Frankincense and myrrh have potential value as important ingredients in cosmetics such as creams, lotions, soaps, and detergents and often in meditation blends (FAO, 1995). In addition to the above, myrrh and frankincense have a wide range of applications in the areas of food, chewing gums, gelatins, beverages, confectioneries, candies, puddings, nut products, and canned vegetables. Typically, myrrh and frankincense are applied as stabilizers, thickeners, adhesive agents and emulsifying agents in the food industry, discharge agents in rubber products and clarification agents in beverages. In the Middle East, especially in Saudi Arabia, roughly 500 tons of olibanum of Somali type are imported for manufacturing chewing gum, and a similar amount of oleo gum resin is used in these countries as incense material at household traditional ceremonies (FAO, 1995). In the market, frankincense products are mainly exported, however considerable amounts of oleo gum resins are also traded in the domestic markets and used locally in households and religious institutions. The demand for frankincense has increased the supply requirement, and the present supply satisfies less than 15% of the domestic demand (Tilahun, 1997; Demel and Mulugeta, 2005).

The demand for frankincense, used to make everything from exportable perfume and lotion to fancy soaps is rising risk on the frankincense producing species leading to their decline in their population globally. The management plans which can be supported by the governmental policies through the forest officials might be helpful in in conservation and can restore the environmental issues such as soil erosion, forest degradation and invasive species (Lemenih and Kassa, 2011). During the last few decades, the abundance of exploitable *Boswellia* trees has decreased mainly owing to intensification of land use practices, such as land clearing for agriculture, grazing by livestock and extraction of resin (Ogbazghi, 2001; Ogbazghi et al., 2006).

The International Union for Conservation of Nature (IUCN) and Natural Resources reported 9 species with a threat status of high vulnerability, and include critically endangered (EN) for *B. ameero* (Miller, 2004a), and Vulnerable for *B. bullata* (Miller 2004b), *B. dioscoridis* (Miller, 2004c), *B. elongata* (Miller, 2004d), *B. nana* (Miller 2004e), *B. popoviana* (Miller, 2004f), *B. socotrana* (Miller, 2004g), *B. ovalifoliolata* (Saha et al., 2015), *B. ogadensis* (Alemu et al., 2018), and *B. pirottae* (Awas et al., 2018), requiring the attention for their conservation.

The plant wealth is dwindling very rapidly, because of nonstop pressure from deforestation also associated with expansion of agriculture, over-grazing, insect attack, improper tapping, etc., as well as insufficient/inadequate research undertakings, lack of master and action plans for the development, defendable utilization, and conservation of plant resources. Given the notable ecological, cultural, and socio-economic importance of natural gum bearing species, viable master plans and realistic action plans for research, growth, defendable utilization and conservation of the species and their habitats have to be developed and implemented in the country before it is too late. Therefore, research and development efforts and international collaborations have strong potentials to the conservation, production, and commercialization of these huge and unexploited renewable natural resources for the well-being of the communities of local, national as well international. There is a growing concern for its increasing global demand and unregulated international trade of this high valued commodity. This might cause a serious threat on the survival of these species in nature.

The main aim of the discussion is to identify the concerns for the existence of the *Boswellia* species in their natural habitats and the pressure posing on the economic thrives for the oleo gum. Understanding the conservation strategies, suitable to the specific species are required to implement their conservation, improvement, and enhancing the tapping sites, sustainable utilization, besides its demographics.

The natural risks, such as habitat destruction, insect infestation, grazing, altitudinal variations, seedling sustenance, inbuilt plant constraints such as lack of dormancy, partial successes in resisting drought during development stage from seedling to adult individual and anthropogenic actions, such as over tapping for domestic or international trade and industrial usage of wood for carving, paper, softwood articles, etc., are jeopardizing the future of the *Boswellia* and major threats across these species' ranges. Among the constraints the important threats were discussed previously by researchers (Kushwaha and Singh, 2005; Singh and Kushwaha, 2006; Sagar and Singh, 2006; Bongers et al., 2019).

In Somalia, the populations of *B. sacra* and *B. frereana* were reduced at alarming rate, because of over and careless tapping and complete debarking during gum collection, resulting decline in the number of species and increasing scarcity. Besides, *B. serrata* has wide distribution and has been a major source of resin for many centuries (DeCarlo and Ali, 2014).

Also, *B. serrata,* in Somalia is facing threats due to grazing, fire, lack of samplings (regeneration stage), adult mortality, etc. (Kushwaha and Singh, 2005; Singh and Kushwaha, 2006; Sagar and Singh, 2006).

The assessment of the extinction risk of *Boswellia* spp., based on the IUCN criteria suggested to move almost all the species to a higher risk category than that ascribed by the 2004 assessment (Miller, 2004a–g; Fabio et al., 2014).

Different species of *Boswellia* have been reported to grow under various types of climatic conditions. For example, Indian frankincense grows in the humid conditions, and *B. sacra* grows in dry tropical conditions. Nevertheless, *B. socotrana* did not grow in a greenhouse environment, but it prefers the winter season. As observations made by Eslamieh (2017), all species of *Boswellia* needs moistness and heat, while there is no broad commonality among these species otherwise, each species has its own genetic makeup for cultivation and adaptation. For example, when *B. nana* grown in rock crevices, totally changed its vegetative characters to fit the rock cavities, while in the natural field, it grows normally as a dwarf shrub. Whereas *B. frereana* forms a bulbous-shaped mass in an extremely dry habitat, while in cultivated form it grows similar to a normal plant.

Boswellia species can withstand extreme environments for long periods of time without any serious adverse effects (Eslamieh, 2017). Hence, this species may be useful for domestication and propagation and relevantly conserved in the social-forestry.

In vitro propagation methods have been used for the conservation of medicinal plants, mainly those species facing problem to propagate through conventional methods. The propagation of *Boswellia* species through micropropagation was reported by plant tissue culturists (Chandrasekhar et al., 2005; Purohit et al., 1995; Vijay, 2014) might be useful in developing the plantlets and introducing them into their habitat for the sake of conservation and restoration as the *in situ* environmental conservation approach. The tissue culture method is one of the important tools that can be used to reach conservation goals with respect to threatened, EN, and rare plant species, including *Boswellia* plants.

B. ameero is a tree with a scattered and somewhat fragmented distribution. Some of the peripheral populations are not regenerating and considered relict in nature, strongly proposed a decline in the extent and nature of its habitat. The tree has been classified as 'Vulnerable' in the IUCN Red List of threatened species (Tropical Plants Database, 2011). Hence maximum attention has to be taken to improve its population to a satisfactory level in connection with conservation.

12.2 CONSERVATION STRATEGIES OF *BOSWELLIA* SPECIES

The scientifically advanced technology utilization on the evaluation of *B. serrata* stands in Indian climatic prediction and the satellite data revealed that about 21.95% area is suitable and predicted that the sustainable area is expected to decrease by 2050 alarming to implement the long-term action plans for its conservation (Radha et al., 2020).

According to Brendler et al. (2018), the majority of the gum oleoresin of Indian frankincense (*B. serrata*) was collected by local tribes inhabiting in the forests. In recent years, attempts have been made to increase populations of frankincense for sustainable utilization. Best example for this is Haryana Forest Department planned to establish *B. serrata* plantations with 1,100 plant samples raised from branch cuttings and seeds in 100 ha in the districts of Rewari, Mohindergarh, and Bhiwani (Haryana State Medicinal Plants Board, 2013).

Sabo et al. (2021) reported that usage of land affects the propagation characteristics such as regeneration of species, size of population, stand structure of *B. dalzielii* in the Sudano-Sahelian region present in Burkina Faso. The analysis revealed that human disturbances like over debarking during gum collection and others contribute to unstable conditions of *B. dalzielii* populations. Environmental factors such as forest fire and wind also prevent high seedling density in natural habitats of *B. dalzielii.* Therefore, it is important to conduct assisted natural regeneration experiments to enhance the *B. dalzielii* populations in their natural habitat. Actions such as the formation of fire lines in the forest area and agroforestry practices are needed to enhance individual species of *B. dalzielii.* In addition to this training in gum tapping and promoting best harvesting practices to local tribes can also minimize the decline in population size of *B. dalzielii.* It is understood that the reforestation and awareness activities are critical to enhance the populations in the fragile and harsh environment (Sabo et al., 2021).

As per the spatial analysis, three ground-rooted *Boswellia* species namely *B. socotrana*, *B. ameero* and *B. elongata* were abundant and wide spread in Socotra Island and these species were characterized by geo-altitudinal zonation (Fabio et al., 2014). The progressive decline in the populations of these species is because of uncontrolled grazing by cattle and the absence of particular conservation strategies. The two species *B. dioscorides* and *B. popoviana* are numerically abundant and need lower conservation priority. The other two species like *B. bullata* and *B. nana* may be under threat condition by uncertain reasons due to a reduction in population size in a very smaller number of locations needs to be given more priority. Specific conservation plans such as the establishment of a seed bank, nurseries, and conservation through use programs are assigned to each species based on their economic importance, ecological features, and environmental condition.

Overgrazing is one of risk of population enhancement of *B. elongata* and an action plan for its protection from over grazing, namely "Socotra Conservation and Development Program" has been implemented by GEF-UNDP in 2008 in the Homil plateau. The action plan includes a collection of seeds by conducting intensive field trips, the growing of plant saplings through branch cuttings by following international standard methods. The project resulted high seed production and germination rate in the rainy season only. But the seedings survival capacity was very poor during the summer season. This may be due to the excessive hardness of the eroded soil and very poor availability of nutrients in the dry season. Similar type of observations was recorded in well-established enclosures (Negussie et al., 2008). They suggested that it required additional types of intervention methods to develop appropriate soil preparation to shading of seedlings.

The two-ground rooted *Boswellia* species, i.e., *B. socotrana* and *B. elongata* producing oleo gum resin with high market value, which is the extra income source to the local rural communities (Newton, 2008). Because of over tapping and collection of gum resin leads to the decline of populations of these species. Effective conservation plants are to be undertaken by giving scientific training in gum tapping and collection, chemical analysis of the gum resins, development of agroforestry methods such as collection time of seeds, nursery establishment, selection of suitable planting sites and care of plants after plantation and good processing practices, national, and international marketing (Gebrehiwot et al., 2003).

Moreover, utilization of the traditional system of land management among local communities need to be involved in forest protection and management activities. People's participation can be ensured by providing incentives and including in the form of ownership/user rights and benefit-sharing mechanisms might improve social responsibility and involvement in the conservation.

12.3 STRATEGIC EVENTS IN CONSERVATION OF *B. SERRATA* FROM THE INDIAN CONTEXT

The anthropogenic activities such as cutting, grazing, lopping, habitation, and infrastructure changed the distribution of *B. serrata* tree species from clumped to uniform with lower stem density (Sagar et al., 2003). Shahabuddin and Prasad (2004) observed that anthropogenic activities such as tree cutting, grazing, and forest fires along with environmental factors, soil nutrients composition and micro-climatic conditions showed greater impact on regeneration of *B. serrata* samplings in natural conditions. In India, *B. serrata* is distributed in dry deciduous forest. These forests are highly threatened due to expansion of cities, over grazing, commercial crop farming and firewood collection. The major forest areas of *B. serrata* populations have been lost due to over mining, constructions of dams and power projects and population resettlements (WWF, 2015a, b). Further, the introduction of invasive species *Lantana camara* L., which covers the understory of forests, can be considered as threat to the survival of *Boswellia* and other forest tree species of dry woodlands of Northern India (Sharma and Raghubanshi, 2007).

Indian frankincense (*B. serrata*) has not yet been listed as threatened taxa in IUCN categories. Several taxonomists based on intensive field trips assessed different threat categories to *Boswellia serrata,* such as Vulnerable (VU) in Jabalpur of Madhya Pradesh by the State Research Institute and in Orissa by ENVIS center at Wildlife Institute of India (Chaubey et al., 2015; Biswal and Nair, 2008). Further, Conservation Assessment and Management Plan workshops conducted in 2003 and 2007 in Bhopal and Rajasthan ranked *B. serrata* as VU and EN, respectively (Ved, 2013). Indian Red Data Book assessed Rare (R) category to *B. serrata,* but based on field study carried out in Jawahar Sagar Sanctuary by Joshi and Shringi (2014) assigned least concern (LC) category. National Red List of Sri Lanka listed *B. serrata* as critically EN to possibly extinct (MOE, 2012).

The government of Madhya Pradesh state (India), restricted the collection of oleogum of *B. serrata* since tapping damaged heavily on the trees and restricted its collection and notified under the Non-Wood Forest Products as specific forest produce in the act–1969 (M. P. Van Uapj, Vyapar Viniyaman). This act enables only an authorized person allowed to gum collection through cooperative societies. Further, in the 1980s, state government-imposed ban, which completely stopped the gum collection, later in 1995 the ban has been lifted and allowed controlled extraction in selected districts. Subsequently, the restrictions were completely withdrawn and removed from the notified list allowed for gum collection (Yogi et al., 2014).

B. serrata trees are well grown in forested areas where the tribal people's lives are associated with nature and wild life. The tribals are trained for gum collection methods and allowed as a part of conservation of in the protected sites (Jathar and Patil, 2011; Dular, 2015).

Soumya et al. (2019) examined the oleo gum extraction methods in Karnataka state in the three Protected Areas namely, Biligiri Rangaswamy Temple Tiger Reserve (BRT) of the Western Ghats, Malai Mahadeshwara Wildlife Sanctuary (MMH) and Cauvery Wildlife Sanctuary (CWS) with special an emphasis on the tapping, collection, and grading types of *B. serrata* gum, extracted through following the conservation strategies using efficient methods.

12.4 THREATS

12.4.1 HABITAT DESTRUCTION

Throughout the Central African Republic, Cameroon, Chad, Ethiopia, Eritrea, Nigeria, Uganda, and Sudan (Volleson, 1989, as summarized in Abdoul-Latif et al., 2012), threats include farming, overgrazing, fire, and land clearing for shifting cultivation (Oqbazghi, 2001, as cited in Abdoul-Latif et al., 2012). Overgrazing and the use of wood for fuel are also threats (Eshete, 2002). In Oman, gravel mining erodes soil, increases evaporation of water, decreases nutrient availability – all of which directly affect plant mortality. Adam and El-Tayeb (2008) noted that in Sudan, more than half of the seeds produced in three populations were infested and therefore not viable. Increased plant stress reduces seed production, which in turn reduces or eliminates the seed bank, greatly diminishing regeneration potential (Eshete, 2002; Eshete et al., 2005; Hassan, 2012).

12.4.2 INSECT INFESTATION

Boswellia species are subject to attack from several species of beetles that bore into live trees. Termites and other insects infest the trees (Oqbazghi, 2001; Abdoul-Latif et al., 2012). In Oman, at least three species of beetles, including *Sphenoptera chalcichroa*, lay eggs in the wounds of the bark and the larvae tunnel under the bark and bore into the trunks of the living trees (Strumia et al., 2001, as cited in Hassan, 2012). Wood-boring insects reduce the health of the tree, increasing their susceptibility to disease and contributing to high adult mortality observed in *Boswellia* populations (Groenendijk et al., 2012). *Sphenoptera chalcichroa* beetles are also found in Sudan and since the 1950s have decimated concomitant *Acacia nilotica* forests (FAO, 2007). Infestations have also been reported on *B. papyrifera* in Ethiopia (Eshete, 2002). In India, poor fruit set in *B. ovalifoliolata* has been linked to three predators: a weevil that feeds on buds and flowers; the palm squirrel, which eats the flowers and fruit; and the garden lizard that feeds on the pollinators of the tree, especially bees and wasps (Raju et al., 2012).

12.4.3 OVER-TAPPING

Research indicates that trees are being over-tapped to meet international demand, for example, *B. serrata* in India (Bhattacharya and Hayat, 2004); *B. papyrifera* and *B. sacra* in Djibouoti (Abdoul-Latif et al., 2012); *B. papyrifera* in Eritrea and Ethiopia (Eshete, 2002; Abiyu et al., 2010; Mengistu, 2011). Ebuen (2016) in Oman, monitored several years on the *B. sacra* and found its tendency on the decline in population and attributed to inappropriate tapping methods, little attention on the tapping, and easily accessible trees are excessively tapped and harvested throughout the year (Hassan, 2012). Inappropriate tapping methods, such as making cuts too deep or too long, in effective girdling the trees which causes life risk. Ogbazghi (2001) noted that germination and regeneration are better than in areas that are closed to tapping or that have not been tapped. Since tapping occurs during the dry season when the trees are leafless, Mengistu (2011) reports that tapping the tree depletes carbon stores and forces the tree to make trade-offs in defenses, vegetative growth, flowering, and fruit production. Over-harvested populations are characterized by a lack of young and saplings denoting loss of regeneration to woods. The scientific studies by several scholars on the demographics of *Boswellia* populations and inferred that, the lack of young trees and saplings and high mortality of adult trees in populations across these species range is an indicative of poor regeneration (Abtew et al., 2011; Gessmalla et al., 2012; Nour, 2008; Ogbazghi, 2001; Ogbazghi et al., 2006). In a study, it is studied and notice that the over-tapped trees have a lower pollination rate (16%) as compared to 80% in less tapped healthy trees (Rijkers et al., 2006). Sunnichan et al. (2005) found lower fruit set in tapped trees. Poor harvesting practices are contributing to the decline in these species (Ogbazghi, 2001; Abdoul-Latif et al., 2012) and the method of tree tapping and collection needs to be revised for long-term tree protection (Hassan, 2012).

12.4.4 REGENERATION CONSTRAINTS

One of the major problems associated with *B. papyrifera* is hampered natural regeneration, which could be attributed to several factors. The frankincense production, and even the existence of the species is seriously threatened by accelerated deforestation rate mainly through land clearing for more agricultural land, frequent wildfires, and overgrazing besides low rate of regeneration (Oqbazghi, 2001; Abeje, 2002; Wale et al., 2012).

12.4.5 INACCESSIBILITY/LOCATION OF THE RESOURCE IN REMOTE AREAS

The populations of *Boswellia* trees were extended distribution to remote or inaccessible areas characterized by rugged terrain, lack of pathways and infrastructure facilities contributed much affected the volume and quality of gum collected. In addition, these areas also associated hardship (harsh temperature, thirst, and disease) and difficulties to mobilize human resources, equipment, and transportation also affects collection (Girmay, 2000; Wubalem et al., 2002; Adefires, 2006).

12.4.6 LACK OF QUALITY CONTROL AND ILLEGAL BOARDER TRADES

Although there is no quantitative information on the exact amount of gum resin, registered enterprises are complaining that the major hindrance of the expansion of oleo gum might be lacking control on the illegitimate trade in Ethiopia (Demel and Mulugeta, 2005). Khamis et al. (2016) reported that *B. papyrifera* in Sudan is facing high pressure due to deforestation, fuel wood preparations, conversion into agricultural lands. In addition to above mentioned constrains the following challenges are recorded from the available literature for its survival such as wildfire, grazing, lack of plantations, low rate of regeneration, less number of saplings, political issues in the boarders of Somalia besides migrants destructing the tree species, lack of belongingness or individual tree caring during or after caring, improper maintenance of tapping intervals, high rate of mortality in mature trees, besides susceptibility to insects and pests. In a study, Bongers (2019) noticed that, *B. papyrifera* populations disappeared in more than 40% of villages, alarming to implement necessary measures for its conservation.

12.5 SUSTAINABLE TAPPING THROUGH MANAGEMENT PRACTICES

The studies of Girma et al. (2015) demonstrated that the quantum of secondary metabolites was correlated to the diameter at breadth height (DBH) of *B. papyrifera* in the bark and the study conducted through the PLSR (partial least square regression technique) spectral analysis. This study might be helpful in quick and reliable information for the decision makers to decide the tapping schedules. However, further extensive studies required further confirmation and portable technology to efficient usage of the hyperspectral remote sensing technology. Hence such scientific studies are highly determinative for the selection of correct tapping periods without disturbing the life of trees. Hence these studies have to be fastened and elaborate to other gum yielding *Boswellia* trees.

Four alternative ways were then suggested by Gidey et al. (2020) for the analytical hierarchy process (AHP) such as, free grazing with no tapping resting period (FGNTR), free grazing with a rotational tapping (FGRT), area enclosure with medium tapping resting period (AEMTR) and area enclosure with long tapping resting period (AELTR). The population structure analysis of *B. papyrifera* found unstable and is specified by low density (266 trees/ha) and found an absence of saplings indicating no regeneration (DBH<10 cm) and is attributed to multifactorial interrelated disturbances such as poor management, overgrazing, excessive tapping, agricultural expansion and pests. Further, suggested to conduct training programs to the gum tappers on its sustainable utilization and management, educating them on the importance of environmental and natural resources (Gidey et al., 2020). The outcomes of the above strategies helpful in conservation through sustainability and can be applied to other *Boswellia* species as per the requirements.

12.6 ADDITIONAL PROBABLE WAYS FOR CONSERVATION

In addition to sustainable tapping, certain conservation strategies might be strictly implemented and monitored by the governmental agencies of respective countries to save the heavenly tree species which might have multiple beneficial aspects to human kind.

- The establishment of Nurseries and vigorous plantations and their care up to survival in natural habitat has to be practiced through the involvement of the local people or non-governmental agencies.
- Implementation of Social forestry or Agroforestry in the accessible rural or semi-urban areas to maintain the satisfactory populations of native *Boswellia* species.
- The scientific propagation methods are followed to generate the required saplings and distributed to their localities for plantations

involving the public participation through wide publicity and awareness programs.

- Providing alternate source of revenue generation, the dependent people by declaring tapping holiday for a certain period to enable free sexual reproduction among the trees species. Even it relaxes the stress of carbon starving during the flowering season.
- Establishing/reviewing the progression of conservative works in the Protected Areas, National Parks, or Reserve Forests in the *Boswellia* growing sites.
- Adopting the groups of *Boswellia* or other commercial gum-yielding trees to village people by forming self-help groups or Biodiversity Monitoring Committees at the village level.
- Sustainable harvesting and post-tapping care and proper storage technology have to be followed for reducing gum loss and indirectly frees the stress on Trees from over-tapping.
- Providing awareness on the government programs (National Environmental Management act on Protected areas acts and legislations or Forest policies) and the role of Biodiversity to the local people.

12.7 INTERNATIONAL FOCUS ON THE *BOSWELLIA* CONSERVATION AND ITS FUTURE PROSPECTIVE

The important challenges were put forth on the international platforms for finding the roadmaps to ensure well-being of *Boswellia* and was deeply discussed in an International online conference "Global Frankincense Alliance (GFA)" held in 17th–18th March 2021, involving working groups of UN Convention on International Trade on Endangered species (CITES). The sessions focused on the Frankincense and Myrrh producing trees biology, medicinal claims, tapping, collection, grading, commercialization, and identified the gaps is good initiative for the sustainability and conservation. The convention identified certain crucial gaps and recommended for implementation for the sake of a better future of these trees and the dependent people of in the vicinity of *Boswellia* stands. In addition, the conference also focused on the strategies of action plans, implementation of high through-put techniques, involvement of people participation, quality analysis, supply chain, etc., along with action plans for the improvement of *Boswellia* populations in the wild. The presentations of participants from Ethiopia, Somalia, and Kenya, emphasized that, there was a harmonious relation existing between tappers or harvesters and Trees, which play a pivotal role in sustainability and forest management plans. This event also emphasized that more clinical research on the oleo gum products and genomic studies of the *Boswellia* trees are needed from the Scientific world for better management practices and to "*Save an Herb for Future*." Dr. De Carlo, Advisory Board Member of GFA, stated that, "How can we fix the problem of frankincense when we don't really know what the problem is? How come, after all these years, we know so little about the science and culture of this tree, which acts as a symbol of peace and a bridge between so many religions and cultures" (GFA, 2021)?

12.8 CONCLUSION

Since *Boswellia* is growing in harsh environments in Arabian Countries, the Horn of Africa, and India, certain conservation ambiance sites may develop to save the germplasm for the sake of future generations. The species of *Boswellia* are ecologically sensitive and involve socio-economic vulnerability and producing commercially tradable olio gum resin and might need control and conservation for sustainable utilization of natural resources. Hence the cumulative efforts of the government and local people to save the trees through safe tapping methods. Further, the utilization of safety gum tapping protocols, massive plantations, caring of saplings from the grazing, implementation of species-specific conservation techniques based on the scientific reports, and people's participation yield fruitful results in the conservation of *Boswellia* species.

KEYWORDS

- **analytical hierarchy process**
- ***Boswellia* species**
- **conservation strategies**
- **endangered**
- **oleo gum resin**
- **sustainable utilization**

REFERENCES

Abdoul-Latif, F. M., Obame, L. C., Bassole, I. H. N., & Dicko, M. H., (2012). Antimicrobial activities of essential oil and methanol extract of *Boswellia sacra* Flueck. and *Boswellia papyrifera* (Del.) Hochst from Djibouti. *Intern. J. Management Modern Sci. Technologies, 1*(1), 1–10.

Abeje, E., (2002). *Regeneration Status, Soil Seed Bank and Socio-Economic Importance of Boswellia papyrifera (Del.) Hoschst. in Two Woredas of North Gonder Zone, Northern Ethiopia*. SIDS, Uppsala (Sweden).

Abiyu, A., Bongers, F., Eshete, A., Gebrehiwot, K., Kindu, M., Lemenih, M., Moges, Y., et al., (2010). Chapter 7 – Incense woodlands in Ethiopia and Eritrea: Regeneration problems and restoration possibilities. In: Bongers, F., & Tennigkeit, T., (eds.), *Degraded Forests in Eastern Africa: Management and Restoration* (pp. 133–152). Oxford: Earthscan.

Abtew, A., Pretzsch, J., Mohamoud, T., & Adam, Y., (2011). *Population Structure, Density and Natural Regeneration of Boswellia papyrifera (Del.) Hochst in Dry Woodlands of Nuba Mountains* (p. 245). South Kordofan State, Sudan. DITSL GmbH. Bonn, Germany.

Adam, A. A., & El Tayeb, A. M., (2008). A comparative study of natural regeneration of *B. papyrifera* and other tree species in Jebel Marra Darfur Sudan. *Res. J. Agric. Biol. Sci., 4*(1), 94–102.

Adefires, W., (2006). *Population Status and Socio-Economic Importance of Gum and Resin Bearing Species in Borana Lowlands, Southern Ethiopia* (pp. 30–55). MSc. Thesis. Addis Ababa, Ethiopia: Addis Ababa University.

Alemu, S., Atnafu, H., Awas, T., Belay, B., Demissew, S., Luke, W. R. Q., Mekbib, E., et al., (2018). *Boswellia Ogadensis: The IUCN Red List of Threatened Species 2018*. e.T34385A128140745. http://dx.doi.org/10.2305/IUCN.UK.2004.RLTS.T44814A10950258.en.

Awas, T., Belay, B., Demissew, S., Nemomissa, S., Mekbib, E., Atnafu, H., & Alemu, S., (2018). *Boswellia pirottae: The IUCN Red List of Threatened Species 2018*. e.T34394A128137387. http://dx.doi.org/10.2305/IUCN.UK.2015-2.RLTS.T50126567A50131280.en.

Bhattacharya, P., & Hayat, S. F., (2004). Sustainable NTFP management for rural development: A case from Madhya Pradesh, India. *Intern. Forestry Review, 6*(2), 161–168.

Biswal, A., & Nair, M., (2008). Threatened plants of Orissa and priority species for conservation. Special habitats and threatened plants of India. *ENVIS Bulletin: Wildlife and Protected Areas, 11*(1), 175–186.

Bongers, F., Groenendijk, P., Bekele, T., et al., (2019). Frankincense in peril. *Nat. Sustain., 2*, 602–610. https://doi.org/10.1038/s41893-019-0322-2.

Brendler, T., Brinckmann, J. A., & Schippmann, U., (2018). Sustainable supply, a foundation for natural product development: The case of Indian frankincense (*Boswellia serrata* Roxb. ex Colebr.). *J Ethnopharmacol., 28*, 279–286. doi: 10.1016/j.jep.2018.07.017.

Chandrasekhar, T., Hussain, T. M., & Jayanand, B., (2005). *In vitro* micropropagation of *Boswellia ovalifoliolata. Journal of Natural Science C, 60*(5, 6), 505–507.

Chaubey, O., Sharma, A., & Krishnamurthy, G., (2015). *Ex-situ* conservation of indigenous, threatened and ethno-medicinal diversity of forest species. *Intern. J. Bio-Science and Bio-Technology, 7*(3), 9–22.

DeCarlo, A., & Ali, S. H., (2014). *Sustainable Sourcing of Phytochemicals as a Development Tool: The Case of Somaliland's Frankincense Industry.* Institute for Environmental Diplomacy & Security. 2014.

Demel, T., & Mulugeta, L., (2015). *Role of Forestry in Poverty Alleviation in Ethiopia* (p. 70). Debub University, Wondo Genet College of Forestry, Shashamane, Ethiopia.

Dular, A., (2015). Plant diversity assessment of Sariska tiger reserve in Aravallis with emphasis on minor forest products. *Trop. Plant Res., 2*(1), 30–35.

Ebuen, Y., (2016). *Protecting the Frankincense.* Sultanate of Oman, Ministry of Information. https://omaninfo.om/english/module.php?module=topics-showtopic&CatID=35&ID=2744 (accessed on 10 October 2022).

Eshete, A., (2002). *Regeneration Status, Soil Seed Bank and Socio-Economic Importance of Boswellia papyrifera (Del.) Hoschst. in Two Woredas of North Gondar Zone, Northern Ethiopia.* Master's Thesis, Swedish University of Agricultural Sciences and Wondo Genet College of Forestry. 2002.

Eshete, A., Teketay, D., & Hulten, H., (2005). The socio-economic importance and status of populations of *Boswellia papyrifera* (Del.) Hochst. in Northern Ethiopia: The Case of North Gondar Zone. *Forests Trees and Livelihoods, 15*, 55–74.

Eslamieh, J., (2020). The genus *Boswellia*: Preservation through horticulture. Fort Collins, CO: A Book's Mind. 2017. In: Al-Harrasi, et al., (eds.), *Biology of Genus Boswellia: Propagation and Conservation of Boswellia sacra* (pp. 71–85, 80527).

Fabio, A., Nadim, T., Michele, S., Alessio, F., Alfredo, G., & Marcello, V., (2014). Developing conservation strategies for endemic tree species when faced with time and data constraints: *Boswellia* spp. on Socotra (Yemen). *Biodiversity and Conservation., 20*, 1483–1499. 10.1007/s10531-011-0039-7.

FAO, (2007). *Overview of Forest Pests: Sudan* (p. 18). Forest Health & Biosecurity Working Papers FBS/31E. FAO: Rome, Italy. http://www.fao.org/forestry/12279-0139fe4fdeb6e212bc4c1d24d189d3663.pdf (accessed on 10 October 2022).

FAO. (Food and Agricultural Organization), (1995). Flavors and fragrances of plant origin. *Non-Wood Forest Products, 1*, Rome.

Gebrehiwot, K., Muys, B., Haile, M., & Mitloehner, R., (2003). Introducing *Boswellia papyrifera* (Del.) Hochst and its non-timber forest product, frankincense. *Int. For. Rev., 5*, 348–353.

Gessmalla, A. F., Raddad, E. Y. A., Ibrahim, G. E. A., (2015). Effects of tapping date, tapping direction and elevation on resin yield from *Boswellia papyrifera* in the Blue Nile State, Sudan. *Natural Resources, Agricultural Development and Food Security-International Working Paper Series, 15*(12), 14. http://economia.unipv.it/naf/ (accessed on 10 October 2022).

GFA, (2021). *The Future of Frankincense Online Workshop*. Global Frankincense Alliance (GFA). https://static1.squarespace.com/static/600c3628f983552f0a5535ec/t/6127c7d7b7405d58d8ffa93c/1629997019049/Outcomes+and+Road+Map.pdf (accessed on 10 October 2022).

Gidey, T., Hagos, D., Juhar, H. M. S., Solomon, N., Negussie, A., Crous-Duran, J., Oliveira, T. S., et al., (2020). Population status of *Boswellia papyrifera* woodland and prioritizing its conservation interventions using multi-criteria decision model in Northern Ethiopia, *Heliyon, 6*(10), e05139.

Girma, A., Andrew, K. S., De Bie, C. A. J. M., & Bongers, F., (2015). Understanding *Boswellia papyrifera* tree secondary metabolites through bark spectral analysis. *ISPRS J. Photogrammetry and Remote Sensing, 105*, 30–37.

Girmay, F., (2000). The status of gum Arabic and resins in Ethiopia. In: Chikamai, B. N., (ed.), *J. Proceedings of the Meeting of the Network* (pp. 29–31). Natural Gums and Resins in Africa (NGARA). NGARA, Nairobi. Kenya.

Groenendijk, P., Eshete, A., Sterck, F. J., Zuidema, P. A., & Bongers, F., (2012). Limitations to sustainable frankincense production: Blocked regeneration, high adult mortality and declining populations. *J. Appl. Ecol., 49*, 164–173.

Hassan, A. M. M., (2012). Distribution *Boswellia sacra* in Dhofar Mountains, Sultanate of Oman: Economic value and environmental role. *J. Life Sci., 6*, 632–636.

Jathar, G., & Patil, D., (2011). *Reassessment of the Status of Forest Owlet in its Known Distribution and Evaluation of Conservation Status*. Foundation for Ecological Conservation and Sustainable Development, Pune, India.

Joshi, S., & Shringi, S., (2014). Floristic diversity with special reference to rare and threatened plants of Jawahar Sagar Sanctuary Area near Kota, Rajasthan. *Biological Forum: An International Journal, 6*(1), 84–91.

Khamis, M. A., Siddig, E. N. E., Khalil, A., & Csaplovics, E., (2016). Changes in forest cover composition of *Boswellia papyrifera* (Del.) Hochst. stands and their consequences, South Kordofan, Sudan. *Mediterr. J. Biosci., 1*, 99–108.

Kushwaha, C. P., & Singh, K. P., (2005). Diversity of leaf phenology in a tropical deciduous forest in India. *J. Trop. Ecol., 21*, 47–56.

Lemenih, M., & Kassa, H., (2011). *Management Guide for Sustainable Production of Frankincense*. CIFOR, Bogor, Indonesia.

Mengistu, T. M., (2011). *Physiological Ecology of the Frankincense Tree*. PhD Thesis, Wageningen University and Research Centre, The Netherlands.

Miller, A., (2004a). *Boswellia ameero: The IUCN Red List of Threatened Species*. e.T30414A9546504. http://dx.doi.org/10.2305/IUCN.UK.2004.RLTS.T30414A9546504.en.

Miller, A., (2004b). *Boswellia bullata: The IUCN Red List of Threatened Species*. e.T44812A10950015. http://dx.doi.org/10.2305/IUCN.UK.2004.RLTS.T44812A10950015.en.

Miller, A., (2004c). *Boswellia dioscoridis: The IUCN Red List of Threatened Species*. e.T44813A10950138. http://dx.doi.org/10.2305/IUCN.UK.2004.RLTS.T44813A10950138.en.

Miller, A., (2004d). *Boswellia elongata: The IUCN Red List of Threatened Species*. e.T30415A9546667. http://dx.doi.org/10.2305/IUCN.UK.2004.RLTS.T30415A9546667.en.

Miller, A., (2004e). *Boswellia nana: The IUCN Red List of Threatened Species*. e.T44814A10950258. http://dx.doi.org/10.2305/IUCN.UK.2004.RLTS.T44814A10950258.en.

Miller, A., (2004f). *Boswellia popoviana: The IUCN Red List of Threatened Species*. e.T37866A10082448. http://dx.doi.org/10.2305/IUCN.UK.2004.RLTS.T37866A10082448.en.

Miller, A., (2004g). *Boswellia socotrana: The IUCN Red List of Threatened Species.* e.T30416A9546843. http://dx.doi.org/10.2305/IUCN.UK.2004.RLTS.T30416A9546843.en.

MOE, (2012). *The National Red List 2012 of Sri Lanka; Conservation Status of the Fauna and Flora.* Ministry of Environment, Colombo, Sri Lanka.

Negussie, A., Aertsb, R., Gebrehiwota, K., & Muysb, B., (2008). Seedling mortality causes recruitment limitation of *Boswellia papyrifera* in northern Ethiopia. *J. Arid. Environ., 72*, 378–383.

Newton, A. C., (2008). Conservation of tree species through sustainable use: How can it be achieved in practice? *Oryx., 42*(2), 195–205.

Nour, L. A. M., (2008). *Production and Productivity of Boswellia papyrifera in Jebel Elgarrie Area (Blue Nile State).* Master's Thesis, University of Khartoum, Sudan.

Ogbazghi, W., (2001). *The Distribution and Regeneration of Boswellia Papyrifera (Del.) Hochst. in Eritrea.* PhD thesis, Wageningen University, The Netherlands.

Ogbazghi, W., Rijkers, T., Wessel, M., & Bongers, F., (2006). The distribution of the frankincense tree *Boswellia papyrifera* in Eritrea: The role of environment and land use. *J. Biogeography, 33*, 524–535.

Purohit, S., Tak, K., & Kukda, G., (1995). *In vitro* propagation of *Boswellia serrata* Roxb. *Biologia. Plantarum, 37*(3), 335–340.

Radha, R., Dibyendu, A., Satyam, V., Purabi, S., Amit, K., Grant, K. R., Arun, D., et al., (2020). Climate models predict a divergent future for the medicinal tree *Boswellia serrata* Roxb. in India. *Global Ecology and Conservation, 23*, e01040.

Raju, A. J. S., Lakshmi, P. V., Ramana, K. V., & Chandra, P. H., (2012). Entomophily, ornithophily and anemochory in the self-incompatible *Boswellia ovalifoliolata* Bal. & Henry (Burseraceae), an endemic and endangered medicinally important tree species. *J. Threatened Taxa, 4*(7), 2673–2684.

Rijkers, T., Ogbaschi, W., Wessel, M., & Bongers, F., (2006). The effect of tapping for frankincense on sexual reproduction in *Boswellia papyrifera. J. Appl. Ecol., 43*, 1188–1195. https://besjournals.onlinelibrary.wiley.com/doi/full/10.1111/j.1365-2664.2006.01215.x (accessed on 10 October 2022).

Sabo, P., Ouédraogo, A., Gbemavo, D. S. J. C., Salako, K. V., & Glèlè, K. R., (2021). Land use impacts on *Boswellia dalzielii* Hutch., an African frankincense tree in Burkina Faso. *Bois et Forêts des Tropiques, 349*, 51–63. doi: Https://doi.org/10.19182/bft2021.349.a31960.

Sagar, R., & Singh, J. S., (2006). Tree density, basal area and species diversity in a disturbed dry tropical forest of northern India: Implications for conservation. *Environ. Conserv., 33*, 256.

Sagar, R., Raghubanshi, A., & Singh, J., (2003). Tree species composition, dispersion and diversity along a disturbance gradient in a dry tropical forest region of India. *Forest Ecology and Management, 186*, 61–71.

Saha, D., Ved, D., Ravikumar, K., & Haridasan, K., (2015). *Boswellia ovalifoliolata: The IUCN Red List of Threatened Species.* E.T50126567A50131280. http://dx.doi.org/10.2305/IUCN.UK.2015-2.RLTS.T50126567A50131280.en (accessed on 10 October 2022).

Shahabuddin, G., & Prasad, S., (2004). Assessing ecological sustainability of non-timber forest produce extraction: The Indian scenario. *Conservation and Society, 2*(2), 235–250.

Sharma, G., & Raghubanshi, A., (2007). Effect of *Lantana camara* L. cover on local depletion of tree population in the Vindhyan tropical dry deciduous forest of India. *Appl. Ecol. Environ. Res., 5*(1), 109–121.

Singh, K. P., & Kushwaha, C. P., (2006). Diversity of flowering and fruiting phenology of trees in a tropical deciduous forest in India. *Ann. Bot., 97*, 265–276.

Soumya, K. V., Shackleton, C. M., & Setty, S. R., (2019). Harvesting and local knowledge of a cultural non-timber forest product (NTFP): Gum-Resin from *Boswellia serrata* Roxb. in three protected areas of the Western Ghats, India. *Forests, 10*, 907.

Sunnichan, V. G., Mohan, R. H. Y., & Shivanna, K. R., (2005). Reproductive biology of *Boswellia serrata*, the source of Salai guggul, an important gum-resin. *Bot. J. Linn. Soc., 147*, 73–82.

Tilahun, G., (1997). *Boswellia papyrifera (Del.) Hochst. from Western Tigray: Opportunities, Constraints and Seed Germination Responses*. MSc. Thesis No. 12. Swedish University of Agricultural Sciences, Skinnskatteberg.

Tropical Plants Database. Ken Fern. tropical.theferns.info. tropical.theferns.info/viewtropical.php?id=Boswellia%20ameero (accessed on 10 October 2022).

Ved, D., (2013). Medicinal plants species of conservation concern identified for Andhra Pradesh, Arunachal Pradesh, Madhya Pradesh & Rajasthan. *ENVIS Newsletter on Medicinal Plants, 6*, 1–4.

Vijay, N., (2014). *Studies on in Vitro Cloning of Mature Tree of Capparis decidua and Boswellia serrata*. Forest Research Institute University, Dehradun, India.

Vollesen, K., (1989). Burseraceae. In: Hedberg, I., & Edwards, S., (eds.), *Flora of Ethiopia* (Vol. 3, pp. 442–478). Addis Ababa, Ethiopia.

Wale, H. A., Bekele, T., & Dalle, G., (2012). Floristic diversity, regeneration status, and vegetation structure of woodlands in Metema Area, Amhara National Regional State, Northwestern Ethiopia. *J. Forestry Res., 23*, 391–398. https://doi.org/10.1007/s11676-012-0275-z.

Wubalem, T., Demel, T., Mulugeta, L., & Girmay, F., (2003). Country report for Ethiopia. In: Chikamai, B. N., (ed.), *Review and Symphysis on the State of Knowledge of Boswellia Species and Commercialization of Frankincense in the Drylands of Africa* (pp. 11–35). Association of Forestry Research Institutions in Eastern Africa, Nairobi, Kenya.

WWF, (2015a). *Tropical and Subtropical Dry Broadleaf Forests*. Southern Asia: India. [Online] Available at: https://www.worldwildlife.org/ecoregions/im0209 (accessed on 10 October 2022).

WWF, (2015b). *Tropical and Subtropical Dry Broadleaf Forests.* Southern Asia: Northwest India. [Online] Available at: http://www.worldwildlife.org/ecoregions/im0206 (accessed on 10 October 2022).

Yogi, R., Bhattacharya, A., & Jaiswal, A. L., (2013). Plant resins and gums statistics at a glance 2013. In: *ICAR – Indian Institute of Natural Resins and Gums, Bulletin (Technical) No. 06/2014* (p. 38). Ranchi (Jharkhand), India.

INDEX

C

H

J

K

L

M

N

S

T

For Product Safety Concerns and Information please contact our EU
representative GPSR@taylorandfrancis.com
Taylor & Francis Verlag GmbH, Kaufingerstraße 24, 80331 München, Germany

www.ingramcontent.com/pod-product-compliance
Ingram Content Group UK Ltd.
Pitfield, Milton Keynes, MK11 3LW, UK
UKHW021512250926
475702UK00043B/228
9781774912638